differential equations in Banach spaces

PURE AND APPLIED MATHEMATICS

A Program of Monographs, Textbooks, and Lecture Notes

LECTURE NOTES IN PURE AND APPLIED MATHEMATICS

1. *N. Jacobson,* Exceptional Lie Algebras
2. *L.-Å. Lindahl and F. Poulsen,* Thin Sets in Harmonic Analysis
3. *I. Satake,* Classification Theory of Semi-Simple Algebraic Groups
4. *F. Hirzebruch, W. D. Newmann, and S. S. Koh, Differentiable Manifolds and Quadratic Forms*
5. *I. Chavel,* Riemannian Symmetric Spaces of Rank One
6. *R. B. Burckel,* Characterization of C(X) Among Its Subalgebras
7. *B. R. McDonald, A. R. Magid, and K. C. Smith,* Ring Theory: Proceedings of the Oklahoma Conference
8. *Y.-T. Siu,* Techniques of Extension on Analytic Objects
9. *S. R. Caradus, W. E. Pfaffenberger, and B. Yood,* Calkin Algebras and Algebras of Operators on Banach Spaces
10. *E. O. Roxin, P.-T. Liu, and R. L. Sternberg,* Differential Games and Control Theory
11. *M. Orzech and C. Small,* The Brauer Group of Commutative Rings
12. *S. Thomier,* Topology and Its Applications
13. *J. M. Lopez and K. A. Ross,* Sidon Sets
14. *W. W. Comfort and S. Negrepontis,* Continuous Pseudometrics
15. *K. McKennon and J. M. Robertson,* Locally Convex Spaces
16. *M. Carmeli and S. Malin,* Representations of the Rotation and Lorentz Groups: An Introduction
17. *G. B. Seligman,* Rational Methods in Lie Algebras
18. *D. G. de Figueiredo,* Functional Analysis: Proceedings of the Brazilian Mathematical Society Symposium
19. *L. Cesari, R. Kannan, and J. D. Schuur,* Nonlinear Functional Analysis and Differential Equations: Proceedings of the Michigan State University Conference
20. *J. J. Schäffer,* Geometry of Spheres in Normed Spaces
21. *K. Yano and M. Kon,* Anti-Invariant Submanifolds
22. *W. V. Vasconcelos,* The Rings of Dimension Two
23. *R. E. Chandler,* Hausdorff Compactifications
24. *S. P. Franklin and B. V. S. Thomas,* Topology: Proceedings of the Memphis State University Conference
25. *S. K. Jain,* Ring Theory: Proceedings of the Ohio University Conference
26. *B. R. McDonald and R. A. Morris,* Ring Theory II: Proceedings of the Second Oklahoma Conference
27. *R. B. Mura and A. Rhemtulla,* Orderable Groups
28. *J. R. Graef,* Stability of Dynamical Systems: Theory and Applications
29. *H.-C. Wang,* Homogeneous Branch Algebras
30. *E. O. Roxin, P.-T. Liu, and R. L. Sternberg,* Differential Games and Control Theory II
31. *R. D. Porter,* Introduction to Fibre Bundles
32. *M. Altman,* Contractors and Contractor Directions Theory and Applications
33. *J. S. Golan,* Decomposition and Dimension in Module Categories
34. *G. Fairweather,* Finite Element Galerkin Methods for Differential Equations
35. *J. D. Sally,* Numbers of Generators of Ideals in Local Rings
36. *S. S. Miller,* Complex Analysis: Proceedings of the S.U.N.Y. Brockport Conference
37. *R. Gordon,* Representation Theory of Algebras: Proceedings of the Philadelphia Conference
38. *M. Goto and F. D. Grosshans,* Semisimple Lie Algebras
39. *A. I. Arruda, N. C. A. da Costa, and R. Chuaqui,* Mathematical Logic: Proceedings of the First Brazilian Conference
40. *F. Van Oystaeyen,* Ring Theory: Proceedings of the 1977 Antwerp Conference
41. *F. Van Oystaeyen and A. Verschoren,* Reflectors and Localization: Application to Sheaf Theory
42. *M. Satyanarayana,* Positively Ordered Semigroups
43. *D. L Russell,* Mathematics of Finite-Dimensional Control Systems
44. *P.-T. Liu and E. Roxin,* Differential Games and Control Theory III: Proceedings of the Third Kingston Conference, Part A
45. *A. Geramita and J. Seberry,* Orthogonal Designs: Quadratic Forms and Hadamard Matrices
46. *J. Cigler, V. Losert, and P. Michor,* Banach Modules and Functors on Categories of Banach Spaces

47. *P.-T. Liu and J. G. Sutinen,* Control Theory in Mathematical Economics: Proceedings of the Third Kingston Conference, Part B
48. *C. Byrnes,* Partial Differential Equations and Geometry
49. *G. Klambauer,* Problems and Propositions in Analysis
50. *J. Knopfmacher,* Analytic Arithmetic of Algebraic Function Fields
51. *F. Van Oystaeyen,* Ring Theory: Proceedings of the 1978 Antwerp Conference
52. *B. Kadem,* Binary Time Series
53. *J. Barros-Neto and R. A. Artino,* Hypoelliptic Boundary-Value Problems
54. *R. L. Sternberg, A. J. Kalinowski, and J. S. Papadakis,* Nonlinear Partial Differential Equations in Engineering and Applied Science
55. *B. R. McDonald,* Ring Theory and Algebra III: Proceedngs of the Third Oklahoma Conference
56. *J. S. Golan,* Structure Sheaves Over a Noncommutative Ring
57. *T. V. Narayana, J. G. Williams, and R. M. Mathsen,* Combinatorics, Representation Theory and Statistical Methods in Groups: YOUNG DAY Proceedings
58. *T. A. Burton,* Modeling and Differential Equations in Biology
59. *K. H. Kim and F. W. Roush,* Introduction to Mathematical Consensus Theory
60. *J. Banas and K. Goebel,* Measures of Noncompactness in Banach Spaces
61. *O. A. Nielson,* Direct Integral Theory
62. *J. E. Smith, G. O. Kenny, and R. N. Ball,* Ordered Groups: Proceedings of the Boise State Conference
63. *J. Cronin,* Mathematics of Cell Electrophysiology
64. *J. W. Brewer,* Power Series Over Commutative Rings
65. *P. K. Kamthan and M. Gupta,* Sequence Spaces and Series
66. *T. G. McLaughlin,* Regressive Sets and the Theory of Isols
67. *T. L. Herdman, S. M. Rankin III, and H. W. Stech,* Integral and Functional Differential Equations
68. *R. Draper,* Commutative Algebra: Analytic Methods
69. *W. G. McKay and J. Patera,* Tables of Dimensions, Indices, and Branching Rules for Representations of Simple Lie Algebras
70. *R. L. Devaney and Z. H. Nitecki,* Classical Mechanics and Dynamical Systems
71. *J. Van Geel,* Places and Valuations in Noncommutative Ring Theory
72. *C. Faith,* Injective Modules and Injective Quotient Rings
73. *A. Fiacco,* Mathematical Programming with Data Perturbations I
74. *P. Schultz, C. Praeger, and R. Sullivan,* Algebraic Structures and Applications: Proceedings of the First Western Australian Conference on Algebra
75. *L Bican, T. Kepka, and P. Nemec,* Rings, Modules, and Preradicals
76. *D. C. Kay and M. Breen,* Convexity and Related Combinatorial Geometry: Proceedings of the Second University of Oklahoma Conference
77. *P. Fletcher and W. F. Lindgren,* Quasi-Uniform Spaces
78. *C.-C. Yang,* Factorization Theory of Meromorphic Functions
79. *O. Taussky,* Ternary Quadratic Forms and Norms
80. *S. P. Singh and J. H. Burry,* Nonlinear Analysis and Applications
81. *K. B. Hannsgen, T. L. Herdman, H. W. Stech, and R. L. Wheeler,* Volterra and Functional Differential Equations
82. *N. L. Johnson, M. J. Kallaher, and C. T. Long,* Finite Geometries: Proceedings of a Conference in Honor of T. G. Ostrom
83. *G. I. Zapata,* Functional Analysis, Holomorphy, and Approximation Theory
84. *S. Greco and G. Valla,* Commutative Algebra: Proceedings of the Trento Conference
85. *A. V. Fiacco,* Mathematical Programming with Data Perturbations II
86. *J.-B. Hiriart-Urruty, W. Oettli, and J. Stoer,* Optimization: Theory and Algorithms
87. *A. Figa Talamanca and M. A. Picardello,* Harmonic Analysis on Free Groups
88. *M. Harada,* Factor Categories with Applications to Direct Decomposition of Modules
89. *V. I. Istrăţescu,* Strict Convexity and Complex Strict Convexity
90. *V. Lakshmikantham,* Trends in Theory and Practice of Nonlinear Differential Equations
91. *H. L. Manocha and J. B. Srivastava,* Algebra and Its Applications
92. *D. V. Chudnovsky and G. V. Chudnovsky,* Classical and Quantum Models and Arithmetic Problems
93. *J. W. Longley,* Least Squares Computations Using Orthogonalization Methods
94. *L. P. de Alcantara,* Mathematical Logic and Formal Systems
95. *C. E. Aull,* Rings of Continuous Functions
96. *R. Chuaqui,* Analysis, Geometry, and Probability
97. *L. Fuchs and L. Salce,* Modules Over Valuation Domains

98. *P. Fischer and W. R. Smith,* Chaos, Fractals, and Dynamics
99. *W. B. Powell and C. Tsinakis,* Ordered Algebraic Structures
100. *G. M. Rassias and T. M. Rassias,* Differential Geometry, Calculus of Variations, and Their Applications
101. *R.-E. Hoffmann and K. H. Hofmann,* Continuous Lattices and Their Applications
102. *J. H. Lightbourne III and S. M. Rankin III,* Physical Mathematics and Nonlinear Partial Differential Equations
103. *C. A. Baker and L, M. Batten,* Finite Geometrics
104. *J. W. Brewer, J. W. Bunce, and F. S. Van Vleck,* Linear Systems Over Commutative Rings
105. *C. McCrory and T. Shifrin,* Geometry and Topology: Manifolds, Varieties, and Knots
106. *D. W. Kueker, E. G. K. Lopez-Escobar, and C. H. Smith,* Mathematical Logic and Theoretical Computer Science
107. *B.-L. Lin and S. Simons,* Nonlinear and Convex Analysis: Proceedings in Honor of Ky Fan
108. *S. J. Lee,* Operator Methods for Optimal Control Problems
109. *V. Lakshmikantham,* Nonlinear Analysis and Applications
110. *S. F. McCormick,* Multigrid Methods: Theory, Applications, and Supercomputing
111. *M. C. Tangora,* Computers in Algebra
112. *D. V. Chudnovsky and G. V. Chudnovsky,* Search Theory: Some Recent Developments
113. *D. V. Chudnovsky and R. D. Jenks,* Computer Algebra
114. *M. C. Tangora,* Computers in Geometry and Topology
115. *P. Nelson, V. Faber, T. A. Manteuffel, D. L. Seth, and A. B. White, Jr.,* Transport Theory, Invariant Imbedding, and Integral Equations: Proceedings in Honor of G. M. Wing's 65th Birthday
116. *P. Clément, S. Invernizzi, E. Mitidieri, and I. I. Vrabie,* Semigroup Theory and Applications
117. *J. Vinuesa,* Orthogonal Polynomials and Their Applications: Proceedings of the International Congress
118. *C. M. Dafermos, G. Ladas, and G. Papanicolaou,* Differential Equations: Proceedings of the EQUADIFF Conference
119. *E. O. Roxin,* Modern Optimal Control: A Conference in Honor of Solomon Lefschetz and Joseph P. Lasalle
120. *J. C. Díaz,* Mathematics for Large Scale Computing
121. *P. S. Milojević,* Nonlinear Functional Analysis
122. *C. Sadosky,* Analysis and Partial Differential Equations: A Collection of Papers Dedicated to Mischa Cotlar
123. *R. M. Shortt,* General Topology and Applications: Proceedings of the 1988 Northeast Conference
124. *R. Wong,* Asymptotic and Computational Analysis: Conference in Honor of Frank W. J. Olver's 65th Birthday
125. *D. V. Chudnovsky and R. D. Jenks,* Computers in Mathematics
126. *W. D. Wallis, H. Shen, W. Wei, and L. Zhu,* Combinatorial Designs and Applications
127. *S. Elaydi,* Differential Equations: Stability and Control
128. *G. Chen, E. B. Lee, W. Littman, and L. Markus,* Distributed Parameter Control Systems: New Trends and Applications
129. *W. N. Everitt,* Inequalities: Fifty Years On from Hardy, Littlewood and Pólya
130. *H. G. Kaper and M. Garbey,* Asymptotic Analysis and the Numerical Solution of Partial Differential Equations
131. *O. Arino, D. E. Axelrod, and M. Kimmel,* Mathematical Population Dynamics: Proceedings of the Second International Conference
132. *S. Coen,* Geometry and Complex Variables
133. *J. A. Goldstein, F. Kappel, and W. Schappacher,* Differential Equations with Applications in Biology, Physics, and Engineering
134. *S. J. Andima, R. Kopperman, P. R. Misra, J. Z. Reichman, and A. R. Todd,* General Topology and Applications
135. *P Clément, E. Mitidieri, B. de Pagter,* Semigroup Theory and Evolution Equations: The Second International Conference
136. *K. Jarosz,* Function Spaces
137. *J. M. Bayod, N. De Grande-De Kimpe, and J. Martínez-Maurica,* p-adic Functional Analysis
138. *G. A. Anastassiou,* Approximation Theory: Proceedings of the Sixth Southeastern Approximation Theorists Annual Conference
139. *R. S. Rees,* Graphs, Matrices, and Designs: Festschrift in Honor of Norman J. Pullman
140. *G. Abrams, J. Haefner, and K. M. Rangaswamy,* Methods in Module Theory

141. *G. L. Mullen and P. J.-S. Shiue*, Finite Fields, Coding Theory, and Advances in Communications and Computing
142. *M. C. Joshi and A. V. Balakrishnan*, Mathematical Theory of Control: Proceedings of the International Conference
143. *G. Komatsu and Y. Sakane*, Complex Geometry: Proceedings of the Osaka International Conference
144. *I. J. Bakelman*, Geometric Analysis and Nonlinear Partial Differential Equations
145. *T. Mabuchi and S. Mukai*, Einstein Metrics and Yang–Mills Connections: Proceedings of the 27th Taniguchi International Symposium
146. *L. Fuchs and R. Göbel*, Abelian Groups: Proceedings of the 1991 Curaçao Conference
147. *A. D. Pollington and W. Moran*, Number Theory with an Emphasis on the Markoff Spectrum
148. *G. Dore, A. Favini, E. Obrecht, and A. Venni*, Differential Equations in Banach Spaces
149. *T. West*, Continuum Theory and Dynamical Systems
150. *K. D. Bierstedt, A. Pietsch, W. Ruess, and D. Vogt*, Functional Analysis

Additional Volumes in Preparation

differential equations in Banach spaces

proceedings of the Bologna conference

edited by
Giovanni Dore
Angelo Favini
Enrico Obrecht
Alberto Venni
University of Bologna
Bologna, Italy

Marcel Dekker, Inc. **New York • Basel • Hong Kong**

Library of Congress Cataloging-in-Publication Data

Differential equations in Banach spaces : proceedings of the Bologna conference / edited by
Giovanni Dore ... [et al.].
p. cm. -- (Lecture notes in pure and applied mathematics ; 148)
"The Second Conference on Differential Equations in Banach Spaces was held in
Bologna..."--Preface.
Includes bibliographical references and index.
ISBN 0-8247-9067-7 (acid-free)
1. Differential equations--Congresses. 2. Banach spaces--Congresses. I. Dore,
Giovanni. II. Conference on Differential Equations in Banach Spaces (2nd: 1991 :
Bologna, Italy) III. Series: Lecture notes in pure and applied mathematics : v. 148.
QA370.D525 1993
515'.353--dc20

93-14088
CIP

This book is printed on acid-free paper.

Marcel Dekker, Inc.
270 Madison Avenue, New York, New York 10016

Current printing (last digit):
10 9 8 7 6 5 4 3 2 1

PRINTED IN THE UNITED STATES OF AMERICA

Preface

The Second Conference on Differential Equations in Banach Spaces was held in Bologna exactly six years after the first. The meeting was devoted not only to evolution equations and semigroup theory, but also to those branches of functional analysis that have proven to be useful for the study of evolution problems, such as functional calculus, perturbation theory, interpolation and extrapolation spaces, spaces of vector-valued functions, Volterra operator equations, and operator matrices. Many contributions were made to the abstract theory of (linear or nonlinear) parabolic and hyperbolic evolution equations; a remarkable interest in problems of applied mathematics is witnessed by the important contributions to such topics as von Karmàn systems, semilinear wave equations, Euler–Bernoulli and Kirchhoff equations, and viscoelasticity problems.

The Organizing Committee thanks all the institutions that made this meeting possible: the Consiglio Nazionale delle Ricerche (in particular, the Gruppo Nazionale per l'Analisi Funzionale e Applicazioni), the Ministero della Università e della Ricerca Scientifica e Tecnologica of Italy, the University of Bologna (in particular, its Rector, Professor Fabio Roversi Monaco) for financial support, and the Ente Fiere of Bologna.

In addition, the Organizing Committee wishes to acknowledge the assistance and support of Professors Giuseppe Da Prato, Alessandra Lunardi, Carlo Domenico Pagani, and Eugenio Sinestrari. Further, special thanks are due to Mauro Fabrizio, Head of the Department of Mathematics of the University of Bologna, and all the staff of the department, for their help in the administration of the meeting.

Giovanni Dore
Angelo Favini
Enrico Obrecht
Alberto Venni

Contents

Contributors

PAOLO ACQUISTAPACE University of Pisa, Pisa, Italy

MARCO LUIGI BERNARDI University of Brescia, Brescia, Italy

JULIO E. BOUILLET University of Buenos Aires, Buenos Aires, Argentina

PIERMARCO CANNARSA University of Rome "Tor Vergata", Rome, Italy

GIUSEPPE DA PRATO Scoola Normale Superiore, Pisa, Italy

GABRIELLA DI BLASIO University of Rome "La Sapienza", Rome, Italy

KLAUS-JOCHEN ENGEL Tübingen University, Tübingen, Germany

A. FAVINI University of Bologna, Bologna, Italy

JEROME A. GOLDSTEIN Louisiana State University, Baton Rouge, Louisiana

ALBRECHT HOLDERRIETH Tübingen University, Tübingen, Germany

I. LASIECKA University of Virginia, Charlottesville, Virginia

RALPH deLAUBENFELS Ohio University, Athens, Ohio

FABIO LUTEROTTI University of Brescia, Brescia, Italy

JAN PRÜSS Paderborn University, Paderborn, Germany

HIROKI TANABE Osaka University, Osaka, Japan

D. TATARU University of Virginia, Charlottesville, Virginia

R. TRIGGIANI University of Virginia, Charlottesville, Virginia

VINCENZO VESPRI University of Pavia, Pavia, Italy

G. F. WEBB Vanderbilt University, Nashville, Tennessee

L. WEIS Louisiana State University, Baton Rouge, Louisiana

ATSUSHI YAGI Himeji Institute of Technology, Himeji, Japan

Conference Participants

Paolo ACQUISTAPACE (Università di Pisa)
Wolfgang ARENDT (Université de Franche Comté, Besançon)
Alberto AROSIO (Università di Parma)
Marco Luigi BERNARDI (Università di Brescia)
Julio E. BOUILLET (Universidad de Buenos Aires)
Rosanna BRESSAN VILLELLA (Università di Padova)
Francesca BUCCI (Università di Modena)
Anna BUTTU (Università di Cagliari)
Piermarco CANNARSA (Università di Roma "Tor Vergata")
Gabriella CARISTI (Università di Udine)
Philippe CLÉMENT (Delft University of Technology)
Giuseppe DA PRATO (Scuola Normale Superiore di Pisa)
Giovanni DORE (Università di Bologna)
Janet DYSON (Oxford University, Mansfield College)
Klaus-Jochen ENGEL (Universität Tübingen)
Angelo FAVINI (Università di Bologna)
Bruno FRANCHI (Università di Bologna)
Marco FUHRMAN (Politecnico di Milano)
Jerome A. GOLDSTEIN (Tulane University, New Orleans)
Ronald GRIMMER (Southern Illinois University, Carbondale)
Davide GUIDETTI (Università di Bologna)
Mats GYLLENBERG (Lulea University of Technology)
Albrecht HOLDERRIETH (Universität Tübingen)
Mimmo IANNELLI (Università di Trento)

Vangipuvam LAKSHMIKANTHAM (Florida Institute of Technology, Melbourne)
Irena LASIECKA (University of Virginia, Charlottesville)
Ralph DELAUBENFELS (Ohio University, Athens)
Srinavasa LEELAMMA (State University of New York, Geneseo)
Gunter LUMER (Université de Mons)
Alessandra LUNARDI (Università di Cagliari)
Fabio LUTEROTTI (Università di Brescia)
Enzo MITIDIERI (Università di Udine)
Rainer NAGEL (Universität Tübingen)
Enrico OBRECHT (Università di Bologna)
Shinnosuke OHARU (Hiroshima University)
El-Maati OUHABAZ (Université de Franche Comté, Besançon)
Stefano PANIZZI (Università di Parma)
Sergiei PISKAREV (Ecologos, People's Academy of Human Values, Moscow)
Jan PRÜSS (Universität Paderborn)
Andrea PUGLIESE (Università di Trento)
René RAU (Universität Tübingen)
Gisèle Ruiz RIEDER (Louisiana State University, Baton Rouge)
Silvia ROMANELLI (Università di Bari)
Wolfgang RÜSS (Universität Essen)
Wilhelm SCHAPPACHER (Universität Graz)
Daniela SFORZA (Università di Pisa)
Yourii SILTCHENKO (Voronezh University)
Hiroki TANABE (Osaka University)
Horst THIEME (Arizona State University, Tempe)
Roberto TRIGGIANI (University of Virginia, Charlottesville)
Alberto VENNI (Università di Bologna)
Vincenzo VESPRI (Università di Pavia)
Michiaki WATANABE (Niigata University)
Glenn WEBB (Vanderbilt University, Nashville)
Lutz WEIS (Louisiana State University, Baton Rouge)
Atsushi YAGI (Himeji Institute of Technology)

Abstract Linear Nonautonomous Parabolic Equations: A Survey

PAOLO ACQUISTAPACE Dipartimento di Matematica, Università di Pisa, I-56127 Pisa, Italy

0. Introduction.

Let E be a Banach space. Consider the abstract Cauchy problem

$$\begin{cases} u'(t)-A(t)u(t)=f(t), & t\in]0,T], \quad (0.1) \\ u(0)=x \quad , & \quad (0.2) \end{cases}$$

where $T>0$, $x\in E$, $f\in C([0,T],E)$ and $\{A(t), t\in[0,T]\}$ is a family of generators of analytic semigroups in E; more precisely we assume for each $t\in[0,T]$:

$$\begin{cases} A(t):D_{A(t)}\subseteq E\longrightarrow E \text{ is a closed linear operator,} \\ \rho(A(t))\supseteq S(\vartheta_0):=\{0\}\cup\{z\in\mathbb{C}: |\arg z|<\vartheta_0\} \text{ with } \vartheta_0>\pi/2, \\ \|[\lambda-A(t)]^{-1}\|_{\mathcal{L}(E)}\le \frac{M}{1+|\lambda|} \quad \forall\lambda\in S(\vartheta_0). \end{cases} \quad (0.3)$$

The domains $D_{A(t)}$ may depend on t and be not dense in E (however they are necessarily dense if E is reflexive, see [15]). Further regularity assumptions on the map $t\longrightarrow A(t)$ will be introduced later; there is a lot of different hypotheses of this kind in the literature, generally independent of one another, which lead to existence and regularity results for strict and classical solutions of problem (0.1)-(0.2).

The aim of this paper is to describe the main known results, improve some of them and, above all, show how slight modifications of a unique approach naturally lead to consider such different kinds of assumptions.

1. Notations.

In the sequel we write Au instead of $A(\cdot)u(\cdot)$, and use the spaces

$$C([0,T],D_A):=\{v\in C([0,T],E):\ v(t)\in D_{A(t)}\ \forall t\in[0,T] \text{ and } Av\in C([0,T],E)\}$$

and $C(]0,T],D_A)$, whose definition is similar.

A strict solution of (0.1) is a function $u\in C^1([0,T],E)\cap C([0,T],D_A)$ which satisfies (0.1) in $[0,T]$; a classical solution is a function $u\in C([0,T],E)\cap C^1(]0,T],E)\cap C(]0,T],D_A)$ which satisfies (0.1) in $]0,T]$.

We will deal with the real interpolation spaces

$$D_{A(t)}(\vartheta,\infty):=\left(D_{A(t)},E\right)_{1-\vartheta,\infty}$$

whose use is crucial in regularity questions; such spaces are exactly characterized in several concrete cases (for $E=L^p(\Omega)$ see [14], for $E=C(\overline{\Omega})$ see [21,10,1]). We will also use the space

$$B(0,T;D_A(\vartheta,\infty)):=$$

$$\left\{v:[0,T]\longrightarrow E:\ v(t)\in D_{A(t)}(\vartheta,\infty)\ \forall t\in[0,T] \text{ and } \sup_{t\in[0,T]}\|v(t)\|_{D_{A(t)}(\vartheta,\infty)}<\infty\right\}.$$

Remark 1.1 The notation $L^\infty(0,T;D_A(\vartheta,\infty))$ has no meaning in general. But even in the case $D_{A(t)}(\vartheta,\infty)\equiv D_\vartheta$ (independent of t), an element of $B(0,T;D_\vartheta)$ needs not be Bochner measurable with values in D_ϑ, i.e. $L^\infty(0,T;D_\vartheta)$ is strictly contained in $B(0,T;D_\vartheta)$, as the following example shows.

Example 1.2 This example is due to J.Zabczyk (unpublished). Set $T:=1$, $E:=C([0,1])$, $A:=d^2/dx^2$ with domain $D_A:=\{u\in C^2([0,1]):u'(0)=u'(1)=0\}$; then it is known that $D_A(\alpha/2,\infty)=C^\alpha([0,1])$ for each $\alpha\in]0,1[$ (see [10]). We sketch the construction of a function $f\in C\big([0,1],C([0,1])\big)\cap B\big(0,1;C^\alpha([0,1])\big)$ which is not Bochner measurable as a $C^\alpha([0,1])$-valued function.

Firstly, we remark that there exists a sequence $\{\varphi_n\}\subseteq C([0,1])$ such that $0\le\varphi_n(t)\le1$ for each $t\in[0,1]$, and the map $\Phi:[0,1]\longrightarrow[0,1]^{\mathbb{N}^n}$, defined by $\Phi(t):=\{\varphi_n(t)\}_{n\in\mathbb{N}}$, is onto (Peano's map). Next, we fix a strictly increasing sequence $\{r_n\}\subseteq[0,1[$ such that $r_0=0$, $r_n\longrightarrow1$ as $n\longrightarrow\infty$, and $\sum_{n=0}^{\infty}\left(r_{n+1}-r_n\right)^\alpha<\infty$; now we define

$$\psi_n(x) := \begin{cases} c_n(x-r_n)^\alpha & \text{if } x\in[r_n,(r_n+r_{n+1})/2], \\ c_n(r_{n+1}-x)^\alpha & \text{if } x\in[(r_n+r_{n+1})/2,r_{n+1}], \end{cases}$$

where c_n is such that $[\psi_n]_{C^\alpha([0,1])} = 1$ for each $n\in\mathbb{N}$. Finally, we set

$$[f(t)](x) := \sum_{n=0}^{\infty} \varphi_n(t)\psi_n(x) \qquad (t,x\in[0,1]).$$

It is a straightforward task to show that this function belongs to $C\big([0,1],C([0,1])\big)\cap B\big(0,1;C^\alpha([0,1])\big)$. If f were also Bochner measurable with values in $C^\alpha([0,1])$, then it is well known that its range $f([0,1])$ would be a separable subset of $C^\alpha([0,1])$. But this is not the case: indeed, fix two distinct subsets $A,B\subseteq\mathbb{N}$ and let $e_A,e_B\in[0,1]^{\mathbb{N}}$ be such that

$$(e_A)_n = \begin{cases} 0 & \text{if } n\in\mathbb{N}-A \\ 1 & \text{if } n\in A \end{cases}, \qquad (e_B)_n = \begin{cases} 0 & \text{if } n\in\mathbb{N}-B \\ 1 & \text{if } n\in B \end{cases};$$

as Φ is onto, there exist $t_A,t_B\in[0,1]$ such that $\Phi(t_A)=e_A$, $\Phi(t_B)=e_B$. Hence it is easy to deduce that

$$\|f(t_A)-f(t_B)\|_{C^\alpha([0,1])} \geq 1,$$

and since $\{f(t_A), A\subseteq\mathbb{N}\}$ is uncountable, $f([0,1])$ cannot be separable. □

2. The autonomous case.

Assume $A(t)\equiv A$, with obvious modifications in (0.3). Then the following facts are well known (see [22, Prop.2.4 and Theorems 4.4, 4.5, 5.4, 5.5]):

Theorem 2.1 (i) *If a solution of* (0.1)-(0.2) *exists, then it has the form*

$$u(t) = e^{tA}x + \int_0^t e^{(t-s)A}f(s)ds, \qquad t\in[0,T], \tag{2.1}$$

where the semigroup $\{e^{sA}\}$ *is expressed by the Dunford integral*

$$e^{sA} = \frac{1}{2\pi i}\int_\gamma e^{s\lambda}[\lambda-A]^{-1}d\lambda \qquad \forall s>0, \tag{2.2}$$

γ *being a smooth path contained in* $S(\vartheta_0)$ *and joining* $+\infty e^{-i\vartheta}$ *to* $+\infty e^{i\vartheta}$ $(\pi/2<\vartheta<\vartheta_0)$.

(ii) *Assume either* $f\in C^\varepsilon([0,T],E)$ *or* $f\in C([0,T],E)\cap B(0,T;D_A(\varepsilon,\infty))$ $(0<\varepsilon<1)$;

then:

(a) *the solution is classical provided* $x\in\overline{D_A}$;

(b) *the solution is strict provided* $x\in D_A$ *and* $Ax+f(0)\in\overline{D_A}$;

(c) *the solution has the maximal regularity property, i.e.* u' *and* Au *belong to the same space as* f, *provided* $x\in D_A$ *and* $Ax+f(0)\in D_A(\varepsilon,\infty)$. □

We now wish to generalize Theorem 2.1 to the non-autonomous case, by a suitable perturbation argument.

3. The approach to the non-autonomous case.

We proceed formally: let u be any solution of the non-autonomous problem (0.1)-(0.2). For fixed $t\in]0,T]$ introduce the auxiliary function

$$v(s):=e^{(t-s)B(t,s)}u(s), \qquad s\in[0,t], \tag{3.1}$$

where B(t,s) is some operator to be chosen later, defined for $0\le s\le t$ and satisfying (0.3). Differentiating (3.1) with respect to s we get (using (0.3)):

$$v'(s)=$$

$$=\left[-B(t,s)e^{(t-s)B(t,s)}+\left[\frac{\partial}{\partial s}e^{rB(t,s)}\right]_{r=t-s}\right]u(s) + e^{(t-s)B(t,s)}\left[A(s)u(s)+f(s)\right];$$

taking into account (0.2), an integration from 0 to t yields

$$\begin{aligned} u(t)-e^{tB(t,0)}x &= \int_0^t e^{(t-s)B(t,s)}f(s)ds + \\ &+ \int_0^t \left[e^{(t-s)B(t,s)}\left[A(s)-B(t,s)\right]+\left[\frac{\partial}{\partial s}e^{rB(t,s)}\right]_{r=t-s}\right]u(s)ds. \end{aligned} \tag{3.2}$$

This is a Volterra integral equation for the solution u, with kernel

$$K(t,s):=e^{(t-s)B(t,s)}\left[A(s)-B(t,s)\right]+\left[\frac{\partial}{\partial s}e^{rB(t,s)}\right]_{r=t-s} \tag{3.3}$$

belonging to $\mathcal{L}(E)$. Thus our strategy will be the following:

<u>Step</u> 1 find some assumption on $t\longrightarrow A(t)$, such that (3.2) is solvable for suitable data x,f;

<u>Step</u> 2 show that the solution of (3.2) is in fact the solution of problem (0.1)-(0.2).

As a consequence of Step 1 we will get a representation formula, and hence uniqueness, for the solution of (0.1)-(0.2); from Step 2 we will deduce existence of (strict and classical) solutions and their maximal regularity. According to different possible choices of $B(t,s)$ we will need different kinds of assumptions on the map $t \longrightarrow A(t)$, corresponding to the various, independent ones appeared in the literature.

We remark that using (3.3) and (2.2) we can rewrite the integral term of (3.2) in the following way:

$$\int_0^t K(t,s)u(s)ds = \tag{3.4}$$
$$= \frac{1}{2\pi i}\int_0^t\int_\gamma e^{(t-s)\lambda}[\lambda-B(t,s)]^{-1}\Big[A(s)-B(t,s)+\frac{\partial}{\partial s}B(t,s)[\lambda-B(t,s)]^{-1}\Big]u(s)d\lambda ds$$

or, after an integration by parts,

$$\int_0^t K(t,s)u(s)ds = \frac{1}{2\pi i}\int_0^t\int_\gamma e^{(t-s)\lambda}\Big[[\lambda-B(t,s)]^{-1}\cdot \tag{3.5}$$
$$\cdot\Big[A(s)-B(t,s)+\Big[\frac{\partial}{\partial s}B(t,s),[\lambda-B(t,s)]^{-1}\Big]+(t-s)\frac{\partial}{\partial s}B(t,s)\Big]u(s)d\lambda ds,$$

where $[U,V]$ stands for the commutator $UV-VU$.

We now have to choose $B(t,s)$. Reasonable choices are the following:

I. $B(t,s):=A(s)$, II. $B(t,s):=A(t)$,

III. $B(t,s):=A(0)$, IV. $B(t,s):=\frac{1}{t-s}\int_s^t A(\sigma)d\sigma$.

In the next section we will perform Step 1 with each one of the above choices of $B(t,s)$.

4. Solving the integral equation (3.2).

4.1. The first choice, $B(t,s)=A(s)$.

By (3.3)

$$\int_0^t K(t,s)u(s)ds = \frac{1}{2\pi i}\int_0^t\int_\gamma e^{(t-s)\lambda}\frac{d}{ds}[\lambda-A(s)]^{-1}u(s)d\lambda ds.$$

Hence in order to solve (3.2) one is lead to assume, for some $\alpha\in]0,1[$ and $N,\eta>0$,

$$\left\|\frac{d}{ds}[\lambda-A(s)]^{-1}\right\|_{\mathcal{L}(E)} \le \frac{N}{1+|\lambda|^{\alpha}} \quad \forall s\in[0,T], \tag{4.1}$$

$$\left\|\frac{d}{dt}A(t)^{-1}-\frac{d}{ds}A(s)^{-1}\right\|_{\mathcal{L}(E)} \le N|t-s|^{\eta} \quad \forall t,s\in[0,T]. \tag{4.2}$$

Conditions (4.1)-(4.2) were introduced in [18] and revisited in [3,11]; they - in fact (4.1) alone: (4.2) is needed only in Step 2 - allow to solve the integral equation (3.2) in the space $C([0,T],E)$, provided $x\in\overline{D}_A$ and $f\in C([0,T],E)$: see [3, Proposition 1.9].

4.2. The second choice, $B(t,s)=A(t)$.

The integral term (3.4) becomes

$$\int_0^t K(t,s)u(s)ds = \int_0^t A(t)e^{(t-s)A(t)}\left[A(t)^{-1}-A(s)^{-1}\right]A(s)u(s)ds.$$

Applying $A(t)$ to both sides of (3.2) we get

$$\begin{aligned} A(t)u(t)-A(t)e^{tA(t)}x &= A(t)\int_0^t e^{(t-s)A(t)}f(s)ds + \\ &+\int_0^t A(t)^2e^{(t-s)A(t)}\left[A(t)^{-1}-A(s)^{-1}\right]A(s)u(s)ds, \end{aligned} \tag{4.3}$$

and this is an integral equation in the unknown Au.

If we assume

$$\begin{cases} \|A(t)[\lambda-A(t)]^{-1}\left[A(t)^{-1}-A(s)^{-1}\right]\|_{\mathcal{L}(E)} \le N\dfrac{|t-s|^{\alpha}}{1+|\lambda|^{\rho}} \ \forall t,s\in[0,T],\ \forall\lambda\in S(\vartheta_0) \\ \delta:=\alpha+\rho-1>0, \end{cases} \tag{4.4}$$

then it is easy to check that

$$\|A(t)^2e^{(t-s)A(t)}\left[A(t)^{-1}-A(s)^{-1}\right]\|_{\mathcal{L}(E)} \le \frac{C}{|t-s|^{1-\delta}} \quad \forall t,s\in[0,T], \tag{4.5}$$

so that (4.3) can be solved in $C([0,T],E)$ for suitable data x,f: see [9, Proposition 3.1]. Of course, Step 2 will consist in showing that $A^{-1}v$, where v is the solution of (4.3), is in fact a strict solution of (0.1) - (0.2).

Condition (4.4), in a somewhat weaker form, was introduced in [9] (see also [8,2]) and used in [30,31].

4.3. The third choice, $B(t,s)=A(0)$.

The integral equation (3.2) just reduces to the Korn device applied to the usual variation of parameters formula (compare with (2.1)):

$$u(t) = e^{tA(0)}x + \int_0^t e^{(t-s)A(0)}\big[f(s)+[A(s)-A(0)]u(s)\big]ds. \tag{4.6}$$

Assume:

$$\begin{cases} D_{A(t)}\equiv D\ ; \quad \exists\vartheta\in]0,1[:\ D_{A(t)}(\sigma,\infty)\equiv D_\sigma\ \ \forall\sigma\in]0,\vartheta+1]-\{1\} \\ A\in C([0,T],\mathcal{L}(D,E))\cap C([0,T],\mathcal{L}(D_{\vartheta+1},D_\vartheta)); \end{cases} \tag{4.7}$$

this allows to solve (4.6), by a fixed point argument, in the space

$$X:= C_{1-\vartheta}(]0,T],D)\cap B_1(]0,T],D_{\vartheta+1})$$

consisting of functions $u\in C(]0,T],D)\cap B(0,T,D_{\vartheta+1})$ such that

$$\|u\|_X:= \sup_{t>0}\ t^{1-\vartheta}\|u(t)\|_D + \sup_{t>0}\ t\|u(t)\|_{D_{\vartheta+1}} < \infty.$$

This is proved in [12], generalizing previous results of [13].

4.4. The fourth choice, $B(t,s)=\frac{1}{t-s}\int_s^t A(\sigma)d\sigma$.

Firstly we remark that

$$(t-s)\frac{\partial}{\partial s}B(t,s)=B(t,s)-A(s),$$

so that the integral term (3.5) becomes

$$\int_0^t K(t,s)u(s)ds = \tag{4.8}$$
$$= \frac{1}{2\pi i}\int_0^t\int_\gamma e^{(t-s)\lambda}[\lambda-B(t,s)]^{-1}\frac{1}{t-s}\big[[\lambda-B(t,s)]^{-1},A(s)\big]u(s)d\lambda ds.$$

We are then led to the following assumption:

$$\begin{cases} D_{A(t)} \equiv D, \quad D \text{ dense in } E, \\ A \in C([0,T],\mathcal{L}(D,E)), \\ \|[\lambda-B(t,s)]^{-1}\|_{\mathcal{L}(E)} \le \dfrac{N}{1+|\lambda|}, \qquad \|[\lambda-B(t,s)]^{-1}\|_{\mathcal{L}(E,D)} \le N, \\ \|[[\lambda-B(t,s)]^{-1},A(s)]\|_{\mathcal{L}(D,E)} \le \dfrac{N}{1+|\lambda|^{\beta}} \quad (0<\beta<1). \end{cases} \tag{4.9}$$

Furthermore, in Step 2 we will also need:

$$\begin{cases} \|[\lambda-B(t,s)]^{-1}\|_{\mathcal{L}(D_A(\eta,\infty),D)} \le \dfrac{N}{1+|\lambda|^{\eta}}, \\ \|[[\lambda-B(t,s)]^{-1},A(s)]\|_{\mathcal{L}(D,D_A(\eta,\infty))} \le \dfrac{N}{1+|\lambda|^{1+\beta-\eta}} \end{cases} \quad \forall\eta\in]0,\beta]. \tag{4.10}$$

This set of assumptions was introduced by Sobolevskii [25]. By (4.8) and (4.9) it is easy to deduce that

$$\|K(t,s)\|_{\mathcal{L}(D)} \le \frac{C}{|t-s|^{1-\beta}},$$

so that (3.2) can be solved in the space $C([0,T],D)$.

5. Existence of solutions of (0.1)-(0.2).

For suitable data x,f we now have a solution v of the integral equation (3.2) (cases I,III,IV) or (4.3) (case II), which can be written as the sum of a uniformly convergent Neumann series; the goal of Step 2 is to show that v coincides with u (cases I,III,IV) or Au (case II), u being the strict solution of (0.1)-(0.2). In fact, this goal is achieved in a somewhat indirect method, since some different representation formulas are used.

Theorem 5.1 (case I) *Assume (0.3) and (4.1)-(4.2). Suppose that* $x\in\overline{D_{A(0)}}$ *and either* $f\in C^{\varepsilon}([0,T],E)$ *or* $f\in C([0,T],E)\cap B(0,T;D_A(\varepsilon,\infty))$*; then the solution* v *of (3.2) is the unique classical solution of problem (0.1)-(0.2). If in addition* $x\in D_{A(0)}$ *and* $A(0)x+f(0)-[dA(t)^{-1}/dt]_{t=0}A(0)x \in \overline{D_{A(0)}}$*, then* v *is strict.*

Proof In [3] it is shown that any strict solution of (0.1)-(0.2) can be represented as

$$u(t) = e^{tA(t)}x + \int_0^t e^{(t-s)A(t)}[(1+P)^{-1}(f-P(\cdot,0)x)](s)ds, \tag{5.1}$$

where P is the integral operator

$$(Pg)(t):=\int_0^t P(t,s)g(s)ds,$$

whose kernel P(t,s) is

$$P(t,s):=\frac{1}{2\pi i}\int_\gamma e^{(t-s)\lambda}\frac{d}{dt}[\lambda-A(t)]^{-1}d\lambda.$$

Then if $f\in C^\varepsilon([0,T],E)$ all statements follow by direct computation (see [3, Theorems 4.1-5.1]. On the other hand if $f\in C([0,T],E)\cap B(0,T;D_A(\varepsilon,\infty))$ we can rewrite (5.1) as

$$u(t) = e^{tA(t)}x + \int_0^t e^{(t-s)A(t)}\left[(1+P)^{-1}(-Pf-P(\cdot,0)x)\right](s)ds + \\ + \int_0^t e^{(t-s)A(t)}f(s)ds; \tag{5.2}$$

now the first two terms represent the solution of problem (0.1)-(0.2) with right member $-Pf\in C^\varepsilon([0,T],E)$ (by [3, Proposition 3.5(v)]) and initial datum x, so that the previous argument applies. The third term is in $C^\varepsilon([0,T],D_A)$ $\cap\ C^{1+\varepsilon}([0,T],E)$ by direct computation, as it is easily seen using [11,Lemma 2.5(ii)]. This implies the result. □

__Theorem__ 5.2 (case II) *Assume* (0.3) *and* (4.4). *Suppose that* $x\in\overline{D_{A(0)}}$ *and either* $f\in C^\varepsilon([0,T],E)$ *or* $f\in C([0,T],E)\cap B(0,T;D_A(\varepsilon,\infty))$; *then the function* $A^{-1}v$, *where* v *is the solution of* (4.3), *is the unique classical solution of* (0.1)-(0.2). *If in addition* $x\in D_{A(0)}$ *and* $A(0)x+f(0)\in\overline{D_{A(0)}}$, *then* $A^{-1}v$ *is strict.*

__Proof__ (See [9].) Consider the Yosida approximations $A_n(t)$ of A(t), i.e.

$$A_n(t)=nA(t)[n-A(t)]^{-1},\qquad n\in\mathbb{N}^+. \tag{5.3}$$

Let u_n be the solution of the Cauchy problem

$$\begin{cases} u'_n(t)-A_n(t)u_n(t)=f(t), & t\in[0,T], \\ u_n(0)=x \end{cases} \tag{5.4}$$

(obviously existing, since $A_n\in C([0,T],\mathcal{L}(E))$!). Then A_nu_n is the solution of an integral equation whose kernel is (compare with (4.5))

$$A_n(t)^2e^{(t-s)A_n(t)}\left[A_n(t)^{-1}-A_n(s)^{-1}\right],$$

and converges in $\mathcal{L}(E)$, as $n\to\infty$, to the kernel of (4.3). Then one sees that u_n converges to some u in $C([0,T],E)$ and $A_n u_n \to v$ in $C([0,T],E)$; hence $v\equiv Au$ and $u_n' \equiv A_n u_n + f \to Au+f$ in $C([0,T],E)$, so that $u' \equiv (A^{-1}v)'$ exists and is equal to $Au+f$, i.e. $A^{-1}v$ solves (0.1)-(0.2). □

Theorem 5.3 (case III) *Assume* (0.3) *and* (4.7). *Suppose that* $x\in D_\varepsilon$ *and* $f\in C([0,T],E)\cap B(0,T;D_\varepsilon))$; *then the solution* v *of* (3.2) *(i.e. of* (4.6)*) is the unique classical solution of problem* (0.1)-(0.2). *If in addition* $x\in D$ *and* $A(0)x\in\bar{D}$, *then* v *is strict.*

Proof The first assertion follows by direct computation (see [12, Theorem 2.2]). Let us prove the second one: fix $x\in D$ such that $A(0)x\in\bar{D}$, and select a sequence $\{x_n\}\subseteq D_{\varepsilon+1}$ such that $x_n \to x$ in D. For each n let v_n be the strict solution of (0.1)-(0.2) with data x_n, f; then, by an argument used in [5], we can write a different representation formula for v_n:

$$v_n(t) = A(t)^{-1}\Big[(1-H)^{-1}\big(L(\cdot,0)x_n + Lf\big)\Big](t)$$

where

$$L(t,s):=A(t)e^{(t-s)A(t)}, \qquad \big(Lf\big)(t):=\int_0^t L(t,s)f(s)ds,$$

$$\big(Hg\big)(t):=\int_0^t A(t)e^{(t-s)A(t)}(1-A(t)A(s)^{-1})g(s).$$

Now it can be seen that $L(\cdot,0)\in\mathcal{L}(D,C([0,T],E))$ and $(1-H)^{-1}\in\mathcal{L}(C([0,T],E))$, so that $v_n \to u$ in $C([0,T],D)$, where

$$u(t) = A(t)^{-1}\Big[(1-H)^{-1}\big(L(\cdot,0)x + Lf\big)\Big](t). \tag{5.5}$$

Moreover $v_n' = Av_n + f \to Au+f$ in $C([0,T],E)$, which implies that u is a strict solution of (0.1)-(0.2). □

Theorem 5.4 (case IV) *Assume* (0.3) *and* (4.9)-(4.10). *Suppose that* $x\in E$ *and* $f\in C([0,T],E)\cap B(0,T;D_\varepsilon))$; *then the solution* v *of* (3.2) *is the unique classical solution of problem* (0.1)-(0.2). *If in addition* $x\in D$, *then* v *is strict.*

Proof It is similar to the case II: let u_n be the solution of problem (5.4), with $A_n(t)$ given by (5.3). If $x\in D$, then u_n is the solution, in the space $C([0,T],D)$, of an integral equation whose kernel is (compare with (4.8)):

$$\frac{1}{2\pi i}\int_{\gamma} e^{(t-s)\lambda}[\lambda-B_n(t,s)]^{-1}\frac{1}{t-s}\Big[[\lambda-B_n(t,s)]^{-1},A_n(s)\Big]d\lambda,$$

where

$$B_n(t,s):=\frac{1}{t-s}\int_s^t A_n(\sigma)d\sigma;$$

hence the above kernel converges as $n\longrightarrow\infty$ to the kernel of (4.8). As a consequence we get $u_n\longrightarrow v$ in $C([0,T],D)$ and, in particular, $A_nu_n\longrightarrow Av$ in $C([0,T],E)$. But then $u'_n\equiv A_nu_n+f\longrightarrow Av+f$, so that $v\in C^1([0,T],E)$ and $v'\equiv Av+f$. If $x\in E$ only, the above argument works, with some more technicalities, in the space $X:=C_{1-\eta}(]0,T],D_\eta)\cap C_1(]0,T],D)$. □

Remark 5.5 In cases III and IV we are not able to find classical or strict solutions of problem (0.1)-(0.2) with $f\in C^{\varepsilon}([0,T],E)$, because the $A(t)$'s are just continuous in t.

6. Maximal regularity.

There are two kinds of maximal regularity, namely with respect to time and with respect to space: that is, there are certain subspaces M of $C([0,T],E)$, consisting either of E-valued Hölder continuous functions, or of functions which are bounded with values in some interpolation space, which have the following property: if the right member of (0.1)-(0.2) belongs to M, then the solution u is such that both u' and Au belong to M, provided suitable compatibility conditions hold. Such conditions turn out to be necessary and sufficient for maximal regularity.

Theorem 6.1 (case I) *Assume* (0.3) *and* (4.1)-(4.2), *and fix* $\varepsilon\in]0,\alpha\wedge\eta]$.

(i) *Suppose* $x\in D_{A(0)}$, $f\in C^{\varepsilon}([0,T],E)$ *and* $A(0)x+f(0)-\big[dA(t)^{-1}/dt\big]_{t=0}A(0)x \in \overline{D_{A(0)}}$; *let* u *be the strict solution of problem* (0.1)-(0.2). *Then* u', u', $Au \in C^{\varepsilon}([0,T],E)$ *if and only if* $A(0)x+f(0)-\big[dA(t)^{-1}/dt\big]_{t=0}A(0)x \in D_{A(0)}(\varepsilon,\infty)$; *in this case, one has also* $u'-\big[dA(\cdot)^{-1}/dt\big]Au \in B(0,T;D_A(\varepsilon,\infty))$.

(ii) *Suppose that* $x\in D_{A(0)}$, $f\in C([0,T],E)\cap B(0,T;D_A(\varepsilon,\infty))$ *and* $A(0)x-\big[dA(t)^{-1}/dt\big]_{t=0}A(0)x\in\overline{D_{A(0)}}$; *let* u *be the strict solution of problem* (0.1)-(0.2). *Then* u', $Au\in B(0,T;D_A(\varepsilon,\infty))$ *if and only if* $A(0)x-\big[dA(t)^{-1}/dt\big]_{t=0}A(0)x \in D_{A(0)}(\varepsilon,\infty)$; *in this case, one has also* $Au-\big[dA(\cdot)^{-1}/dt\big]Au \in C^{\varepsilon}([0,T],E)$.

Proof The first assertion of (i) is proved in [3, Theorem 5.3] and [6,

Appendix]; let us prove the second one. To this purpose we need to rewrite the integral equation (3.2) with the choice II instead of I, i.e. taking $B(t,s)=A(t)$: then we get

$$u(t) = \int_0^t A(t)e^{(t-s)A(t)}\left[A(t)^{-1}-A(s)^{-1}\right]A(s)u(s)ds + \\ + e^{tA(t)}x + \int_0^t e^{(t-s)A(t)}f(s)ds. \tag{6.1}$$

Next, we split it conveniently:

$$\begin{aligned} u(t) = &\int_0^t A(t)e^{(t-s)A(t)}\left[A(t)^{-1}-A(s)^{-1}-(t-s)\tfrac{d}{dt}A(t)^{-1}\right]A(s)u(s)ds + \\ &+ \int_0^t A(t)e^{(t-s)A(t)}(t-s)\tfrac{d}{dt}A(t)^{-1}\left[A(s)u(s)-A(t)u(t)\right]ds + \\ &+ \left[te^{tA(t)}-A(t)^{-1}\left[e^{tA(t)}-1\right]\right]\tfrac{d}{dt}A(t)^{-1}\cdot A(t)u(t) + e^{tA(t)}x + \\ &+ \int_0^t e^{(t-s)A(t)}\left[f(s)-f(t)\right]ds + A(t)^{-1}\left[e^{tA(t)}-1\right]f(t). \end{aligned}$$

It is clear that each term belongs to $D_{A(t)}$, so that after a further splitting we easily have:

$$\begin{aligned} u'(t) - \tfrac{d}{dt}A(t)^{-1}\cdot A(t)u(t) &= A(t)u(t) + f(t) - \tfrac{d}{dt}A(t)^{-1}\cdot A(t)u(t) = \\ = &\int_0^t A(t)e^{(t-s)A(t)}\left[A(t)^{-1}-A(s)^{-1}-(t-s)\tfrac{d}{dt}A(t)^{-1}\right]A(s)u(s)ds + \\ &+ \int_0^t A(t)^2e^{(t-s)A(t)}(t-s)\tfrac{d}{dt}A(t)^{-1}\left[A(s)u(s)-A(t)u(t)\right]ds + \\ &+ A(t)e^{tA(t)}\left[t\tfrac{d}{dt}A(t)^{-1}\left[A(t)u(t)-A(0)x\right] + \left[t\tfrac{d}{dt}A(t)^{-1}+A(0)^{-1}-A(t)^{-1}\right]\right]A(0)x + \\ &+ e^{tA(t)}\left[\left[f(t)-f(0)\right] - \tfrac{d}{dt}A(t)^{-1}\left[A(t)u(t)-A(0)x\right] - \right. \\ &\left. - \left[\tfrac{d}{dt}A(t)^{-1}-\left[\tfrac{d}{dt}A(t)^{-1}\right]_{t=0}\right]A(0)x\right] + e^{tA(t)}\left[A(0)x+f(0)-\left[\tfrac{d}{dt}A(t)^{-1}\right]_{t=0}A(0)x\right] - \\ &+ \int_0^t A(t)e^{(t-s)A(t)}f(s)ds. \end{aligned}$$

Now it is a straightforward task to verify that each term in the right-hand side is in $D_{A(t)}(\varepsilon,\infty)$, with bounded norms (using, for the last two, [11, Lemma 2.5(i)-(ii)]). This proves (i).

Let us prove (ii). We start from the last assertion. Using (5.3) we see that

$$A(t)u(t) = A(t)w(t) + A(t)\int_0^t e^{(t-s)A(t)}f(s)ds,$$

where w is the strict solution of problem (0.1)-(0.2) with data -Pf, x. Hence Aw is in $C^{\varepsilon}([0,T],E)$ by (i), whereas the second term is in the same space as remarked in the proof of Theorem 5.1. This proves the last assertion of (ii). Finally, concerning the first one, starting from (6.1) we can repeat the argument used in the proof of the last part of (i), and the result follows. □

Theorem 6.2 (case II) *Assume* (0.3) *and* (4.4), *and fix* $\varepsilon\in]0,\delta]$.

(i) *Suppose* $x\in D_{A(0)}$, $f\in C^{\varepsilon}([0,T],E)$ *and* $A(0)x+f(0) \in \overline{D_{A(0)}}$; *let* u *be the strict solution of problem* (0.1)-(0.2). *Then* u', Au $\in C^{\varepsilon}([0,T],E)$ *if and only if* $A(0)x+f(0) \in D_{A(0)}(\varepsilon,\infty)$; *in this case, one has also* u'$\in B(0,T;D_A(\varepsilon,\infty))$.

(ii) *Suppose that* $x\in D_{A(0)}$, $f \in C([0,T],E) \cap B(0,T;D_A(\varepsilon,\infty))$ *and* $A(0)x \in \overline{D_{A(0)}}$; *let* u *be the strict solution of problem* (0.1)-(0.2). *Then* u', Au $\in B(0,T;D_A(\varepsilon,\infty))$ *if and only if* $A(0)x \in D_{A(0)}(\varepsilon,\infty)$; *in this case, one has also* Au $\in C^{\varepsilon}([0,T],E)$.

Proof See [9, Theorems 6.1(iii)-6.2(iii)]. □

Theorem 6.3 (case III) *Assume* (0.3) *and* (4.7), *and fix* $\varepsilon\in]0,\vartheta]$. *Suppose* $x\in D$, $A(0)x\in\overline{D}$ *and* $f\in C([0,T],E)\cap B(0,T;D_{\varepsilon}))$, *and let* u *be the strict solution of problem* (0.1)-(0.2). *Then* u', Au $\in B(0,T;D_{\varepsilon})$ *if and only if* $A(0)x\in D_{\varepsilon}$.

Proof By direct computation, starting from (5.5) (the original proof is different: see [12, Corollary 2.3]). □

Theorem 6.4 (case IV) *Assume* (0.3) *and* (4.9)-(4.10), *and fix* $\varepsilon\in]0,\beta[$. *Suppose* $x\in D$ *and* $f\in C([0,T],E)\cap B(0,T;D_{\varepsilon}))$, *and let* u *be the strict solution of* (0.1)-(0.2). *Then* u', Au $\in B(0,T;D_{\varepsilon})$ *if and only if* $A(0)x\in D_{\varepsilon}$.

Proof The solution u is in C([0,T],D) and solves the integral equation

$$u(t) - \int_0^t K(t,s)u(s)ds = e^{tB(t,0)}x + \int_0^t e^{(t-s)B(t,s)}f(s)ds, \tag{6.2}$$

with K(t,s) given by (4.8). Now it is easily seen that, since $u\in C([0,T],D)$,

$$t\longrightarrow\int_0^t K(t,s)u(s)ds \in B(0,T;D_{1+\eta}), \quad \forall\eta\in]0,\beta[,$$

whereas it is not difficult to verify that the right member of (6.2) belongs to $B(0,T;D_{1+\varepsilon})$ if and only if $A(0)x\in D_{\varepsilon}$. Thus the result follows at once. □

Remark 6.5 In cases III and IV one cannot expect maximal regularity in time, i.e. that $f\in C^{\varepsilon}([0,T],E)$ implies $u', Au\in C^{\varepsilon}([0,T],E)$: this is false even in the scalar case $E=\mathbb{R}$.

7. The evolution operator.

In all cases I-...-IV one can construct the evolution operator associated to problem (0.1), i.e. an operator $U(t,s)\in\mathcal{L}(E)$ defined for $0\le s\le t\le T$, such that

$$U(t,s) = U(t,\tau)U(\tau,s) \quad \forall\tau\in[s,t], \qquad U(t,t) = I \quad \forall t\in[0,T], \tag{7.1}$$

$$\frac{d}{dt} U(t,s) = A(t)U(t,s) \quad \forall t\in]s,T], \tag{7.2}$$

$$\frac{d}{ds} U(t,s) = - U(t,s)A(s) \quad \forall s\in[0,t[. \tag{7.3}$$

More precisely:

Theorem 7.1 (case I) *Assume* (0.3) *and* (4.1)-(4.2). *Then there exists a unique operator* $U(t,s)$ *satisfying* (7.1)-(7.2); *moreover* $dU(t,s)/ds$ *exists in* $\mathcal{L}(E)$ *for each* $s\in[0,t[$ *and satisfies* (7.3) *pointwise for each* $x\in D_{A(s)}$.

Proof The result is classical [18] in the case of dense domains; for the general case see [11]. □

Theorem 7.2 (case II) *Assume* (0.3) *and* (4.4). *Then there exists a unique operator* $U(t,s)$ *satisfying* (7.1)-(7.2); *if in addition the domains* $D_{A(t)}$ *are dense in* E *and the operators* $\{A(t)^*\}$ *satisfy* (0.3) *and* (4.4) *in the space* E^*, *then* $dU(t,s)/ds$ *exists in* $\mathcal{L}(E)$ *for each* $s\in[0,t[$ *and satisfies* (7.3) *pointwise for each* $x\in D_{A(s)}$.

Proof See [2] and [11]. □

Theorem 7.3 (case III) *Assume* (0.3) *and* (4.7). *Then there exists a unique operator* $U(t,s)$ *satisfying* (7.1)-(7.2); *moreover, setting* $X:=\{x\in D: A(0)x\in\overline{D}\}$, $dU(t,s)/ds$ *exists in* $\mathcal{L}(X,E)$ *for each* $s\in[0,t[$ *and satisfies* (7.3).

Proof First of all, we remark that if $x\in X$, then $A(s)x\in\overline{D}$ for each $s\in[0,T]$. Now, the first assertion is essentially proved in [12]. Let us show the second one. Fix $x\in D_{1+\varepsilon}$: by (5.5) we can write

$$U(t,s)x = A(t)^{-1}\Big[(1-H)^{-1}\big(L(\cdot,0)x\big)\Big](t) =$$

$$= e^{(t-s)A(t)}x + \sum_{n=1}^{\infty} A(t)^{-1}\left[H^n\left(L(\cdot,0)x\right)\right](t) =$$

$$= e^{(t-s)A(t)}x + \sum_{n=1}^{\infty} A(t)^{-1}\int_s^t H_n(t,\sigma)A(\sigma)e^{(\sigma-s)A(\sigma)}x d\sigma ,$$

where $H_n(t,\sigma)$ is the iterated kernel of the integral operator H^n, i.e.

$$H_1(t,\sigma):=H(t,\sigma); \qquad H_{n+1}(t,\sigma):=\int_\sigma^t H_n(t,r)H(r,\sigma)dr \quad \forall n\in\mathbb{N}.$$

Now differentiating with respect to s we get

$$\frac{d}{ds}U(t,s)x = - A(t)e^{(t-s)A(t)}x - \sum_{n=1}^{\infty} A(t)^{-1}H_n(t,s)A(s)x -$$

$$- \sum_{n=1}^{\infty} A(t)^{-1}\int_s^t H_n(t,\sigma)A(\sigma)^2e^{(\sigma-s)A(\sigma)}x d\sigma =$$

$$= e^{(t-s)A(t)}\left[1-A(t)A(s)^{-1}\right]A(s)x - e^{(t-s)A(t)}A(s)x - \sum_{n=1}^{\infty} A(t)^{-1}H_n(t,s)A(s)x -$$

$$+ \sum_{n=1}^{\infty} A(t)^{-1}\int_s^t H_n(t,\sigma)A(\sigma)e^{(\sigma-s)A(\sigma)}\left[1-A(\sigma)A(s)^{-1}\right]A(s)x d\sigma -$$

$$- \sum_{n=1}^{\infty} A(t)^{-1}\left[H^n\left(L(\cdot,0)A(s)x\right)\right](t) =$$

$$= A(t)^{-1}H_1(t,s)A(s)x - e^{(t-s)A(t)}A(s)x - \sum_{n=1}^{\infty} A(t)^{-1}H_n(t,s)A(s)x +$$

$$+ \sum_{n=1}^{\infty} A(t)^{-1}H_{n+1}(t,s)A(s)x - \sum_{n=1}^{\infty} A(t)^{-1}\left[H^n\left(L(\cdot,0)A(s)x\right)\right](t) =$$

$$= - U(t,s)A(s)x .$$

Thus (7.3) is established when $x\in D_{1+\varepsilon}$. Finally we extend (7.3) to the whole X by approaching any $x\in X$ by a suitable sequence $\{x_n\}\subseteq D_{1+\varepsilon}$. □

Theorem 7.4 (case IV) *Assume* (0.3) *and* (4.9)-(4.10). *Then there exists a unique operator* U(t,s) *satisfying* (7.1)-(7.2); *moreover* dU(t,s)/ds *exists in* $\mathcal{L}(D,E)$ *for each* $s\in[0,t[$ *and satisfies* (7.3).

Proof See [25]. □

Remark 7.5 In all cases the operators U(t,s) fulfill further regularity properties with respect to t and s. □

8. Examples and remarks.

Consider a general linear, non-autonomous parabolic initial-boundary value problem. The various assumptions of cases I-...-IV correspond to different concrete situations: case II means moderate regularity in t, with a strong parabolic structure, whereas case I allows a less stringent parabolicity (the boundary operators may reduce their order for some t) provided there is a very good dependence on t; on the other hand cases III and IV concern very restrictive boundary conditions (independent of t), requiring however just continuity with respect to t. Thus it is not surprising that the assumptions of the four cases are independent of one another. To verify this independence it is sufficient to take a one space dimension example. Consider a second order operator

$$A(t,x,D)u := u'' + a(t,x)u' + b(t,x)u, \quad t\in[0,T],\ x\in[0,1],$$

equipped with endpoint conditions

$$u(0) = 0, \quad u(1) + c(t)u'(1) = 0;$$

here $a,b\colon [0,T]\times[0,1]\to\mathbb{R}$ and $c\colon [0,T]\to\mathbb{R}$ are continuous functions. Next, set

$$\begin{cases} D_{A(t)} := \{u\in E:\ A(t,\cdot,D)u\in E,\ u(0)=0,\ u(1)+c(t)u'(1)=0\} \\ A(t)u := A(t,\cdot,D)u \end{cases}$$

with $E:=C([0,1])$.

Assume now $a\equiv b\equiv 0$, $c(t)=1+t^{2/3}$; then it is easy to see that assumptions of case II hold and assumptions of case I do not, and conversely, if we take again $a\equiv b\equiv 0$ and $c(t)=t$, then assumptions of case I hold and those of case II do not (see [9, §7] . Moreover in both situations above the assumptions of cases III and IV are not fulfilled (since $D_{A(t)}$ is not constant).

On the other hand, take $a\equiv 0$, $b(t,x)=\omega(t)\gamma(x)$ and $c\equiv 0$, with $\gamma\in C^{\varepsilon}([0,1])$ $(0<\varepsilon<1)$ and $\omega\in C([0,T])$ (but not Hölder continuous for any $\alpha\in]0,1[$): then an easy inspection shows that assumptions of case III hold but those of case IV do not (since the constant domain is not dense), and assumptions of cases I-II do not too (since ω is not Hölder continuous).

Finally it does not seem easy to construct an example where assumptions of case IV hold but assumptions of case III do not: in fact the hypotheses of case IV, and in particular the existence in $\mathcal{L}(D,E)$ of the commutator $\left[[\lambda-B(t,s)]^{-1},A(s)\right]$, seem to imply that $D_{A(t)}(\vartheta+1,\infty)$ is independent of t in all "reasonable" examples.

The assumptions of cases I-II-III-IV cover most part of the available literature. Case II generalizes the papers [23,4,5,24,16,17,7]; case I contains [18,3,26,27,28], case III corresponds to [12,13] and finally case IV is introduced in [25]. It is to be noted that in the paper [29] there is an assumption which is related to and weaker than that of case I: it implies Theorem 7.1 and existence of classical solutions if $f \in C^{\varepsilon}([0,T],E)$, but no maximal regularity results seem to hold in this case; in fact we are not able even to include it in our unitary approach.

We also have to mention the results of [19,20] where assumptions and methods of case II are generalized to the more abstract setting of the sum of two closed linear operators, with applications to elliptic as well as parabolic problems.

References

[1] - P.Acquistapace, Zygmund classes with boundary conditions as interpolation spaces, in "Semigroup theory and applications", proceedings Trieste 1987, Ph.Clément, S.Invernizzi, E.Mitidieri & I.I.Vrabie editors, Lect.Notes in Pure Appl.Math.n°116, M.Dekker, New York 1989, 1-19.

[2] - P.Acquistapace, Evolution operators and strong solutions of abstract linear parabolic equations, Diff.Int.Eq.**1** (1988) 433-457.

[3] - P.Acquistapace, B.Terreni, Some existence and regularity results for abstract non-autonomous parabolic equations, J.Math.Anal.Appl. **99** (1984) 9-64.

[4] - P.Acquistapace, B.Terreni, On the abstract non-autonomous Cauchy problem in the case of constant domains, Ann.Mat.Pura Appl.(4) **140** (1985) 1-55.

[5] - P.Acquistapace, B.Terreni, Maximal space regularity for abstract non-autonomous parabolic equations, J.Funct.Anal.**60** (1985) 168-210.

[6] - P.Acquistapace, B.Terreni, Existence and sharp regularity results for linear parabolic non-autonomous integro-differential equations, Israel J.Math.**53** (1986) 257-303.

[7] - P.Acquistapace, B.Terreni, Linear parabolic equations in Banach spaces with variable domains but constant interpolation spaces, Ann. Sc.Norm.Sup.Pisa (4) **13** (1986) 75-107.

[8] - P.Acquistapace, B.Terreni, On fundamental solutions for abstract parabolic equations, in "Differential equations in Banach spaces",

proceedings Bologna 1985, A.Favini & E.Obrecht editors, Lect.Notes in Math.n°1223, Springer-Verlag, Berlin Heidelberg 1986, 1-11.

[9] - P.Acquistapace, B.Terreni, A unified approach to abstract linear non-autonomous parabolic equations, Rend.Sem.Mat.Univ.Padova **78** (1987) 47-107.

[10] - P.Acquistapace, B.Terreni, Hölder classes with boundary conditions as interpolation spaces, Math.Z.**195** (1987) 451-471.

[11] - P.Acquistapace, B.Terreni, Regularity properties of the evolution operator for abstract linear parabolic equations, Diff.Int.Eq.**5** (1992) 1151-1184.

[12] - A.Buttu, On the evolution operator for a class of non-autonomous abstract parabolic equations, preprint Univ. Cagliari (1990).

[13] - G.Da Prato, P.Grisvard, Équations d'évolution abstraites non linéaires de type parabolique, Ann.Mat.Pura Appl.(4) **120** (1979) 329-396.

[14] - P.Grisvard, Équations différentielles abstraites, Ann.Sci.École Norm. Sup.(4) **2** (1969) 311-395.

[15] - T.Kato, Remarks on pseudo-resolvents and infinitesimal generators of semigroups, Proc.Japan Acad.Ser.A Math.Sci.**35** (1959) 467-468.

[16] - T.Kato, Abstract evolution equations of parabolic type in Banach and Hilbert spaces, Nagoya Math.J.**19** (1961) 93-125.

[17] - T.Kato, Semigroups and temporally inhomogeneous evolution equations, C.I.M.E.(1°ciclo), Varenna 1963.

[18] - T.Kato, H.Tanabe, On the abstract evolution equations, Osaka Math.J. **14** (1962) 107-133.

[19] - R.Labbas, B.Terreni, Somme d'opérateurs linéaires de type parabolique. 1re partie, Boll.Un.Mat.Ital.(7) **1B** (1987) 545-569.

[20] - R.Labbas, B.Terreni, Sommes d'opérateurs de type elliptique et parabolique. 2me partie: applications, Boll.Un.Mat.Ital.(7) **2B** (1988) 141-162.

[21] - A.Lunardi, Interpolation between domains of elliptic operators and spaces of continuous functions with applications to nonlinear parabolic equations, Math.Nachr.**121** (1985) 295-318.

[22] - E.Sinestrari, On the abstract Cauchy problem of parabolic type in spaces of continuous functions, J.Math.Anal.Appl.**107** (1985) 16-66.

[23] - P.E.Sobolevskii, Equations of parabolic type in a Banach space, Trudy Moscow Mat.Obsc.**10** (1961) 297-350 (Russian); English transl.: Amer. Math.Soc.Transl.**49** (1965) 1-62.

[24] - P.E.Sobolevskii, Parabolic equations in a Banach space with an unbounded variable operator, a fractional power of which has a constant domain of definition, Dokl.Akad.Nauk **138** (1961) 59-62 (Russian); English transl.: Soviet Math.Dokl.**2** (1961) 545-548.

[25] - P.E.Sobolevskii, The average principle for differential equations with a variable main member, manuscript, Voronezh 1984.

[26] - H.Tanabe, On the equations of evolution in a Banach space, Osaka Math.J.**12** (1960) 363-376.

[27] - H.Tanabe, Note on singular perturbations for abstract differential equations, Osaka J.Math.**1** (1964) 239-252.

[28] - A.Yagi, On the abstract evolution equations in Banach spaces, J.Math. Soc.Japan **28** (1976) 290-303.

[29] - A.Yagi, On the abstract evolution equations of parabolic type, Osaka J.Math.**14** (1977) 557-568.

[30] - A.Yagi, Fractional powers of operators and evolution equations of parabolic type, Proc.Japan Acad.**64A** (1988) 227-230.

[31] - A.Yagi, Abstract quasilinear evolution equations of parabolic type in Banach spaces, Boll.Un.Mat.Ital.**5B** (1991) 341-368.

On Some Classes of Singular Variational Inequalities

MARCO LUIGI BERNARDI Dipartimento di Elettronica per l'Automazione, Università degli Studi di Brescia, Via Valotti 9, I-25133 Brescia, Italy

FABIO LUTEROTTI Dipartimento di Elettronica per l'Automazione, Università degli Studi di Brescia, Via Valotti 9, I-25133 Brescia, Italy

1 - INTRODUCTION.

The subject of singular or degenerate evolution equations was widely and extensively studied by many authors. In addition to the book by CARROLL and SHOWALTER [11] (with its bibliography), we can refer e.g. to the following papers (in particular, for abstract evolution equations with singularities or degeneracies with respect to the time variable): SOBOLEVSKIĬ [27], BAIOCCHI-BAOUENDI [1], DA PRATO-GRISVARD [13], BERNARDI [3], [4], LEWIS-PARENTI [21], COPPOLETTA [12], FAVINI [16], DORE-VENNI [15], DORE-GUIDETTI [14], GUIDETTI [19], POVOAS [26], KUTTLER [20], FAVINI-PLAZZI [17], FAVINI-YAGI [18], LUTEROTTI [23], [24], BUTTU [10].

On the other hand, many results are well known on various kinds of evolution variational inequalities: see e.g., as general references, LIONS [22], BREZIS [8], [9], BARBU [2].

Then, a natural problem is the study of evolution variational inequalities containing singularities or degeneracies with respect to the time varia-

---

This work was supported in part by the "G.N.A.F.A. del C.N.R." (Italy), and by the "Ministero dell'Università e della Ricerca Scientifica e Tecnologica" (through 40% and 60% grants).

ble. (Remark, in particular, that LIONS [22] (chap.3; problem 11.7) proposed, as an open problem, the study of the singular hyperbolic variational inequality connected, "in a natural way", with the Euler-Poisson-Darboux operator).

First, let us consider the case of singular or degenerate variational inequalities of "parabolic" type, where the unilateral constraints concern the unknown function. In this case, some existence and uniqueness results were proved in BERNARDI-POZZI [6], [7], in a natural framework of suitable weighted spaces.

Moreover, let us consider the case of singular or degenerate evolution variational inequalities, where now the unilateral constraints concern the time derivative of the unknown function. Recently, in this case, some existence and uniqueness results (also in a framework of suitable weighted spaces) were proved by BERNARDI, LUTEROTTI and POZZI [5] for inequalities of "hyperbolic" type, and by LUTEROTTI [25] for inequalities of "parabolic" type. Remark that (see the following Section 2), in all of the results contained in [5] and [25], the solutions u(t) of the inequalities under consideration vanish at the initial time t=0.

The outline of this paper is as follows. Section 2 deals with the notation, the general assumptions, and with a very brief (but necessary) survey of the main results contained in [5] and [25]. In Sections 3 and 4, we prove some new results, which are related to those obtained in [5] and [25]. In particular, Section 3 deals with a singular perturbations theorem, while Section 4 concerns other results, where suitable non-zero initial data can be prescribed for the solutions u(t) of some of the (singular or degenerate) inequalities under consideration.

2 - NOTATION AND ASSUMPTIONS. SOME PREVIOUS RESULTS.

2.1. Throughout the paper, let T be given, with $0<T<+\infty$. Let moreover $p\in[1,+\infty]$ and $\mu\in\mathbb{R}$ be given; let $\mathcal{X}$ be a Banach space. We define:

$$L^p_\mu(\mathcal{X})=\{v(t)\,|\,t^\mu v(t)\in L^p(0,T;\mathcal{X})\}, \tag{2.1}$$

which is a Banach space with respect to its natural norm. (Clearly, if $\mathcal{X}$ is a Hilbert space, $L^2_\mu(\mathcal{X})$ is a Hilbert space too). We also adopt the following notation:

$$\begin{cases} z(t)\in {}_0C^0(\mathcal{X}) \text{ means that} \\ z(t)\in C^0([0,T];\mathcal{X}), \text{ and vanishes at } t=0. \end{cases} \tag{2.2}$$

Let now

$$V \subseteq H \equiv H^* \subseteq V^*, \text{ with } V \text{ separable,} \tag{2.3}$$

be the standard <u>real</u> Hilbert triplet. $(\cdot,\cdot)$ denotes both the scalar product in H and the duality pairing between V^* and V. $\|\cdot\|$ and $|\cdot|$ denote respectively the norms in V and H.

To recall the main results in [5] and [25], we have to take into account the functions f(t) and u(t) with the following properties. First, <u>let some $\alpha \in \mathbb{R}$ be given</u>. Consider the functions f(t) such that:

$$f(t)=f_1(t)+f_2(t), \text{ where, for some } \mu \in \mathbb{R}, \tag{2.4}$$

$$f_1(t) \in L^2_{\mu+1/2}(H),\ f_1'(t) \in L^2_{\mu+3/2}(H) \text{ (so that } t^{\mu+1} f_1(t) \in {}_0C^0(H)); \tag{2.5}$$

$$\begin{cases} f_2(t) \in L^1_{\mu-\alpha}(V^*),\ f_2'(t) \in L^1_{\mu-\alpha+1}(V^*),\ f_2''(t) \in L^1_{\mu-\alpha+2}(V^*) \\ \text{(so that: } t^{\mu-\alpha+1} f_2(t) \in {}_0C^0(V^*),\ t^{\mu-\alpha+2} f_2'(t) \in {}_0C^0(V^*)). \end{cases} \tag{2.6}$$

Moreover, we will consider the functions u(t) such that, for some $\mu \in \mathbb{R}$:

$$\begin{cases} u(t) \in L^2_{\mu+\alpha+1/2}(V),\ u'(t) \in L^2_{\mu+\alpha+3/2}(V) \\ \text{(so that } t^{\mu+\alpha+1} u(t) \in {}_0C^0(V)),\ u'(t) \in L^\infty_{\mu+\alpha+2}(V); \end{cases} \tag{2.7}$$

$$u'(t) \in L^2_{\mu+3/2}(H),\ u''(t) \in L^2_{\mu+5/2}(H) \text{ (so that } t^{\mu+2} u'(t) \in {}_0C^0(H)). \tag{2.8}$$

Now, we introduce some other notation and assumptions. First, let

$$\mathcal{K} \text{ be a closed convex subset of V, with } 0 \in \mathcal{K}. \tag{2.9}$$

Sometimes, in the sequel, we shall also assume that:

$$\exists \text{ a closed convex subset } \mathcal{K}_1 \text{ of H, such that } \mathcal{K}=\mathcal{K}_1 \cap V. \tag{2.10}$$

(For example, (2.10) holds, when (Ω being an open bounded subset of $\mathbb{R}^n$): $V=H^1_0(\Omega)$, $H=L^2(\Omega)$, $\mathcal{K}=\{v \in H^1_0(\Omega) \mid v \ge 0 \text{ a.e. in } \Omega\}$, $\mathcal{K}_1=\{v \in L^2(\Omega) \mid v \ge 0 \text{ a.e. in } \Omega\}$).

Let moreover A be an operator such that:

$$\begin{cases} A \in \mathcal{L}(V,V^*);\ (Au,v)=(Av,u),\quad \forall u,v \in V; \\ \exists c>0 \text{ such that } (Av,v) \ge c\|v\|^2,\ \forall v \in V. \end{cases} \tag{2.11}$$

2.2. Now, we recall the main results proved in BERNARDI-LUTEROTTI-POZZI [5], which concern a class of singular or degenerate inequalities of "hyperbolic" type.

<u>Theorem 2.1</u>. Let (2.3), (2.9), and (2.11) hold. Let $a \in \mathbb{R}$, and $\alpha \in \mathbb{R}$ be given. Define:

$$\mu_0 \equiv \mu_0(a,\alpha) \equiv \min[-\alpha-1; a-2; a-2\alpha-2]. \tag{2.12}$$

Fix any $\mu<\mu_0$. Then, for every f(t) satisfying (2.4), (2.5), (2.6), there exists a unique u(t), which satisfies (2.7), (2.8), and

$$u'(t)\in\mathcal{K}, \text{ for a.e. } t\in]0,T[; \tag{2.13}$$

$$\begin{cases} (t^2u''(t)+atu'(t)+t^{2\alpha}Au(t)-f(t),v-u'(t))\geq 0, \\ \forall v\in\mathcal{K}, \text{ for a.e. } t\in]0,T[. \end{cases} \tag{2.14}$$

Moreover, it results also that $u''(t)\in L^\infty_{\mu+3}(H)$.

Remark 2.1. Note (see [5]) that Theorem 2.1 also holds, by assuming that $f_1(t)\in L^1_\mu(H)$ and $f_1'(t)\in L^1_{\mu+1}(H)$, instead of (2.5). Moreover, the uniqueness result in Theorem 2.1 holds true more generally, by taking $\mu\leq\mu_0^*\equiv\min[-\alpha-1;a-2]$, instead of $\mu<\mu_0$, and with some weaker regularity assumptions on f(t). On the other hand, we recall that the existence result in Theorem 2.1 was proved, by using, as a main tool, a convenient procedure of penalization.

When the convex $\mathcal{K}$ also satisfies (2.10), the existence result in Theorem 2.1 can be improved (as far as the "threshold" exponent μ_0 is concerned), for suitable values of a and α. To this aim, define:

$$\hat{\mu}_0\equiv\hat{\mu}_0(a,\alpha)\equiv\max_{r\geq 1}\{\min[-\alpha-1;a-2;a-1-r;-3\alpha-2+r]\}. \tag{2.15}$$

Then (also see [5]), the following conclusions hold (where $\mu_0^*\equiv\mu_0^*(a,\alpha)$ is defined in Remark 2.1).

$$\mu_0(a,\alpha)\leq\hat{\mu}_0(a,\alpha)\leq\mu_0^*(a,\alpha), \ \forall a,\alpha\in\mathbb{R}; \tag{2.16}$$

$$\mu_0(a,\alpha)<\hat{\mu}_0(a,\alpha) \text{ (strictly), iff } \alpha>\max[0;a-1]; \tag{2.17}$$

$$\begin{cases} \text{if } \alpha>\max[0;a-1], \text{ it results that } \hat{\mu}_0(a,\alpha)=a-2, \\ \text{when } a+3\alpha\leq 1, \\ \text{while } \hat{\mu}_0(a,\alpha)=2^{-1}(a-3\alpha-3), \text{ when } a+3\alpha\geq 1. \end{cases} \tag{2.18}$$

Theorem 2.2. Let (2.3), (2.9), (2.10), and (2.11) hold. Let moreover $a\in\mathbb{R}$ and $\alpha\in\mathbb{R}$ satisfy: $\alpha>\max[0;a-1]$ (and hence $\mu_0(a,\alpha)<\hat{\mu}_0(a,\alpha)$). Fix any $\mu<\hat{\mu}_0(a,\alpha)$. Then, for every f(t) satisfying (2.4), (2.5), (2.6), there exists a (unique) u(t) satisfying (2.7), (2.8) (with also $u''(t)\in L^\infty_{\mu+3}(H)$), (2.13), (2.14).

Remark 2.2. Observe that, in Theorems 2.1 and 2.2, no initial condition (at t=0) on the solution u(t) of (2.13)-(2.14) appears explicitly. However, since u(t) satisfies (2.7) and (2.8), u(t) (resp. u'(t)) has a "weighted initial trend", i.e. $t^{\mu+\alpha+1}u(t)\in{}_0C^0(V)$ (resp. $t^{\mu+2}u'(t)\in{}_0C^0(H)$). Moreover, since we are taking (in Theorems 2.1 and 2.2) some μ such that $\mu<-\alpha-1$ (in any case), it results that u(t) vanishes at t=0 with an order greater than $-(\mu+\alpha+1)$. (The fact that $u'(t)$ vanishes or not at $t=0$ depends on the values of a, α, and μ).

Example 2.1. Take, in (2.14), $\alpha=0$ and any $a\in\mathbb{R}$. Then, (2.14) concerns an operator of Euler-Fuchs type. It results here that $\mu_0(a,0)=\mu_0^*(a,0)=\hat{\mu}_0(a,0)=$ $=\min[-1;a-2]$. It must be noted that, as far as the "threshold" exponent $\mu_0(a,0)$ is concerned, the result given by Theorem 2.1 is optimal, even in the very special case of o.d.e. of Euler-Fuchs type. Indeed (see [5]), the existence (resp. the uniqueness) of $u(t)$, in our weighted spaces framework, fails, in general, when $\mu\geq\mu_0(a,0)$ (resp. when $\mu>\mu_0(a,0)$). (We also recall that several new results on abstract differential equations of Euler-Fuchs type were recently obtained by LUTEROTTI [23], [24]).

Example 2.2. By dividing both sides of (2.14) by t^2, and putting $g(t)\equiv$ $\equiv t^{-2}f(t)$, we obtain the following inequality:

$$\begin{cases} (u''(t)+at^{-1}u'(t)+t^{2\alpha-2}Au(t)-g(t),v-u'(t))\geq 0, \\ \forall v\in\mathcal{K}, \text{ for a.e. } t\in]0,T[. \end{cases} \tag{2.19}$$

If we take $\alpha=1$, and any $a\in\mathbb{R}$, (2.19) concerns an operator of Euler-Poisson-Darboux type (in particular, when moreover $a=0$, (2.19) is, in fact, a "regular" (i.e. neither singular nor degenerate) inequality). On the other hand, if we take $\alpha=3/2$ and $a=0$, (2.19) concerns an operator of Tricomi type. Here, we limit ourselves to point out that, according to Remark 2.2, we can obtain only solutions $u(t)$ such that $u(0)=0$. The possibility of taking $u(0)=u_0\neq 0$, with some suitable u_0, will be investigated in Section 4.

2.3. Now, we recall the main result proved in LUTEROTTI [25], which concerns a class of singular or degenerate evolution inequalities of "parabolic" type.

Theorem 2.3. Let (2.3), (2.9), and (2.11) hold. Let $\alpha\in\mathbb{R}$ be given. Fix any $\mu<-1-\alpha$. Then, for every $f(t)$ satisfying (2.4), (2.5), (2.6), there exists a unique $u(t)$, which satisfies (2.7), (2.8), and

$$u'(t)\in\mathcal{K}, \text{ for a.e. } t\in]0,T[; \tag{2.20}$$

$$\begin{cases} (tu'(t)+t^{2\alpha}Au(t)-f(t),v-u'(t))\geq 0, \\ \forall v\in\mathcal{K}, \text{ for a.e. } t\in]0,T[. \end{cases} \tag{2.21}$$

Remark 2.3. Note (see [25]) that the uniqueness result in Theorem 2.3 holds true more generally, by taking some weaker regularity assumptions on $f(t)$, and with also $\mu\leq-1-\alpha$. On the other hand, we recall that the existence result in Theorem 2.3 was proved, by using the penalization method as well. In particular, the starting point was the approximation of (2.20)-(2.21) through the equation

$$tu_k'(t)+t^{2\alpha}Au_k(t)+kt^{2\alpha+1}\beta(u_k'(t))=f(t),\ 0<t<T, \tag{2.22}$$

where $\beta(\cdot)$ is the "natural" penalty operator related to the convex $\mathcal{K}$, and k is any positive integer. On the other hand, the singular or degenerate nonlinear implicit evolution equation (2.22) was handled through the following "approximating" equations (as $\eta\to 0^+$):

$$\eta t^2u_{k\eta}''(t)+tu_{k\eta}'(t)+t^{2\alpha}Au_{k\eta}(t)+kt^{2\alpha+1}\beta(u_{k\eta}'(t))=f(t),\ 0<t<T. \tag{2.23}$$

Remark 2.4. In Theorem 2.3, no initial condition (at t=0) on the solution u(t) of (2.20)-(2.21) appears explicitly. However, since u(t) satisfies (2.7) and (2.8), u(t) (resp. u'(t)) has a "weighted initial trend", i.e. $t^{\mu+\alpha+1}u(t)\in {}_0C^0(V)$ (resp. $t^{\mu+2}u'(t)\in {}_0C^0(H)$). Moreover, since $\mu<-\alpha-1$ is taken in Theorem 2.3, it results that u(t) vanishes at t=0 with an order greater than $-(\mu+\alpha+1)$. (The fact that u'(t) vanishes or not at t=0 depends on the values of α, and μ. For example, it also results that u'(0)=0, when $\alpha\geq 1$).

Example 2.3. Take $\alpha=0$. Then, in such a case, (2.21) concerns an operator of Euler-Fuchs type. The "threshold" exponent is equal to -1. As far as the "threshold" exponent is concerned, the result given by Theorem 2.3 is here "quasi-optimal", even in the very special case of o.d.e. of Euler-Fuchs type (see [25] for the details).

Example 2.4. By dividing both sides of (2.21) by t, and putting $g(t)\equiv t^{-1}f(t)$, we obtain the following inequality:

$$\begin{cases} (u'(t)+t^{2\alpha-1}Au(t)-g(t),v-u'(t))\geq 0, \\ \forall v\in\mathcal{K},\ \text{for a.e. } t\in]0,T[. \end{cases} \tag{2.24}$$

Note, in particular, that, when $\alpha=1/2$, (2.24) is a "regular" (i.e. neither singular nor degenerate) inequality. However, by using Theorem 2.3 (recall Remark 2.4), we can obtain only solutions u(t) such that u(0)=0. We shall investigate, in Section 4, the possibility of taking $u(0)=u_0\neq 0$, with some suitable u_0 (and we shall obtain a positive answer, when $\alpha>0$, as we could expect).

3 - A SINGULAR PERTURBATIONS RESULT.

We prove here a theorem of "singular perturbations type", which concerns the problems (2.13)-(2.14) and (2.20)-(2.21) (see e.g. BARBU [2], chap.5, for the standard results in this direction, in the case of neither singular nor degenerate evolution inequalities). First, we have the following

Proposition 3.1. Let (2.3), (2.9), and (2.11) hold. Let $a>0$ and $\alpha\in\mathbb{R}$ be given.

Take any ε such that $0<\varepsilon\leq a(1+|\alpha|)^{-1}$. Fix any $\mu<-\alpha-1$. Then, for every $f(t)$ satisfying (2.4), (2.5), (2.6), there exists a unique $u_\varepsilon(t)$, which satisfies (2.7), (2.8), and

$$u'_\varepsilon(t)\in\mathcal{K}, \text{ for a.e. } t\in]0,T[; \tag{3.1}$$

$$\begin{cases} (\varepsilon t^2 u''_\varepsilon(t)+atu'_\varepsilon(t)+t^{2\alpha}Au_\varepsilon(t)-f(t),v-u'_\varepsilon(t))\geq 0, \\ \forall v\in\mathcal{K}, \text{ for a.e. } t\in]0,T[. \end{cases} \tag{3.2}$$

Proof. It suffices to review and to remake carefully the proof of Theorem 2.1 above (see [5]). Since a and ε are taken, with $a>0$ and $0<\varepsilon\leq a(1+|\alpha|)^{-1}$, it results here that the "true" "threshold exponent" is, in fact, $-\alpha-1$.
Now, we can prove the following

Theorem 3.1. Let (2.3), (2.9), and (2.11) hold. Let $\alpha\in\mathbb{R}$ be given. Take any ε such that $0<\varepsilon\leq(2+2|\alpha|)^{-1}$. Fix any $\mu<-\alpha-1$, and any $f(t)$ satisfying (2.4), (2.5), (2.6). Let $u(t)$ (resp. $u_\varepsilon(t)$) the corresponding solution of (2.20)-(2.21) (resp. of (3.1)-(3.2), where $a=1$), satisfying (2.7) and (2.8) (thanks to Theorem 2.3 and to Proposition 3.1). Then, it results that:

$$\begin{cases} \|u'_\varepsilon(t)-u'(t)\|_{L^2_{\mu+3/2}(H)}+\|u_\varepsilon(t)-u(t)\|_{L^\infty_{\mu+\alpha+1}(V)}+ \\ +\|u_\varepsilon(t)-u(t)\|_{L^2_{\mu+\alpha+1/2}(V)}\leq c_1\varepsilon, \end{cases} \tag{3.3}$$

$$\|u'_\varepsilon(t)-u'(t)\|_{L^\infty_{\mu+2}(H)}\leq c_2\varepsilon^{1/2}, \tag{3.4}$$

where neither c_1 nor c_2 depends on ε.

Proof. Take $v=u'(t)$ in (3.2) (where $a=1$), and $v=u'_\varepsilon(t)$ in (2.21). Then, by adding the resulting inequalities and by putting $w_\varepsilon(t)\equiv u_\varepsilon(t)-u(t)$, we get that:

$$(\varepsilon t^2 w''_\varepsilon(t)+tw'_\varepsilon(t)+t^{2\alpha}Aw_\varepsilon(t)+\varepsilon t^2 u''(t),w'_\varepsilon(t))\leq 0 \text{ (a.e. on }]0,T[), \tag{3.5}$$

that is (thanks also to (2.11)):

$$\frac{1}{2}\varepsilon t^2\frac{d}{dt}|w'_\varepsilon(t)|^2+t|w'_\varepsilon(t)|^2+\frac{1}{2}t^{2\alpha}\frac{d}{dt}(Aw_\varepsilon(t),w_\varepsilon(t))\leq-\varepsilon t^2(u''(t),w'_\varepsilon(t)). \tag{3.6}$$

Next, we multiply both sides of (3.6) by $t^{2\mu+2}$, and we integrate from 0 to t ($0<t\leq T$). Taking into account the properties (2.7), (2.8) of ($u(t)$, $u_\varepsilon(t)$ and) $w_\varepsilon(t)$ (in particular, the "weighted initial trends" at $t=0$), we obtain that

$$\begin{cases} \frac{1}{2}\varepsilon t^{2\mu+4}|w'_\varepsilon(t)|^2+[1-\varepsilon(\mu+2)]\int_0^t s^{2\mu+3}|w'_\varepsilon(s)|^2 ds+\frac{1}{2}t^{2\mu+2\alpha+2}\cdot \\ \cdot(Aw_\varepsilon(t),w_\varepsilon(t))-[\mu+\alpha+1]\int_0^t s^{2\mu+2\alpha+1}(Aw_\varepsilon(s),w_\varepsilon(s))ds\le \\ \le\varepsilon\int_0^t s^{2\mu+4}|u''(s)|\cdot|w'_\varepsilon(s)|ds\le \\ \le\frac{1}{2}\varepsilon^2\int_0^T s^{2\mu+5}|u''(s)|^2 ds+\frac{1}{2}\int_0^t s^{2\mu+3}|w'_\varepsilon(s)|^2 ds, \\ 0<t\le T. \end{cases} \tag{3.7}$$

Now, since $\mu<-\alpha-1$ and $0<\varepsilon\le(2+2|\alpha|)^{-1}$, the quantity $[2^{-1}-\varepsilon(\mu+2)]$ is strictly positive. Hence, thanks to (2.11) and to the fact that $u''(t)\in \in L^2_{\mu+5/2}(H)$, (3.3) and (3.4) follow from (3.7). Thus, the theorem is completely proved.

4 - SOME RESULTS CONCERNING NON-ZERO INITIAL DATA.

As we observed in Remark 2.2 (resp. Remark 2.4), the solutions $u(t)$ of (2.13)-(2.14) (resp. (2.20)-(2.21)), obtained through Theorems 2.1 and 2.2 (resp. Theorem 2.3), satisfy $u(0)=0$. Here, we study the possibility of having some results where $u(0)=u_0\neq 0$, with some suitable u_0 (and also, for (2.13)-(2.14), $u'(0)=u_1\neq 0$, with some suitable u_1).

4.1. We consider firstly the problem (2.20)-(2.21). We have the following

Theorem 4.1. Let (2.3), (2.9), and (2.11) hold. Let $\alpha>0$ (strictly) and $u_0\in D(A)=\{z\in V\,|\,Az\in H\}$ be given. Fix any $\mu\in\mathbb{R}$, such that: $-1-2\alpha<\mu<-1-\alpha$. Take any $f(t)$ satisfying (2.4), (2.5), (2.6). Then, there exists a unique $u(t)\in \in C^0([0,T];V)$, solution of (2.20)-(2.21), with the following properties:

$$\begin{cases} u(0)=u_0;\ (u(t)-u_0)\in L^2_{\mu+\alpha+1/2}(V),\ u'(t)\in L^2_{\mu+\alpha+3/2}(V), \text{ so that} \\ t^{\mu+\alpha+1}(u(t)-u_0)\in {}_0C^0(V);\ u'(t)\in L^\infty_{\mu+\alpha+2}(V);\ u(t) \text{ satisfies (2.8).} \end{cases} \tag{4.1}$$

Proof. Let us consider the problem of finding $w(t)$ such that:

$$\begin{cases} (tw'(t)+t^{2\alpha}Aw(t)-(f(t)-t^{2\alpha}Au_0),v-w'(t))\ge 0, \\ \forall v\in\mathcal{K}, \text{ for a.e. } t\in]0,T[; \\ w'(t)\in\mathcal{K}, \text{ for a.e. } t\in]0,T[. \end{cases} \tag{4.2}$$

Since $\mu>-1-2\alpha$, $\tilde{f}_1(t)\equiv t^{2\alpha}Au_0$ satisfies (2.5). Hence, being also $\mu<-1-\alpha$, Theorem 2.3 applies. Then, there exists a (unique) $w(t)$, solution of (4.2), with the properties (2.7) and (2.8) (stated there for $u(t)$, instead of $w(t)$). Now, define $u(t)\equiv w(t)+u_0$ (and hence $u'(t)=w'(t)$). It is clear that such $u(t)$ belongs to $C^0([0,T];V)$, and satisfies (2.20)-(2.21) and (4.1). On the other hand, the uniqueness of such u(t) results from the following argument. Let $u(t)=u_1(t)\in C^0([0,T];V)$ and $u(t)=u_2(t)\in C^0([0,T];V)$ both satisfy (2.20)-(2.21) and (4.1) (with the same f(t) and u_0). Define $w_i(t)\equiv u_i(t)-u_0$, $i=1,2$ (and hence $w_i'(t)=u_i'(t)$, $i=1,2$). Then, $w(t)=w_1(t)$ and $w(t)=w_2(t)$ both satisfy (4.2). Hence, by proceeding as in the proof of Theorem 3.1 in LUTEROTTI [25], it can be proved that $w_1(t)=w_2(t)$, i.e. $u_1(t)=u_2(t)$.

Remark 4.1. The proof of Theorem 4.1 is based upon the result given by Theorem 2.3, where the condition $\mu<-1-\alpha$ is crucial. On the other hand, $\tilde{f}_1(t)\equiv t^{2\alpha}Au_0$ ($u_0\neq 0$) satisfies (2.5) if and only if $\mu>-1-2\alpha$. Hence, such conditions on μ can be fulfilled if and only if $\alpha>0$. Moreover, the condition $u_0\in D(A)$ cannot be weakened, in the context of our proof. In fact, if we take $u_0\in V$, with $u_0\notin D(A)$, we have that $\tilde{f}_2(t)\equiv t^{2\alpha}Au_0$ satisfies (2.6) if and only if $\mu>-1-\alpha$.

Example 4.1. The outcome given by Theorem 4.1 is also more evident, if we rewrite (2.21) in the form (2.24), where $g(t)=t^{-1}f(t)$. By putting $\lambda\equiv 2\alpha-1$, we see that our result applies to problems of the type

$$\begin{cases} (u'(t)+t^{\lambda}Au(t)-g(t),v-u'(t))\geq 0, \\ \forall v\in\mathcal{K}, \text{ for a.e. } t\in]0,T[; \\ u'(t)\in\mathcal{K}, \text{ for a.e. } t\in]0,T[, \end{cases} \tag{4.3}$$

provided that $\lambda>-1$. The case of "regular" (i.e. neither singular nor degenerate) inequalities corresponds to the choice $\lambda=0$ (i.e. $\alpha=1/2$). In such case, Theorem 4.1 gives the result which is detailed in the following corollary.

Corollary 4.1. Let (2.3), (2.9), and (2.11) hold. Let $u_0\in D(A)$ be given. Fix any $\mu\in\mathbb{R}$ such that: $-2<\mu<-3/2$. Take any $g(t)=g_1(t)+g_2(t)$ such that: $g_1(t)\in L^2_{\mu+3/2}(H)$, $g_1'(t)\in L^2_{\mu+5/2}(H)$ (so that: $t^{\mu+2}g_1(t)\in{}_0C^0(H)$); $g_2(t)\in L^1_{\mu+1/2}(V^*)$, $g_2'(t)\in L^1_{\mu+3/2}(V^*)$, $g_2''(t)\in L^1_{\mu+5/2}(V^*)$ (so that: $t^{\mu+3/2}g_2(t)\in{}_0C^0(V^*)$, $t^{\mu+5/2}g_2'(t)\in{}_0C^0(V^*)$). Then, there exixts a unique $u(t)\in C^0([0,T];V)$, solution of (4.3) (where $\lambda=0$), with the following properties:

$$\begin{cases} u(0)=u_0;\ (u(t)-u_0)\in L^2_{\mu+1}(V),\ u'(t)\in L^2_{\mu+2}(V), \text{ so that} \\ t^{\mu+3/2}(u(t)-u_0)\in{}_0C^0(V);\ u'(t)\in L^{\infty}_{\mu+5/2}(V);\ u'(t)\in L^2_{\mu+3/2}(H), \\ u''(t)\in L^2_{\mu+5/2}(H) \text{ (so that } t^{\mu+2}u'(t)\in{}_0C^0(H)). \end{cases} \tag{4.4}$$

(Remark that the result given by Corollary 4.1 is close (although not identical) to the "standard" outcomes for "regular" inequalities, as e.g. Proposition II.10 in the chapter 2 of BREZIS [8]).

4.2. Now, we consider the problem (2.13)-(2.14). We try to proceed as we did in subsection 4.1, for the problem (2.20)-(2.21). So, we look for solutions u(t) of (2.13)-(2.14), which have the form $u(t)=u_0+tu_1+w(t)$, where $u_1\in\mathcal{K}$ and u_0 are suitably chosen, while $w(0)=w'(0)=0$. By putting such u(t) in (2.13)-(2.14), we obtain formally that

$$\begin{cases} (t^2w''(t)+atw'(t)+t^{2\alpha}Aw(t)-[f(t)-atu_1-t^{2\alpha}Au_0-t^{2\alpha+1}Au_1], \\ v-u_1-w'(t))\geq 0, \\ \forall v\in\mathcal{K}, \text{ for a.e. } t\in]0,T[; \end{cases} \tag{4.5}$$

$$[u_1+w'(t)]\in\mathcal{K}, \text{ for a.e. } t\in]0,T[. \tag{4.6}$$

Define now $\tilde{\mathcal{K}}\equiv[\mathcal{K}\backslash u_1]=\{z=v-u_1 \mid v\in\mathcal{K}\}$ (which is also a closed convex subset of V, with $0\in\tilde{\mathcal{K}}$). Then, (4.5)-(4.6) can be rewritten in the following form:

$$\begin{cases} (t^2w''(t)+atw'(t)+t^{2\alpha}Aw(t)-h(t),z-w'(t))\geq 0, \\ \forall z\in\tilde{\mathcal{K}}, \text{ for a.e. } t\in]0,T[, \end{cases} \tag{4.7}$$

$$w'(t)\in\tilde{\mathcal{K}}, \text{ a.e. on }]0,T[, \tag{4.8}$$

where we have set $h(t)\equiv f(t)-atu_1-t^{2\alpha}Au_0-t^{2\alpha+1}Au_1$. Hence, when the outcomes of Theorems 2.1 and 2.2 apply to (4.7)-(4.8) (with also $\mu+2\leq 0$, so that $w'(0)=0$ in addition to $w(0)=0$), we can solve (2.13)-(2.14), with $u(0)=u_0$ and $u'(0)=u_1$. However, the situation is here much more complicated than in the case of the "parabolic" inequalities considered in the previous subsections 2.3 and 4.1. In fact, the "threshold" exponent was there $-1-\alpha$, while the "threshold" exponents $\mu_0(a,\alpha)$ and $\hat{\mu}_0(a,\alpha)$ of Theorems 2.1 and 2.2 have a quite less simple form. Moreover, in the case of "parabolic" inequalities (see subsection 4.1), we had to see when the term $t^{2\alpha}Au_0$ could be treated as $f_1(t)$ or $f_2(t)$, while here we have to consider the behaviour of each of the terms atu_1, $t^{2\alpha}Au_0$, $t^{2\alpha+1}Au_1$. For example, we remark that a necessary condition for solving (2.13)-(2.14) with $u(0)=u_0\neq 0$ (through the above device and Theorem 2.1 or 2.2) is that $u_0\in D(A)$ and $\alpha>0$: this can be easily proved, by considering (2.5), (2.6), (2.12), (2.15).

We could perform here a complete study of the various cases: however, for the sake of brevity, we detail only the results we can obtain for some of the main examples considered in subsection 2.2.

Example 4.2. Take $\alpha=3/2$ and $a=0$. In this case, the problem (2.14)-(2.13) (or, more evidently, (2.19), with $g(t)=t^{-2}f(t)$) concerns an operator of Tricomi type. It results here that $\mu_0=-5$, and $\hat{\mu}_0=-15/4$. By using the above procedure, we can obtain the following Proposition.

Proposition 4.1. Let (2.3), (2.9), (2.10), (2.11) hold. Let $u_0 \in D(A)$ and $u_1 \in \mathcal{K} \cap D(A)$ be given. Fix any $\mu \in \mathbb{R}$, such that: $-4<\mu<-15/4$. Take any $g(t)$ of the form (2.4), (2.5), (2.6), where μ is replaced by $\mu+2$ and $\alpha=3/2$. Then there exists a unique $u(t) \in C^0([0,T];V) \cap C^1([0,T];H)$, solution of (2.14)-(2.19) (where $\alpha=3/2$ and $a=0$), with the following properties: $u(0)=u_0$; $u'(0)=u_1$; $(u(t)-u_0-tu_1) \equiv \tilde{u}(t)$ satisfies (2.7)-(2.8) (also with $\alpha=3/2$). (We also remark, that, if we take $u_0=0$ (and $u_1 \in D(A) \cap \mathcal{K}$), the above result holds true, more generally, when $-5<\mu<-15/4$).

Example 4.3. Take $\alpha=1$ and any $a \in \mathbb{R}$. In this case, the problem (2.14)-(2.19) (with $g(t)=t^{-2}f(t)$) concerns an operator of Euler-Poisson-Darboux type. It results here that:

$$\begin{cases} \mu_0=a-4, \forall a\leq 2;\ \mu_0=-2,\ \forall a\geq 2; \\ \hat{\mu}_0=a-2, \forall a\leq -2;\ \hat{\mu}_0=a/2-3,\ \forall a\in[-2,2];\ \hat{\mu}_0=-2, \forall a\geq 2. \end{cases} \tag{4.9}$$

By using the above procedure, we can obtain the following Proposition.

Proposition 4.2. Let (2.3), (2.9), and (2.11) hold. Let $u_0 \in D(A)$ (and $u_1=0$) be given. Let $a>1$. Fix any $\mu \in \mathbb{R}$, such that: $-3<\mu<\min[-2;a-4]$. Take any $g(t)$ of the form (2.4), (2.5), (2.6), where μ is replaced by $\mu+2$ and $\alpha=1$. Then, there exists a unique $u(t) \in C^0([0,T];V) \cap C^1([0,T];H)$, solution of (2.14)-(2.19) (where $\alpha=1$), with the following properties: $u(0)=u_0$; $u'(0)=0$; $(u(t)-u_0) \equiv \tilde{u}(t)$ satisfies (2.7)-(2.8) (also with $\alpha=1$). If $\mathcal{K}$ also satisfies (2.10), such result holds when $a>0$ too, by fixing any $\mu \in \mathbb{R}$ such that $-3<\mu<\min[-2;a/2-3]$.

Acknowledgment. These authors would like to thank Gianni Gilardi for some useful discussions.

- REFERENCES -

[1] C.BAIOCCHI-M.S.BAOUENDI, Singular evolution equations, J. Funct. Anal., 25 (1977), 103-120.

[2] V.BARBU, "Nonlinear Semigroups and Differential Equations in Banach Spaces", Noordhoff, Leiden, 1976.

[3] M.L.BERNARDI, Su alcune equazioni d'evoluzione singolari, Boll. Un. Mat. Ital., (5) B-13 (1976), 498-517.

[4] M.L.BERNARDI, Second order abstract differential equations with singular coefficients, Ann. Mat. Pura Appl., (4) 130 (1982), 257-286.

[5] M.L.BERNARDI-F.LUTEROTTI-G.A.POZZI, On a class of singular or degenerate hyperbolic variational inequalities, Differential Integral Equations, 4 (1991), 953-975.

[6] M.L.BERNARDI-G.A.POZZI, On some singular or degenerate parabolic variational inequalities, Houston J. Math., 15 (1989), 163-192.

[7] M.L.BERNARDI-G.A.POZZI, On a class of singular nonlinear parabolic variational inequalities, Ann. Mat. Pura Appl., 159 (1991), 117-131.

[8] H.BREZIS, Problèmes unilatéraux, J. Math. Pures Appl., 51 (1972), 1-168.

[9] H.BREZIS, "Operateurs Maximaux Monotones et Semi-groupes de Contractions dans les Espaces de Hilbert", North-Holland, Amsterdam, 1973.

[10] A.BUTTU, On a class of abstract degenerate parabolic equations, Differential Integral Equations, 3 (1990), 663-681.

[11] R.W.CARROLL-R.E.SHOWALTER, "Singular and Degenerate Cauchy Problems", Academic Press, New York, 1976.

[12] G.COPPOLETTA, Abstract singular evolution equations of "hyperbolic" type, J. Funct. Anal., 50 (1983), 50-66.

[13] G.DA PRATO-P.GRISVARD, On an abstract singular Cauchy problem, Comm. in P.D.E., 3 (1978), 1077-1082.

[14] G.DORE-D.GUIDETTI, On a singular non-autonomous equation in Banach spaces, in Proc. Conf. on Differential Equations in Banach Spaces (Bologna, 1985) (ed. by A.Favini and E.Obrecht), Lecture Notes in Mathematics, vol.1223, Springer, Berlin, 1986.

[15] G.DORE-A.VENNI, On a singular evolution equation in Banach spaces, J. Funct. Anal., 64 (1985), 227-250.

[16] A.FAVINI, Degenerate and singular evolution equations in Banach spaces, Math. Annalen, 273 (1985), 17-44.

[17] A.FAVINI-P.PLAZZI, On some abstract degenerate problems of parabolic type. 1: The linear case, Nonlinear Analysis, T.M.A., 12 (1988), 1017-1027. 2: The nonlinear case, ibidem, 13 (1989), 23-31.

[18] A.FAVINI-A.YAGI, Multivalued linear operators and degenerate evolution problems, to appear in Ann. Mat. Pura Appl..

[19] D.GUIDETTI, Linear singular parabolic equations in Banach spaces, Math. Z., 195 (1987), 487-504.

[20] K.L.KUTTLER, Time-dependent implicit evolution equations, Nonlinear Analysis, T.M.A., 10 (1986), 447-463.

[21] J.E.LEWIS-C.PARENTI, Abstract singular parabolic equations, Comm. in P.D.E., 7 (1982), 279-324.

[22] J.L.LIONS, "Quelques Méthodes de Résolution des Problèmes aux Limites Non-Linéaires", Dunod-Gauthier Villars, Paris, 1969.

[23] F.LUTEROTTI, On a class of abstract differential equations of Fuchs' type, Boll. Un. Mat. Ital., (7) B-4 (1990), 83-99.

[24] F.LUTEROTTI, Some regularity results for a class of Fuchs' abstract differential equations, to appear in Atti Sem. Mat. Fis. Univ. Modena.

[25] F.LUTEROTTI, On some degenerate evolution variational inequalities, to appear in Ann. Mat. Pura Appl..

[26] M.POVOAS, On some hyperbolic evolution equations with time-dependent singularity, J. Differential Equations, 59 (1985), 396-419.

[27] P.E.SOBOLEVSKIĬ, On degenerate parabolic operators, Dokl. Akad. Nauk SSSR, 196 (1971), 302-304 (and Sov. Math. Dokl., 12 (1971), 129-132).

Nonuniqueness in L^∞: An Example

JULIO E. BOUILLET Departamento de Matemática, FCEyN, Universidad de Buenos Aires, and IAM (CONICET), Viamonte 1636, (1055) Buenos Aires, Argentina

1. Introduction.

In [BC] the following result is obtained:

THEOREM. Let $\varphi : R \mapsto R$ be nondecreasing, continuous, $\varphi(0) = 0$; let $Q = R^N \times (0,T)$, $T > 0$, and assume

(1.1) $u_1, u_2 \in L^\infty(Q)$,

(1.2) $u_{1t} - \Delta\varphi(u_1) = u_{2t} - \Delta\varphi(u_2)$ in $\mathcal{D}'(Q)$,

(1.3) $u_1 - u_2 \in L^1(Q)$,

(1.4) $\operatorname{ess}\lim_{t\downarrow 0} \int_{R^N} |u_1(x,t) - u_2(x,t)|dx = 0.$

Then $u_1 = u_2$ a.e. on Q.

Continuity or even single–valuedness of φ is used in [BC] only to establish that

(1.5) measure $\{(x,t) \in Q : |\varphi(u_1(x,t)) - \varphi(u_2(x,t))| > \varepsilon\} < \infty$ for every $\varepsilon > 0$, so an $L^p(Q)$ condition on u_i, $1 \le p < \infty$, and continuity of φ at zero would suffice.

We present here an equation with nondecreasing but discontinuous – at $u = 1$ – function φ, that admits two $L^\infty(R^N \times (0,T))$ solutions satisfying (1.1) to (1.4). Moreover, (1.4) could be replaced with $\lim_{t\downarrow 0} \|u_1(x,t) - u_2(x,t)\|_{L^\infty(R^N)} = 0$. A variant of the equation with φ discontinuous at $u = 0$ admits two different solutions, integrable on Q.

We present also a variant of the uniqueness result in [BC].

2. An Example.

Let $Q = R^N \times (0,\infty)$, $N \geq 3$, φ be the *graph* $\varphi(u) := \operatorname{sgn}(u-1)^+$. A solution to $u_t + \Delta\varphi(u) = 0$ in $\mathcal{D}'(Q)$ is a locally integrable function $u(x,t)$ for which there is a measurable function $U(x,t) \in \operatorname{sgn}(u(x,t)-1)^+$ a.e. such that $\int_Q (u\psi_t + U\Delta\psi) = 0$ for every $\psi \in C_0^\infty(Q)$.

Let $u_I(x) = u_I(|x|)$ > 1 in $B(0,R_I) \subset R^N$, $R_I > 0$
$\equiv 1$ in $R^N \setminus B(0,R_I)$;

this function will be assumed as smooth as needed.

We define

$$\begin{array}{lll} u_1(x,t) = u_I(|x|), & U_1(x,t) \equiv 1, & \text{for every } t>0; \\ u_2(x,t) = u_I(|x|), & U_2(x,t) = 1, & \text{for } 0<|x|<R(t),\ t>0, \\ u_2(x,t) = 1, & U_2(x,t) = \big(R(t)/|x|\big)^{N-2}, & \text{for } |x| \geq R(t),\ t>0, \end{array}$$

where $R(t)$ is defined by $0 < t = (N-2)^{-1}\int_{R(t)}^{R_I} r(u_I(r)-1)dr$.

It can be shown that

$R'(t) =$ (jump of $(-DU_2 \cdot x/|x|)$ across $|x| = R(t)$)/ (jump of u_2 across $|x| = R(t)$),

whence the verification of the equation in the $\mathcal{D}'(Q)$-sense. Clearly

(2.1) $u_2(x,t)$ is discontinuous, support (u_2-1) *shrinks* as t increases;

(2.2) If $u_I(|x|)$ is continuous at $|x| = R_I$, $\lim_{t\downarrow 0} \|u_1(\cdot,t) - u_2(\cdot,t)\|_{L^\infty(R^N)} = 0$.

(2.3) If $ru_I(r)$ is integrable at $r = 0$, then the evolution stops at time $t^* = (N-2)^{-1}\int_0^{R_I} r(u_2(r)-1)dr$ yielding $u_2(x,t) \equiv 1$, $U_2(x,t) \equiv 0$ for $t \geq t^*$.

(2.4) $u_2(x,t)$ is the monotone limit of the $C([0,\infty); L^1(R^N))$–solutions $u^n(x,t)$ of the equation with initial data $u^n(x,0) = u_I^n(|x|) := u_I(|x|)\cdot\chi_{[0,n]}(|x|)$, χ_E being the characteristic function of E (in $R^1 \times (0,\infty)$, the monotone limit of $u^n(x,t)$ is $u_1(x,t) \equiv u_I(|x|)$).

(2.5) For smooth u_I, the functions

$$\begin{array}{lll} v_1(x,t) := u_I(|x|) - 1, & V_1(x,t) \equiv 1, & t>0, \text{ and} \\ v_2(x,t) := u_2(x,t) - 1, & V_2(x,t) = U_2(x,t), & t>0, \end{array}$$

are $\mathcal{D}'$ solutions to $u_t = \Delta\operatorname{sgn}(u^+)$, integrable in $Q = R^N \times (0,T)$, $T>0$.

(2.6) Solutions with prescribed $\lim_{|x|\to\infty} U(x,t) \in [0,1]$ are also easily obtained.

This suggests as a crucial point the behavior of the functions U when $U(x,t) \in [0,1]$, the jump of the monotone graph φ, for $|x|$ large.

3. A Uniqueness Theorem.

THEOREM. Let φ be a monotone graph, $0 \in \varphi(0)$, $T > 0$. For $x \in R^N$, $N \geq 2$, $i = 1,2$, let $u_i(x,t)$, $U_i(x,t) \in \varphi(u_i(x,t))$ a.e. be such that

(3.1) $u_{1t} - \Delta U_1 = u_{2t} - \Delta U_2$ in $\mathcal{D}'(R^N \times (0,\mathcal{T}))$,

(3.2) $u_1 - u_2,\ U_1 - U_2 \in L^2_{\text{loc}}(R^N \times (0,\mathcal{T}))$,

(3.3) $\operatorname*{ess\,lim}_{t\downarrow 0} \int_{B(R)} (u_1(x,t) - u_2(x,t))^+ dx = 0$ for every ball $B(R) \subset R^N$, and

(3.4) $\operatorname*{ess\,lim}_{R\to\infty} R^{1-N} \int_\tau^T \int_{\partial B(R)} |U_1 - U_2| dS\, dt = 0$ uniformly in τ, $0 < \tau < T < \mathcal{T}$.

Then $u_1(x,t) \le u_2(x,t)$ a.e. in $R^N \times (0,\mathcal{T})$.

Proof. Put $[u] := u_1 - u_2$, $[U] := U_1 - U_2$. Consider the regularizations $[u]^\rho = [u] * \rho(x,t)$, $[U]^\rho = [U] * \rho(x,t)$, $0 \le \rho \in C_0^\infty(R^N \times R^1)$. By employing $\rho(x - \cdot)$ as a test function in the definition of (3.1) one obtains

$$-[u]^\rho_t(x,t) + \Delta[U]^\rho(x,t) = 0 \text{ in } R^N \times (\delta, \mathcal{T} - \delta)$$

for certain $\delta > 0$. Standard manipulations, coupled to [AC, Proposition 1.2] give for arbitrarily small δ, $0 < \delta < \tau < T < \mathcal{T} - \delta$,

$$\text{(3.5)} \quad \int_{B(R)} [u]^\rho(x,T)\psi(x,T)dx = \int_{B(R)} [u]^\rho(x,\tau)\psi(x,\tau)dx + \\ + \int_\tau^T \int_{B(R)} ([u]^\rho \psi_t + [U]^\rho \Delta\psi)dx\, dt - \int_\tau^T dt \int_{\partial B(R)} [U]^\rho D\psi \cdot \frac{x}{R} dS(x)\,.$$

Therefore for almost all τ, T and R with $0 < \tau < T$ and $R > 0$ one gets the identity (3.5) for $[u]$, $[U] \in L^1_{\text{loc}}(R^N \times (0,\mathcal{T}))$. In (3.5) the test functions $\psi(x,t) \in C^{1,0}(\overline{B(R)} \times (0,T]) \cap C^2(B(R) \times (0,T])$ and $\psi(x,t) = 0$ on $\partial B(R) \times (0,T]$ (cf. [AC]).

We now proceed as follows: define

$$\text{(3.6)} \quad \begin{aligned} E_1 &:= \{(x,t) \in R^N \times (0,\mathcal{T}) : |[U](x,t)| \le |[u](x,t)| \text{ and } [u](x,t) \ne 0\}, \\ E_2 &:= \{(x,t) \in R^N \times (0,\mathcal{T}) : |[u](x,t)| \le |[U](x,t)| \text{ and } [U](x,t) \ne 0\}. \end{aligned}$$

We study the integrals in the r.h.s. of (3.5) for $[u]$, $[U]$.

$$\text{(3.7)} \quad \int_\tau^T \int_{B(R)} ([u]\psi_t + [U]\Delta\psi)dx\, dt = \int_\tau^T \int_{B(R)} \chi_{E_1}[u]\Big(\psi_t + \frac{[U]}{[u]}\Delta\psi\Big)dx\, dt + \\ + \int_\tau^T \int_{B(R)} \chi_{E_2}[U]\Big(\frac{[u]}{[U]}\psi_t + \Delta\psi\Big)dx\, dt\,.$$

Put $c(x,t) := [U]/[u]$,

$$\text{(3.8)} \qquad c_n(x,t) := \max\{1/n, \min\{n, [U]/[u]\}\} * \rho_{j_n}\,,$$

$\rho_{j_n} \ge 0$ a mollifier as before, the subsequence j_n to be determined. It is clear that $1/n \le c_n \le n$.

For each $n \in \mathbf{N}$, let $\psi = \psi_n(x,t) \in C^{1,0}(\overline{B(R)} \times (0,T]) \cap C^2(B(R) \times (0,T])$ be the solution to the parabolic equation

$$(3.9)\qquad \psi_{nt} + c_n(x,t)\Delta\psi_n = 0 \left(= \frac{1}{c_n(x,t)}\psi_{nt} + \Delta\psi_n \right)$$

in $B(R) \times (0,T]$, with boundary data

(3.10) $\psi_n(x,T) = \theta(x) \in C_0^\infty(B(R))$, support $\theta \subset B(R_1)$, $R_1 < R$, $0 \le \theta \le 1$, $\psi = 0$ on $\partial B(R) \times (0,T]$.

REMARK. These solutions ψ_n satisfy $\left| D\psi_n \cdot \frac{x}{|x|} \right| \le c(N-2)R^{1-N}$ on $\partial B(R) \times (0,T]$: this is obtained by comparison with the solution $v(x,t) := c\big(|x|^{2-N} - R^{2-N}\big)$ of equation $\psi_t + c_n\Delta\psi = 0$ in $(B(R) \setminus \{0\}) \times (0,T]$. Here c satisfies $c\big(R_1^{2-N} - R^{2-N}\big) = 1$, hence $c \le 2R_1^{N-2}$ if $R > 2^{\frac{1}{N-2}} R_1$.

By the maximum principle it is also clear that $0 \le \psi_n \le 1$. These estimates are independent of c_n and θ.

We observe further that $\frac{1}{c_n}(\psi_{nt})^2 = c_n(\Delta\psi_n)^2$, and that integration by parts in (3.9) gives

$$(3.11)\qquad \int_\tau^T \int_{B(R)} c_n |\Delta\psi_n|^2 dx\, dt = \int_\tau^T \int_{B(R)} \frac{1}{c_n} |\psi_{nt}|^2 dx\, dt \le \frac{1}{2} \int_{B(R_1)} |D\theta|^2 dx \,.$$

With these test functions ψ_n the r.h.s. of (3.7) is

$$\int_\tau^T \int_{B(R)} \chi_{E_1}[u](c - c_n)\Delta\psi_n dx\, dt + \int_\tau^T \int_{B(R)} \chi_{E_2}[U]\Big(\frac{1}{c} - \frac{1}{c_n}\Big)\psi_{nt} dx\, dt \,,$$

and together with (3.8) and (3.11) give

$$\le \Big(\frac{1}{2}\int_{B(R_1)} |D\theta|^2 dx\Big)^{1/2} \Big\{ \Big(n \int_\tau^T \int_{B(R)} \chi_{E_1}[u]^2 (c - c_n)^2 dx\, dt\Big)^{1/2} + \Big(n \int_\tau^T \int_{B(R)} \chi_{E_2}[U]^2 \Big(\frac{1}{c} - \frac{1}{c_n}\Big)^2 dx\, dt\Big)^{1/2} \Big\} .$$

Following a well-known technique (cf.[O]), for each fixed $n \in \mathbf{N}$ we select j_n in (3.8) so that

$$\int_\tau^T \int_{B(R) \cap \{(x,t): c > 1/n\}} \chi_{E_1}[u]^2 (c - c_n)^2 dx\, dt \le n^{-2}, \text{ and likewise}$$

$$\int_\tau^T \int_{B(R) \cap \{(x,t): 1/c \ge 1/n\}} \chi_{E_2}[U]^2 \Big(\frac{1}{c} - \frac{1}{c_n}\Big)^2 dx\, dt \le n^{-2}$$

(here we used the dominated convergence theorem, the fact that $[u], [U] \in L^2_{\text{loc}}(R^N \times (0,T])$, and that $c \le 1$ on E_1, $1/c \le 1$ on E_2).

We finally obtain

$$(3.12)\qquad \left|\int_\tau^T\int_{B(R)}([u]\psi_{nt}+[U]\Delta\psi_n)dx\,dt\right|\le\left(\frac{1}{2}\int_{B(R_1)}|D\theta|^2dx\right)^{1/2}\cdot C\cdot n^{-1/2}\,,$$

where C depends on $B(R)$ and $L^2(B(R)\times(\tau,T))$–norms of $[u]$ and $[U]$.

For the boundary term in (3.5) we have, after the Remark,

$$(3.13)\qquad -\int_\tau^T dt\int_{\partial B(R)}[U]\Big(D\psi\cdot\frac{x}{R}\Big)dS(x)dt\le C\cdot R_1^{N-2}R^{1-N}\int_\tau^T\int_{\partial B(R)}|[U]|dS\,dt$$
$$\text{for } R\ge 2^{\frac{1}{N-2}}R_1\,.$$

To conclude the proof, select $T,R_1>0$, and let $\theta\in C_0^\infty(B(R_1))$, $0\le\theta\le 1$. For every $\varepsilon>0$, choose $R\ge 2^{1/(N-2)}R_1$ so that the bound in 3.13 be $\le\varepsilon$ for all $0<\tau<T$, by (3.4). Then pick $\tau>0$ so that $\int_{B(R)}[u]^+(x,\tau)dx<\varepsilon$.

We now solve (3.9), (3.10) for test functions ψ_n that satisfy the estimates in the Remark, in particular (3.13), independently of n. With T, R and $\tau>0$ already fixed, we can find n to make the bound (3.12) less than ε, and reverting to identity (3.5) for $[u]$, $[U]$, we have for a given $0\le\theta\le 1$,

$$\int_{B(R_1)}[u](x,T)\theta(x)dx\le 3\varepsilon\quad\text{for every }\varepsilon>0\,.$$

Therefore $[u]^+(x,T)=0$, a.e. $x\in B(R_1)$ and $T>0$, as desired.

COROLLARY 1. Under the assumption (1.5):

$$\text{measure}\big\{(x,t):|(U_1-U_2)(x,t)|>\varepsilon\big\}<\infty$$

for every $\varepsilon>0$, we obtain (3.4) if $[U]=U_1-U_2$ is integrable on sets of finite measure, as in the case of our example, cf. Section 2.

COROLLARY 2. In the one–dimensional case ($N=1$) boundedness of $[U]=U_1-U_2$ replaces condition (3.4), as the boundary term in (2) tends to zero, $R\to\infty$ (use $v(x,t):=(R-x)/(R-R_1)$ in Remark). Therefore $u(x,t)\equiv u_I(x)$ is the only solution to our example, in R^1, cf. Section 2.

COROLLARY 3. The Theorem is valid with $(U_1-U_2)^+$ replacing $|U_1-U_2|$ in (3.4).

REMARK. Conditions (1.5) or (3.4) take care of the behavior of the functions U when $U(x,t)$ belongs to a vertical segment of the graph φ for $|x|\to\infty$. They are too stringent when φ is a function of u (cf. [P], [DK1], [DK2] where initial data are taken as measures, and [B]).

Acknowledgments. This work was done while visiting the Equipe de Mathématiques, Université de Besançon, on leave from the U. de Buenos Aires and with partial support of CONICET, Argentina. I wish to thank Ph. Bénilan and C.E. Kenig for many stimulating conversations on the subject.

References.

[AC] D.G. Aronson, L.A. Caffarelli, The initial trace of a solution of the porous medium equation, Trans. A.M.S. 280(1) (1983), 351–366.

[BC] H. Brézis, M.G. Crandall, Uniqueness of solutions of the initial–value problem for $u_t - \Delta\varphi(u) = 0$, J. Math. pures et appl. 58 (1979), 153–163.

[B] J.E. Bouillet, Signed solutions to diffusion–heat conduction equations, in Free–Boundary Problems: Theory and Applications, Proc. Int. Colloq. Irsee/Germany 1987. Pitman Res. Notes Math. Ser. 186 (1990), 480–485.

[DK1] B.E.J. Dahlberg, C.E. Kenig, Non–negative solutions of the porous medium equation, Comm. in P.D.E. 9(5) (1984), 409–437.

[DK2] B.E.J. Dahlberg, C.E. Kenig, Non–negative solutions of generalized porous medium equations, Revista Matemática Iberoamericana 2(3) (1986), 267–305.

[O] O.A. Oleinik, A method of solution of the general Stefan problem, Soviet Math. Doklady (1960), 1350–1353.

[P] M. Pierre, Uniqueness of the solutions of $u_t - \Delta\varphi(u) = 0$ with initial datum a measure, Nonl. Analysis T.M.A. 6 (1982), 175–187.

Some Results on Abstract Evolution Equations of Hyperbolic Type

PIERMARCO CANNARSA Dipartimento di Matematica, Università di Roma "Tor Vergata", Via O. Raimondo, 00173, Roma, Italy

GIUSEPPE DA PRATO Scuola Normale Superiore di Pisa, 56126 Pisa, Italy

1 Introduction

The theory of non–autonomous abstract evolution equations

$$\begin{cases} u'(t) = A(t)u(t) + f(t), t \in [0,T] \\ u(0) = x \end{cases} \tag{1.1}$$

in a Banach space X has a long history starting from [5], in which the domains of $A(t)$, $D(A(t))$, are assumed to be independent of t. The case of non–constant domains is much harder, especially when the family $\{A(t)\}_{t\in[0,T]}$ is of "hyperbolic" type. The basic works on this subject are [6] and [7], in which the notion of stable family of generators plays an essential role (see §2 for details). Roughly speaking, in [6] and [7], a fundamental solution of problem (1.1) is constructed under the assumption that $\{A(t)\}_{t\in[0,T]}$ is a stable family of generators in X and that there exists another Banach space Y, continuously and densely embedded in X, such that the part of $A(t)$ in Y[1] is a stable family of generators in Y and, moreover,

$$A(\cdot) \in L^\infty(0,T;\mathcal{L}(Y,X)). \tag{1.2}$$

Partially supported by the Italian National Project MURST "Equazioni di Evoluzione e Applicazioni Fisico-Matematiche"

[1]If $L : D(L) \subset X \to X$ is a linear operator, the *part* of L in Y is the linear operator L_Y defined as $D(L_Y) = \{y \in D(L) \cap Y \ : \ Ly \in Y\}$, $L_Y y = Ly, \forall y \in D(L_Y)$.

Afterwords, in [4], assumption (1.2) has been weakened to the following:

$$A(\cdot)y \in L^\infty(0,T;Y)\ , \quad \forall\, y \in Y. \tag{1.3}$$

On the other hand, in [4] one has to assume that Y is contained in $D(A^2(t))$ (or in a suitable interpolation space between $D(A(t))$ and $D(A^2(t))$).

In the present paper, we will weaken (1.3) assuming a suitable sommability hypothesis on $A(t)y$, see §2. This generalization is motivated by the analysis of Kolmogoroff equations in infinitely many variables, as we explain in §3.

2 The Main Result

Let X, $|\cdot|$ and Y, $\|\cdot\|$ be Banach spaces such that $Y \subset X$ with a continuous and dense inclusion.

Let us consider a family of linear operators $\{A(t)\}_{t\in[0,T]}$, $A(t) : D(A(t)) \subset X \to X$.

Definition 2.1 *Given numbers $\omega, M \in \mathbf{R}$ we say that $\{A(t)\}_{t\in[0,T]}$ is a (M,ω)–stable family of generators in X if $\rho(A(t)) \supset]\omega, +\infty[$ ([2]) for any $t \in [0,T]$, and for all $n \in \mathbf{N}$, $0 \le t_n \le \ldots \le t_1 \le T$ and $\lambda > \omega$ we have*

$$\|R(\lambda, t_1, \ldots, t_n)\|_{\mathcal{L}(E)} \le \frac{M}{(\lambda-\omega)^n}, \tag{2.1}$$

where

$$R(\lambda; t_1, \ldots, t_n) = R(\lambda, A(t_1)) \cdots R(\lambda, A(t_n)).$$

We will study problem (1.1) under the following assumptions on $\{A(t)\}_{t\in[0,T]}$.

Hypothesis 2.2 *There exist numbers $\omega, \eta, M, N \in \mathbf{R}$ such that:*

(i) *$\{A(t)\}_{t\in[0,T]}$ is a (M,ω)–stable family of generators in X;*

(ii) *$R(\lambda, A(\cdot))x$ is Bochner measurable as a map $[0,T] \to X$ for all $x \in X$;*

(iii) *$Y \subset D(A(t)), \forall\, t \in [0,T]$ with continuous and dense inclusion;*

(iv) *$\{A_Y(t)\}_{t\in[0,T]}$ is a (N,η)–stable family of generators in Y;*

(v) *$R(\lambda, A_Y(\cdot))y$ is Bochner measurable as a map $[0,T] \to Y$ for all $y \in Y$;*

(vi) *there exists $\alpha \in]0,1[$ and a positive function $c \in L^1(0,T)$ such that*

$$|A(t)y|_{D_{A(t)}(\alpha,\infty)} \le c(t)\|y\|, \forall\, y \in Y, \forall\, t \in [0,T]. \tag{2.2}$$

[2]If $L : D(L) \subset X \to X$ is a linear operator, we denote by $\rho(L)$ and $\sigma(L)$ the resolvent set and the spectrum of L, respectively.

In the assumption (vi) above we have denoted by $D_{A(t)}(\alpha,\infty)$ the real interpolation space between X and $D(A(t))$ which consists of all $x \in X$ such that

$$|x|_{D_{A(t)}(\alpha,\infty)} := |x| + \sup_{n \in \mathbf{N}} |n^\alpha A(t)R(n, A(t))x| < \infty.$$

We now turn to problem (1.1), i.e.

$$\begin{cases} u'(t) = A(t)u(t) + f(t), t \in [0,T] \\ u(0) = x \end{cases} \tag{2.3}$$

assuming $f \in L^1(0,T;X)$ and $x \in X$.

We denote by $L^1(0,T;D(A(t)))$ the space of all functions $u \in L^1(0,T;X)$ such that $A(\cdot)u(\cdot) \in L^1(0,T;X)$.

Definition 2.3 *A* strict solution *of problem* (2.3) *is a function*

$$u \in W^{1,1}(0,T;X) \cap L^1(0,T;D(A(t)))$$

fullfilling (2.3).

A function $u \in C([0,T];X)$ *is said to be a* strong solution *of problem* (2.3) *if there exist sequences* $\{u_n\}_{n\in\mathbf{N}} \subset W^{1,1}(0,T;X) \cap L^1(0,T;D(A(t)))$ *and* $\{f_n\}_{n\in\mathbf{N}} \subset L^1(0,T;X)$ *such that* $u_n(0) \to x$ *and*

$$u_n \to u \text{ in } C([0,T];X) \ \& \ f_n \to f \text{ in } L^1(0,T;X)$$

as $n \to \infty$, *and*

$$u_n'(t) = A(t)u_n(t) + f_n(t)$$

for all $n \in \mathbf{N}$ *and a.e.* $t \in [0,T]$.

Remark 2.4 An equivalent way of introducing strict and strong solutions of (2.3) is the following. Define the operator

$$\begin{cases} \gamma_0 : D(\gamma_0) \subset L^1(0,T;X) \to L^1(0,T;X) \oplus X \\ D(\gamma_0) = W^{1,1}(0,T;X) \cap L^1(0,T;D(A(t))) \\ \gamma_0(u) = \{u'(\cdot) - A(\cdot)u(\cdot), u(0)\} \end{cases} \tag{2.4}$$

Clearly, u is a *strict* solution of equation (2.3) if and only if

$$u \in D(\gamma_0) \ \& \ \gamma_0(u) = \{f, x\}$$

Moreover, denote by γ the closure of γ_0, that is γ the (possibly multi–valued) operator whose graph is the closure of the graph of γ_0. Then, it is easy to show that u is a *strong* solution of equation (2.3) if and only if

$$u \in D(\gamma) \ \& \ \gamma(u) \ni \{f, x\}.$$

As a matter of fact under Hypotheses 2.2(i)–(ii), operator γ_0 is closable, so that γ is single valued and the inclusion above becomes an equality (see [4]). However, we will not use this fact in the sequel. ∎

The lemma below is proved in [4]. However, we recall its proof for completeness.

Lemma 2.5 *Assume Hypothesis 2.2(i)–(ii). Then the range of γ is closed, γ is one–to–one and*

$$D(\gamma) \subset \{u \in C([0,T];X) \; : \; u(0) = x\}.\text{[3]}$$

Proof — First of all, let $v_n(\cdot; s, x)$ be the (classical) solution of the homogeneous equation

$$v'(t) \; = \; A_n(t)v(t) \; , \quad v(s) \; = \; x \in X \tag{2.5}$$

where $A_n(t) = nA(t)R(n, A(t))$ is the usual Yosida approximation of $A(t)$. Equation (2.5) is equivalent to

$$v'(t) \; = \; -nv(t) + n^2 R(n, A(t))v(t) \; , \quad v(s) \; = \; x$$

and in turn to the integral equation

$$v(t) \; = \; e^{-nt}x + \int_0^t e^{-n(t-s)} n^2 R(n, A(s))v(s)ds.$$

By a standard iteration argument one can easily get

$$v_n(t; s, x) \; = \; \sum_{k=0}^{\infty} n^{2k} e^{-n(t-s)} \int_{\Delta_k(t,s)} R(n, t_1, ..., t_k)x \; dt_1 ... dt_k \tag{2.6}$$

where

$$\Delta_k(t, s) \; = \; \{(t_1, ..., t_k) \; : \; s \le t_1 \le t, ..., s \le t_k \le t_{k-1}\}.$$

Since $\text{meas}(\Delta_k(t, s)) = \frac{(t-s)^k}{k!}$, from (2.1) and (2.6) we obtain the estimate

$$|v_n(t)| \; \le \; M e^{\frac{n\omega t}{n-\omega}} |x| \; , \; \forall \, n \in \mathbf{N}. \tag{2.7}$$

Next, we consider the non–homogeneous equation

$$u'(t) \; = \; A_n(t)u(t) + f(t) \; , \quad u(0) \; = \; x \tag{2.8}$$

where $x \in X$ and $f \in L^1(0, T; X)$. By the standard variation of constants formula and (2.7), on the solution u_n of the above problem we easily get the estimate

$$|u_n(t)| \; \le \; M e^{\frac{n\omega t}{n-\omega}} |x| + M \int_0^t e^{\frac{n\omega(t-s)}{n-\omega}} |f(s)| ds \; , \; \forall \, n \in \mathbf{N}. \tag{2.9}$$

Let now $u \in D(\gamma_0)$ and $\gamma_0(u) = \{f, x\}$. We claim that

$$|u(t)| \; \le \; M e^{\omega t} |x| + M \int_0^t e^{\omega(t-s)} |f(s)| ds. \tag{2.10}$$

Indeed, we have

$$u'(t) \; = \; A_n(t)u(t) + f(t) + [A(t) - A_n(t)]u(t) \; , \quad u(0) \; = \; x$$

for all $n \in \mathbf{N}$. Hence, (2.10) follows applying (2.9) and letting $n \to \infty$.

Since γ is the closure of γ_0, estimate (2.10) also holds for any $y \in D(\gamma)$ with $\gamma(u) \ni \{f, x\}$.

We are now in a position to prove the conclusions of the lemma. Indeed, (2.10) immediately implies that γ is one–to–one and also, after some standard deductions, that the range of γ is closed. Finally, (2.10) yields that any $u \in D(\gamma)$ belongs to $L^\infty(0, T; X)$ and the continuity of such elements follows by a standard density argument. ∎

The main result of this paper is the following.

[3]The inclusion should be interpreted in the sense that each element of $D(\gamma)$ (which belongs to $L^1(0, T; X)$) possesses a continuous representative.

Theorem 2.6 *Assume Hypothesis 2.2 and let $x \in X$, $f \in L^1(0,T;X)$. Then there exists a unique strong solution u of* (1.1).

Proof — Since the above lemma states that the range of γ is closed, in order to prove existence of strong solutions it will suffice to show that the range of γ is also dense. For this purpose, let us consider the approximating problem (2.8) taking $x \in Y$ and $f \in L^1(0,T;Y)$. Obviously, problem (2.8) has a unique solution $u_n \in C([0,T];Y)$ and the following estimate is proved as estimate (2.10):

$$\|u_n(t)\| \leq Ne^{\eta t}\|x\| + N\int_0^t e^{\eta(t-s)}\|f(s)\|ds, \quad \forall\, t \in [0,T]. \tag{2.11}$$

Since $Y \subset D(A(t))$, we also have

$$\begin{cases} u_n'(t) - A(t)u_n(t) = [A_n(t) - A(t)]u_n(t) + f(t), & t \in [0,T] \\ u_n(0) = x \end{cases}$$

We now claim that

$$\lim_{n\to\infty}\int_0^T |[A_n(t) - A(t)]u_n(t)|dt \;=\; 0, \tag{2.12}$$

which will in turn imply that the closure of the range of γ contains $L^1(0,T;Y) \oplus Y$. The conclusion will follow as $L^1(0,T;Y) \oplus Y$ is dense in $L^1(0,T;X) \oplus X$. To prove (2.12) we note that

$$\begin{aligned}[A_n(t) - A(t)]u_n(t) &= A^2(t)(nI - A(t))^{-1}u_n(t) \\ &= n^{-\alpha}[n^\alpha A(t)(nI - A(t))^{-1}A(t)]u_n(t).\end{aligned}$$

Hence, recalling Hypothesis (2.2)(vi), we get

$$|[A_n(t) - A(t)]u_n(t)| \leq n^{-\alpha}|A(t)u_n(t)|_{D_{A(t)}(\alpha,\infty)} \leq n^{-\alpha}c(t)\|u_n(t)\|.$$

Therefore, recalling (2.11)

$$\begin{aligned}&\textstyle\int_0^T |[A_n(t) - A(t)]u_n(t)|dt \leq \\ &\qquad \leq Nn^{-\alpha}\textstyle\int_0^T c(t)dt\left[C_\eta\|y\| + \int_0^T e^{\eta(t-s)}\|f(s)\|ds\right].\end{aligned}$$

where $C_\eta = \max\{e^{\eta T}, 1\}$. The last inequality yields (2.12).

As the uniqueness of strong solutions is an immediate consequence of the fact that γ is one–to–one, the proof is now complete. ∎

We now study existence of strict solutions. To this purpose, in addition to Hypothesis 2.2 we need the following assumptions.

Hypothesis 2.7 *There exists a Banach space Z, with norm $|\!|\!| \cdot |\!|\!|$, continuosly and densely embedded in Y, and numbers $L, \zeta \in \mathbf{R}$ such that*

(i) *$\{A_Z(t)\}_{t\in[0,T]}$ is a (L,ζ)–stable family of generators in Z;*

(ii) *$R(\lambda, A_Z(\cdot))z$ is Bochner measurable as a map $[0,T] \to Z$ for all $z \in Z$;*

(iii) *$Z \subset D(A_Y(t)), \forall\, t \in [0,T]$ with continuous and dense inclusion;*

(iv) *there exists* $\beta \in]0,1[$ *and a positive function* $k \in L^1(0,T)$ *such that*

$$\|A_Y(t)z\|_{D_{A_Y(t)}(\beta,\infty)} \leq k(t)\|z\|, \forall\, z \in Z, \forall\, t \in [0,T]. \tag{2.13}$$

In the assumption (iii) above we have used the notation

$$\|z\|_{D_{A_Y(t)}(\beta,\infty)} := \|z\| + \sup_{n\in\mathbf{N}} \|n^\alpha A(t)R(n, A(t))z\| < \infty.$$

Theorem 2.8 *Assume Hypotheses 2.2 and 2.7 and let* $f \in L^1(0,T;Y)$ *and* $x \in Y$. *Then problem (2.3) has a unique strict solution* u. *Moreover,*

$$u \in W^{1,1}(0,T;X) \cap C([0,T];Y).$$

Proof — Applying theorem 2.6 to the family $\{A_Y(t)\}$, we conclude that there exists a function $u \in C([0,T];Y)$ and sequences $\{u_n\}_{n\in\mathbf{N}} \subset W^{1,1}(0,T;Y) \cap L^1(0,T;D(A_Y(t)))$ and $\{f_n\}_{n\in\mathbf{N}} \subset L^1(0,T;Y)$ such that $u_n(0) \to x$ in Y and

$$u_n \to u \text{ in } C([0,T];Y) \ \ \& \ \ f_n \to f \text{ in } L^1(0,T;Y)$$

as $n \to \infty$, and

$$u_n'(t) = A_Y(t)u_n(t) + f_n(t)$$

for all $n \in \mathbf{N}$ and a.e. $t \in [0,T]$.

Since Y is continuously embedded in $D(A(t))$ and all operators $A(t)$ are closed, we conclude that

$$A(t)u_n(t) \to A(t)u(t), \forall\, t \in [0,T] \tag{2.14}$$

as $n \to \infty$. Moreover, using (2.2), we deduce that the convergence in (2.14) actually occurs in $L^1(0,T;X)$. Therefore,

$$u_n'(t) \to A_Y(t)u(t) + f(t)$$

in $L^1(0,T;X)$ and $u \in W^{1,1}(0,T;X)$. ∎

3 Application to Kolmogoroff Equations

Let H be a separable Hilbert space and let $C_b^\infty(H)$ be the set of all mappings from H into $\mathbf{R}$ that are continuously differentiable and bounded togheter with all their derivatives. We denote by $\widetilde{C}_b(H)$ the closure of $C_b^\infty(H)$ with respect to the norm

$$\|\varphi\|_0 = \sup_{x\in H} |\varphi(x)|.$$

So $\widetilde{C}_b(H)$ is a Banach space which however is a proper closed subset of the space $C_b(H)$ of all mappings from H into $\mathbf{R}$ that are uniformly continuous and bounded see [10]. Analogously, for any positive integer K, we define the space $\widetilde{C}_b^k(H)$ to be the closure of $C_b^\infty(H)$ with respect to the norm

$$\|\varphi\|_k = \sum_{h=0}^{k} \sup_{x\in H} |\varphi^h(x)|,$$

where φ^h denotes the derivative of order h of φ.

We will apply our abstract result of §2 to problem

$$\begin{cases} \frac{\partial v}{\partial t}(t,x) = \frac{1}{2} \text{ Tr } [e^{tB}Qe^{tB^*}v_{xx}(t,x)] \text{ in }]0,T] \times H \\ \\ v(0,x) = \varphi(x), \quad \varphi \in \widetilde{C}_b(H), \end{cases} \tag{3.1}$$

where B is the infinitesimal generator of a strongly continuous semigroup of contractions e^{tB} on H, e^{tB^*} is the adjoint of e^{tB}, Q is a non– negative bounded self–adjoint operator on H, (possibly equal to the identity) and Tr represents the trace. We remark that equation (3.1) arises in studying some Kolmogoroff equations in Hilbert space, see [2].

In order to set problem (3.1) in the abstract form (2.3), let us introduce the linear operator in $\widetilde{C}_b(H)$

$$\Delta(t)\varphi = \frac{1}{2} \text{ Tr } (e^{tB}Qe^{tB^*}\varphi_{xx}); \ \forall\, \varphi \in \widetilde{C}_b^2(H),\ t \in [0,T] \tag{3.2}$$

We shall assume

Hypothesis 3.1 $\text{Tr}e^{tB}Qe^{tB^*} < +\infty,\ \forall\, t > 0$.

Remark 3.2 In particular, Hypothesis 3.1 implies that, for any $t \in]0,T]$ there exists a complete orthonormal system, $\{e_k(t)\}_{k\in\mathbf{N}}$ in H of eigenvectors of the self–adjoint operator $e^{tB}Qe^{tB^*}$, with corresponding eigenvalues $\{\lambda_k(t)\}_{k\in\mathbf{N}}$, so that $\lambda_k(\cdot)$ and $e_k(\cdot)$ are measurable mappings for all $k \in \mathbf{N}$.

Such a system can be constructed in the following way. Set

$$\lambda_1(t) \ := \ \max_{|x|\le 1} < e^{tB}Qe^{tB^*}x, x >$$

Clearly, $\lambda_1(t)$ is an eigenvalue of $e^{tB}Qe^{tB^*}$ which is measurable in t. Moreover, due to a well known selector theorem (see [8]), there exists a corresponding eigenvector $e_1(t)$ which is also measurable. Similarly, setting

$$\begin{aligned} \lambda_2(t) \ : \ &= \ \max_{|x|\le 1}\left\{< e^{tB}Qe^{tB^*}x, x > - | < x, e_1(t) > |^2\right\} \\ &= \ :< e^{tB}Qe^{tB^*}e_2(t), e_2(t) > . \end{aligned}$$

one constructs the second measurable eigenvalue– eigenvector pair $\{\lambda_2(t), e_2(t)\}$. Then, iterating this procedure easily yields the conclusion. ∎

Therefore, for any fixed $t \in]0,T]$, every function $\varphi \in \widetilde{C}_b(H)$ can be represented as a function of infinitely many variables

$$\varphi(x) \ = \ \varphi^{[t]}(x_1, x_2, ..., x_n, ..),\ x_n =< x, e_n(t) > .$$

The following result is proved in [2].

Proposition 3.3 *Assume Hypothesis* 3.1, *then, for all* $t > 0$, *the linear operator* $\Delta(t)$ *in* $\widetilde{C}_b(H)$ *is closable and its closure* $A(t)$ *is the infinitesimal generator of a strongly continuous semigroup of contractions,* $e^{sA(t)}, s \ge 0$, *in* $\widetilde{C}_b(H)$. *Moreover, for any positive integer k, the* *part of $A(t)$ in $\widetilde{C}_b^k(H)$ is the infinitesimal generator of a strongly continuous semigroup of contractions in $\widetilde{C}_b^k(H)$. Furthermore, for all $\varphi \in \widetilde{C}_b(H)$, we have that*

$$e^{sA(t)}\varphi := \lim_{n\to\infty} \prod_{h=1}^{n} T_{h,t}(s)\varphi, \tag{3.3}$$

uniformly on the bounded sets of $[0, \infty[$, where the above limit is to be understood in $C_b(H)$ and

$$(T_{h,t}(s)\varphi)(x) = \frac{1}{\sqrt{2\pi\lambda_h(t)s}} \int_{-\infty}^{+\infty} e^{-\frac{(x_h-\xi)^2}{2s\lambda_h(t)}} \varphi^{[t]}(x_1, ..., x_{h-1}, \xi, x_{h+1}, ...)d\xi, \tag{3.4}$$

for all $x \in X$.

To apply Theorem 2.6 we need a stronger hypothesis

Hypothesis 3.4 *There exists $\alpha \in]0,1[$ such that*

$$\int_0^T \left[\mathrm{Tr}(e^{tB}Qe^{tB^*})\right]^{1+\alpha} dt < \infty. \tag{3.5}$$

The main result of this section is the following

Theorem 3.5 *Assume Hypothesis 3.4 and let $\varphi \in \widetilde{C}_b(H)$. Then (3.1) has a unique strong solution.*

If moreover $\varphi \in \widetilde{C}_b^3(H)$ the solution is strict.

Proof — Let us set $X = \widetilde{C}_b(H)$,
$Y = (\widetilde{C}_b^4(H), \widetilde{C}_b^2(H))_{1-\alpha,\infty}$([4]). Let $A(t)$ be given by Proposition 3.2 for $t \in]0,T]$ and set $A(0) = 0$. We will check that assumptions of Hypothesis 2.2 hold true. Indeed, conditions (i) and (iv) follow from the property, stated in proposition 3.2, that $A(t)$ generates a contraction semigroup in all spaces $\widetilde{C}_b^k(H)$. Moreover, (iii) is clear from the construction. We now prove (ii) (the proof of (v) is analogous). Recalling Remark 3.2, we have that $\{\lambda_k, e_k\}$ is measurable for all $k \in \mathbf{N}$. Hence, by (3.4), $T_{h,\cdot}$ is measurable for all $k \in \mathbf{N}$ and so is $e^{sA(\cdot)}$ is view of (3.3). Since the mesasurability of the resolvent is equivalent to the measurability of the corresponding semigroup, (iii) follows. Finally, let us prove (vi). For any $\varphi \in \widetilde{C}_b^2(H)$ we have

$$\|\Delta(t)\varphi\|_0 \;\le\; \frac{1}{2}\mathrm{Tr}(Q)\;\|\varphi\|_2.$$

Moreover, for any $\varphi \in \widetilde{C}_b^4(H)$ we have

$$\|\Delta^2(t)\varphi\|_0 \;\le \frac{1}{4}[\mathrm{Tr}(Q)]^2\;\|\varphi\|_4.$$

Hence, the interpolation inequality (see [9]) yields

$$|A(t)\varphi|_{D_{A(t)}(\alpha,\infty)} \;\le\; \frac{1}{2^{1+\alpha}}[\mathrm{Tr}(Q)]^{1+\alpha}\|\varphi\|_{(\widetilde{C}_b^4(H),\widetilde{C}_b^2(H))_{1-\alpha,\infty}}$$

[4] We recall that, given Banach spaces $E_0 \subset E_1$, the real interpolation space $(E_0, E_1)_{1-\alpha,\infty}$ is defined as the space of the traces $u(0)$ of all functions $u : [0,\infty[\to E_1$ such that $t^{1-\alpha}u \in L^\infty(0,\infty; E_0)$ and $t^{1-\alpha}u' \in L^\infty(0,\infty; E_1)$, see [9]

and so Theorem 2.6 applies with

$$c(t) = \frac{1}{2^{1+\alpha}}\left[\mathrm{Tr}(e^{tB}Qe^{tB^*})\right]^{1+\alpha}.$$

The last statement is proved similarly by applying Theorem 2.2 with

$$Z = (\widetilde{C}_b^4(H), \widetilde{C}_b^2(H))_{1-\alpha,\infty}. \ ■$$

Example 3.6 Consider the case of $Q = I$ and $Be_k = -\mu_k e_k$, with $\mu_k \uparrow \infty$, for a given complete orthonormal system in H, $\{e_k\}_{k\in\mathbf{N}} \subset D(B)$. Then Hypothesis 3.4 holds provided

$$\sum_{k=1}^{\infty} \mu_k^{-\frac{1}{1+\alpha}} < \infty. \tag{3.6}$$

Indeed, let $f : [0,\infty[\to \mathbf{R}$ be a smooth non- decreasing function such that $f(k) = \mu_k$. Then,

$$\int_0^T \left[\ \mathrm{Tr}\ (e^{2tB})\right]^{1+\alpha} dt = \int_0^T \left[\sum_{k=1}^{\infty} e^{-2tf(k)}\right]^{1+\alpha} dt$$

$$\leq \int_0^T \left[\int_0^\infty e^{-2tf(x)}dx\right]^{1+\alpha} dt.$$

On the other hand,

$$\int_0^T \left[\int_0^\infty e^{-2tf(x)}dx\right]^{1+\alpha} dt = \sup_{\|\psi\|_{1+1/\alpha}\leq 1} \int_0^T \psi(t)dt \int_0^\infty e^{-2tf(x)}dx,$$

where $\|\psi\|_{1+1/\alpha} = \left\{\int_0^T |\psi(t)|^{1+1/\alpha}dt\right\}^{\frac{\alpha}{1+\alpha}}$. Let $\psi \in L^{1+1/\alpha}(0,T)$ with unit norm, then

$$\int_0^T \psi(t)dt \quad \int_0^\infty e^{-2tf(x)}dx$$

$$\leq \int_0^\infty dx \left(\int_0^T e^{-2(1+\alpha)tf(x)}dt\right)^{\frac{1}{1+\alpha}} \left(\int_0^T |\psi(t)|^{1+1/\alpha}dt\right)^{\frac{\alpha}{1+\alpha}}$$

$$\leq \int_0^\infty \left(\frac{1}{2(1+\alpha)f(x)}\right)^{\frac{1}{1+\alpha}} dx.$$

The conclusion follows, since ψ is artbiray.

In particular, we note that, if $\mu_k = k^\theta, \theta > 1$, then condition (3.6) holds if and only if $\alpha \in]0, \theta - 1[$. ■

References

[1] CANNARSA P. & DA PRATO G. (1991) *A semigroup approach to Kolmogoroff equations in Hilbert spaces*, Appl. Math. Lett. **4**, 49–52.

[2] CANNARSA P. & DA PRATO G. (1991)*On a functional analysis approach to parabolic equations in infinite dimensions*, Universitá di Roma Tor Vergata, Preprint Dipartimento di Matematica, $n^0 1$.

[3] DALECKII YU. (1966) *Differential equations with functional derivatives and stochastic equations for generalized random processes*, Dokl. Akad. Nauk SSSR **166**, 1035–1038.

[4] DA PRATO G. & IANNELLI M. (1976) *On a method for studying abstract evolution equation in the hyperbolic case*, Comm. in Partial Differential Equations **1** (6), 586–608.

[5] KATO T. (1953) *Integration of the equation of evolution in Banach space*, J. Math. Soc. Japan **5**, 208–234.

[6] KATO T. (1970)*Linear evolution equations of "hyperbolic" type*, J. Fac. Sci. Univ. Tokyo, Sect I **17**, 241–258.

[7] KATO T. (1973) *Linear evolution equations of "hyperbolic type" II*, J. Math. Soc. Japan **25**, 648–666.

[8] KURATOWSKI K. & RYLL- NARDZEWSKI C. (1965), *A general theorem on selectors*, Bull. Acad. Pol. Sc. **13**, 397–403.

[9] LIONS J.L. & PEETRE J. (1964),*Sur une classe d'espaces d'interpolation*, Publ. Math de l'I.H.E.S. **19**, 5-68.

[10] NEMIROVSKI A.S. & SEMENOV S.M. (1973), *The polynomial approximation of functions in Hilbert spaces*, Mat. Sb. (N.S), **92**, 134, 257-281.

[11] PAZY A. (1983) SEMIGROUPS OF LINEAR OPERATORS AND APPLICATIONS TO PARTIAL DIFFERENTIAL EQUATIONS, Springer-Verlag,Berlin, New-York.

Interpolation and Extrapolation Spaces and Parabolic Equations

GABRIELLA DI BLASIO Dipartimento di Matematica, Università di Roma "La Sapienza", Piazzale Aldo Moro, I-00185 Roma, Italy

1 Introduction

Let X be a Banach space with norm $\|\cdot\|$ and let $A : D(A) \subset X \to X$ be a linear operator which generates an analytic semigroup $S(t)$. In this note we want to discuss a new approach for the study of the parabolic problem:

$$\begin{cases} u'(t) = Au(t) + f(t), \ t \in]0,T] \\ u(0) = x \ . \end{cases} \tag{1.1}$$

More specifically we are interested in those spaces Y such that if $f \in L^p(0,T;Y)$ (and x satisfies a compatibility condition) then (1.1) admits a solution u satisfying

$$u', \ Au \in L^p(0,T;Y) \tag{1.2}$$

As it is known (see [5]) if Y is a Hilbert space or Y is a ζ-convex space (see [8]) then (1.2) is satisfied. As usual we call a space of *maximal regularity* any space Y satisfying this property.

Example of Banach spaces of maximal regularity are the subspaces $X_{\theta,p}$ (for $0 < \theta < 1$ and $1 \le p < \infty$) of all elements $x \in X$ satisfying

$$\|x\|_{\theta,p} := \left(\int_0^{+\infty} (t^{1-\theta}\|AS(t)x\|)^p t^{-1} dt\right)^{1/p} < +\infty.$$

It is known (see [2]) that we have

$$D(\theta,p) = (X, D(A))_{\theta,p}$$

where $(X, D(A))_{\theta,p}$ denote the real interpolation spaces between $D(A)$ and X. We refer to [3] and [6] for the use of the spaces $D(\theta,p)$ in connection with (1.1).

In this paper we introduce the spaces $D(\theta,p)$, for $\theta = 0$. For these spaces the following inclusions hold

$$D(A) \subset D(\theta,p) \subset D(0,p+s)\ ,\ \theta \geq 0,\ s \geq 0$$

whereas the inclusion $D(0,p) \subset X$ is true for small p and may be false for large p. Therefore for large p the spaces $D(0,p)$ could be called extrapolation spaces, although this definition is different from that introduced by Da Prato and Grisvard [4] and Nagel [10] (see also Amann [1]). This motivates us to find a semigroup $\overline{S}(t)$ on $D(0,p)$ which coincides with $S(t)$ on $X_{0,p}$ and to characterize its generator $\overline{A}$.

The interest in studying the case $\theta = 0$ is due to the fact that also $D(0,p)$ are spaces of maximal regularity for the semigroup $\overline{S}(t)$. This property enables us to find new regularity results for (1.1) if p is small or to find generalized solutions if p is large.

2 The spaces D(0,p) and the semigroup $\overline{S}_p(t)$

Let X be a Banach space with norm $\|\cdot\|$ and let $A : D(A) \subset X \to X$ be a linear, closed and densely defined operator generating an analytic semigroup $S(t)$ on X. Without essential loss of generality we may assume that there exist $M, \alpha > 0$, satisfying

$$\|S(t)x\| \leq M\, e^{-\alpha t}\|x\| \tag{2.1}$$

and

$$\|tA\,S(t)x\| \leq M\, e^{-\alpha t}\|x\| \tag{2.2}$$

In what follows we denote by $X_{0,p}$, $1 \leq p < \infty$, the normed space of all elements $x \in X$ such that

$$\|x\|_{0,p} = \left(\int_0^{+\infty} t^{p-1}\|AS(t)x\|^p dt\right)^{1/p} \tag{2.3}$$

is finite. Using (2.1) and (2.2) we can prove that for each $x \in X_{0,p}$ we have:

$$\|S(t)x\|_{0,p} \leq M\, e^{-\alpha t}\|x\|_{0,p} \tag{2.4}$$

$$\|tA\,S(t)x\|_{0,p} \leq M\, e^{-\alpha t}\|x\|_{0,p} \tag{2.5}$$

$$\|S(t)x\| \leq 4\,M\, e^{-\alpha\frac{t}{2}}(1+|logt|)^{\frac{p-1}{p}}\|x\|_{0,p} \tag{2.6}$$

$$\|tA\,S(t)x\| \leq 4M\, e^{-\alpha/2t}\|x\|_{0,p} \tag{2.7}$$

where M and α satisfy (2.1) and (2.2).

Further let us denote by $D(0,p)$ the completion of $X_{0,p}$. Then we have the following results:

Lemma 2.1 *The following properties hold:*

(i) $D(0,1) = X_{0,1}$

(ii) $D(\theta,p) \subset D(0,p) \subset D(0,p+s)$, *for each* $s \geq 0$

Proof. See [7], section 2. ∎

By Lemma 2.1 (i) we see that $D(0,1) \subset X$. If p is large the inclusion $D(0,p) \subset X$ may be false as the following example shows.

Example 2.2 Let $X = L^p(\mathbf{R}^n), p > 1$, and let A be the operator defined as $D(A) = W^{2,p}(\mathbf{R}^n), Au = \Delta u$, (Δ = laplacian operator). Then we have (see [11])

$$D(0,p) \subset L^p(\mathbf{R}^n),\ if\ p < 2$$

$$D(0,2) = L^2(\mathbf{R}^n),$$

$$D(0,p) \supset L^p(\mathbf{R}^n),\ if\ p > 2$$

For a more detailed description of the properties of these spaces we refer to [11].

Therefore, using (2.4), we introduce the family of operators $\overline{S}_p(t)$ on $D(0,p)$ defined as

$$\overline{S}_p(t)x = \lim_{n\to\infty} S(t)x_n$$

where $\{x_n\}$ is a sequence in $X_{0,p}$ converging to x in $D(0,p)$. Moreover we denote by $A_p : D(A_p) \subset D(0,p) \to D(0,p)$ the operator defined as

$$D(A_p) = \{x \in D(A) : Ax \in X_{0,p}\}$$

$$A_p(x) = Ax$$

and denote by $\overline{A}_p$ the closure of A_p. Then we have:

Theorem 2.3 $\overline{S}_p(t)$ *is an analytic semigroup on* $D(0,p)$ *whose generator is* $\overline{A}_p$ *and we have*

(i) $\|\overline{S}_p(t)x\|_{0,p} \leq M' exp(-\alpha t)\|x\|_{0,p}$

(ii) $t\|\overline{A}_p\overline{S}_p(t)x\|_{0,p} \leq exp(-\alpha t)\|x\|_{0,p}$

Proof. The results follow from (2.4), (2.5) and the definition of $\overline{S}_p(t)$. ∎

Finally we have the following property concerning $\overline{A}_p$:

Theorem 2.4 *For each* $\theta \in [0,1[$ *we have*

$$D(\theta,p) \subset D(\overline{A}_p)$$

Proof. Let $\{x_n\} \subset D(A)$ be such that $\{Ax_n\}$ is Cauchy in $D(0,p)$. From (2.6) we have

$$\begin{aligned}
&\|x_n - x_m\|_{\theta,p}^p \\
&\quad = \int_0^{+\infty} \sigma^{p-1-p\theta} \|S(\sigma)A(x_n - x_m)\|^p \, d\sigma \\
&\quad \leq const \ \|A(x_n - x_m)\|_{0,p}^p
\end{aligned}$$

Hence if $\{x_n\}$ and $\{A(x_n)\}$ are Cauchy in $D(0,p)$ we have that $\{x_n\}$ is Cauchy in $D(\theta,p)$, and the theorem is proved. ∎

We now study the regularizing properties of the semigroup $\overline{S}_p(t)$. As we shall see these properties will provide regularity results for the parabolic equation associated with $\overline{A}_p$.

We use the following notation for spaces of functions from $[a,b]$ into a Banach space E where E is the space X or $D(\theta,p)$

- $L^p(a,b;E)$: is the Banach space of measurable functions u such that $\|u(\cdot)\|_E^p$ is integrable on $]a,b[$
- $C(a,b;E)$: is the Banach space of continuous functions on $[a,b]$

Given $x \in D(0,p)$ and $f \in L^p(0,T;D(0,p))$ we want to study the regularity properties of the functions

$$u_0(t) := \overline{S}_p(t)$$

and

$$u_1(t) := \int_0^t \overline{S}_p(t-s)f(s) \, ds \ .$$

We begin with the following result concerning u_0.

Theorem 2.5 *Let $x \in D(0,p)$. Then for each $\theta \in [0.1[$ we have*

$$u_0 \in L^p(0,T;D(\theta,p))$$

Proof. By density arguments it suffices to take $x \in X_{0,p}$. From (2.7) we get

$$\begin{aligned}
&\int_0^T |u_0(t)|_{\theta,p}^p \\
&\quad = \int_0^T dt \int_0^{+\infty} s^{p-1-p\theta} \|AS(t+s)x\|^p ds \\
&\quad \leq const \|x\|_{0,p} \int_0^T dt \int_0^{+\infty} s^{p-1-p\theta} \frac{e^{\frac{a}{2}(t+s)}}{(s+t)^p} ds
\end{aligned}$$

and the result follows. ∎

Theorem 2.6 *Let $x \in D(1-1/p,p)$. Then*

$$u_0', \ \overline{A}_p u_0 \in L^p(0,T;D(0,p)).$$

Proof. The assertion follows from

$$\int_0^T \|\overline{A}_p u_0(t)\|^p dt = \|x\|_{1-1/p,p}.$$

∎

Concerning u_1 we have:

Theorem 2.7 *Let $f \in L^p(0,T;D(0,p))$. Then*

(i) $u_1 \in C(0,T;D(1-1/p,p))$

(ii) $u_1', \ \overline{A}_p u_1 \in L^p(0,T;D(0,p))$

Proof. The assertions are proved in [7], sections 4 and 5. ∎

Remark 2.8 From Theorems 2.6 and 2.7 we have that $u_0', \overline{A}_p u_0 \in L^p(0,T;D(0,p))$ if and only if $x \in D(1-1/p,p)$.

3 Parabolic equations

In this section we see how the previous results can be used to obtain new regularity properties for the solutions of the parabolic problem

$$\begin{cases} u'(t) = Au(t) + f(t), \ t \in]0,T] \\ u(0) = x\,. \end{cases} \tag{3.1}$$

Together with problem (3.1) we consider the following

$$\begin{cases} u'(t) = \overline{A}_p u(t) + f(t), \ \ t \in]0,T] \\ u(0) = 0 \end{cases} \tag{3.2}$$

where, as defined in section 2, $\overline{A}_p$ is the closure of A in $D(0,p)$. If $D(0,p) \subset X$ it turns out that a solution of (3.2) is also a solution of (3.1). If $D(0,p) \supset X$ a solution of (3.2) is a *generalized solution* of (3.1). We have:

Theorem 3.1 *Let $x \in D(0,p)$ and let $f \in L^p(0,T;D(0,p))$. Then there exist u such that*

(i) $u \in C(0,T;D(0,p)) \cap C(\epsilon,T;D(1-1/p,p))$. *for each* $\epsilon > 0$

(ii) $u', \ \overline{A}_p u, \ \in L^p(\epsilon,T;D(0,p))$, *for each* $\epsilon > 0$

(iii) $u \in L^p(0,T;D(\theta,p))$, *for each* $\theta \in [0,1[$

moreover u satisfies (3.2).

Proof. Using Theorems 2.3 (ii), 2.5, 2.7 and 2.4 it is readily seen that the function

$$u := \overline{S}_p(t) + \int_0^t \overline{S}_p(t-s)f(s)ds$$

satisfies the required properties. ∎

Theorem 3.2 *Let $x \in D(1-1/p,p)$ and let $f \in L^p(0,T;D(0,p))$. Then there exists a unique u such that*

(i) $u \in C(0,T;D(1-1/p,p))$

(ii) $u',\ \overline{A}_p u,\ \in L^p(0,T;D(0,p))$

(iii) $u \in L^p(0,T;D(\theta,p))$, *for each* $\theta \in [0,1[$.

Moreover u satisfies (3.2).

Proof. Using Theorems (2.5), (2.6) and (2.4) it can be checked that the function

$$u := \overline{S}_p(t) + \int_0^t \overline{S}_p(t-s)f(s)ds$$

satisfies the required properties. ∎

Let us apply Theorems (3.1) and (3.2) in the case where $X = L^p(\mathbf{R}^n)$ and A is the realization of the laplacian operator in $L^p(\mathbf{R}^n)$. In this case we have (see [11])

$$D(0,p) \subset L^p(\mathbf{R}^n),\ if\ 1 < p < 2 \tag{3.3}$$
$$D(0,2) = L^2(\mathbf{R}^n), \tag{3.4}$$
$$D(0,p) \supset L^p(\mathbf{R}^n),\ if\ p > 2\ . \tag{3.5}$$

We shall consider the following problem

$$\begin{cases} u_t(t,\xi) = \Delta u(t,\xi) + f(t,\xi),\ t \in]0,T],\ \xi \in \mathbf{R}^n \\ u(0,\xi) = u_0(\xi),\ \xi \in \mathbf{R}^n\ . \end{cases} \tag{3.6}$$

If $1 < p < 2$ we therefore have the following regularity result for solutions of (3.6)

Theorem 3.3 *Let $u_0 \in D(0,p)$ and let $f \in L^p(0,T;D(0,p))$, for some $1 < p < 2$. Then there exists a solution u of problem (3.6). Moreover we have for each $\theta \in]0,1[$*

(i) $u \in C(0,T;L^p(\mathbf{R}^n)) \cap C(\epsilon,T;W^{2-\frac{2}{p},p}(\mathbf{R}^n))$, *for each* $\epsilon > 0$

(ii) $u \in L^p(0,T;W^{2\theta,p}(\mathbf{R}^n))$

(iii) $u_t, \Delta u \in L^p(]\epsilon,T[\times\mathbf{R}^n)$, *for each* $\epsilon > 0$

where $W^{\alpha,p}(\mathbf{R}^n)$ are the Sobolev spaces of non integer order.

If in addition $u_0 \in W^{2-\frac{2}{p},p}(\mathbf{R}^n)$ then we have:

(iii) $u \in C(0,T;W^{2-\frac{2}{p},p}(\mathbf{R}^n))$

(iv) $u_t,\ \Delta u \in L^p(]0,T[\times\mathbf{R}^n)$

Proof. The result follows from Theorems 3.1, 3.2, from (3.3) and the well known characterization of the interpolation spaces $D(\theta,p)$ between $D(\Delta) = W^{2,p}(\mathbf{R}^n)$ and $X = L^p(\mathbf{R}^n)$, (see e.g. [12]). ■

If $p = 2$ we get the well known maximal regularity result in Hilbert spaces (see [5]):

Theorem 3.4 *Let $u_0 \in L^2(\mathbf{R}^n)$ and let $f \in L^2(]0,T[\times\mathbf{R}^n)$. Then there exists a solution u of problem (3.6) and we have:*

(i) $u \in C(0,T;L^2(\mathbf{R}^n)) \cap C(\epsilon,T;W^{1,2}(\mathbf{R}^n))$, *for each* $\epsilon > 0$

(ii) $u_t,\ \Delta u \in L^2(]\epsilon,T[\times\mathbf{R}^n)$, *for each* $\epsilon > 0$

If in addition $u_0 \in W^{1,2}(\mathbf{R}^n)$, then:

(iii) $u \in C(0,T;W^{1,2}(\mathbf{R}^n))$

(iv) $u_t,\ \Delta u \in L^2(]0,T[\times\mathbf{R}^n)$

Proof. The results follow from Theorems 3.1 and 3.2 and from (3.4). ■

Finally if $p > 2$ we get the following existence result for generalized solutions to problem (3.6)

Theorem 3.5 *Let $f \in L^p(0,T;D(0,p))$, for some $p > 2$. Then there exists a function u satisfying in a generalized sense problem (3.6). Moreover we have for each $\theta \in]0,1[$*

(i) $u \in C(\epsilon,T;W^{2-\frac{2}{p},p}(\mathbf{R}^n)) \cap L^p(0,T;W^{2\theta,p}(\mathbf{R}^n))$, *for each* $\epsilon > 0$

If in addition $u_0 \in W^{2-\frac{2}{p},p}(\mathbf{R}^n)$, then

(ii) $u \in C(0,T;W^{2-\frac{2}{p},p}(\mathbf{R}^n))$

Proof. The result follows from Theorems 3.1 and 3.2 and from (3.5). ■

References

[1] Amann H., Parabolic evolution equations in interpolation and extrapolation spaces, J. Funct. Anal., 78, (1988), 233-270.

[2] Butzer P.L. & Berens H. (1967) Semigroups of Operators and Approximation, Springer, Berlin.

[3] Da Prato G. & Grisvard P., Somme d' opérateurs linéaires et équations différentielles opérationelles, J. Math. Pures Appl., 54, (1975), 305-387.

[4] Da Prato G. & Grisvard P., Maximal regularity for evolution equations by interpolation and extrapolation, J. Funct. Anal., 58, (1984), 107-124.

[5] De Simon L., Un' applicazione della teoria degli integrali singolari allo studio delle equazioni differenziali lineari astratte del primo ordine, Rend. Sem. Mat. Padova, 34, (1964), 205-232.

[6] Di Blasio G., Linear parabolic equations in L^p-Spaces, Ann. Mat. Pura e Appl., IV, (1984),55-104.

[7] Di Blasio G., Holomorphic semigroups in interpolation and extrapolation spaces, Semigroup Forum, (to appear).

[8] Dore G. & Venni A., On the closedness of the sum of two closed operators, Math. Z., 196, (1987), 189-201.

[9] Lions J. L.& Magenes E., Problèmes aux limites non homogènes et applications, Ann.Sc. Norm. Pisa, 13, (1968), 389-403.

[10] Nagel R., Sobolev spaces and semigroups, Semesterbericht Funktionalanalysis, Tübingen, Sommersemester, (1983), 1-20.

[11] Taibleson M. H., On the theory of Lipschitz spaces of distributions on Euclidean n-spaces , J. Math. Mech., 13, (1964), 407-479.

[12] Triebel H. (1978) Interpolation Theory, Functions Spaces, Differential Operators, North-Holland, Amsterdam.

On the Diagonalization of Certain Operator Matrices Related to Volterra Equations

KLAUS-JOCHEN ENGEL Mathematisches Institut, Universität Tübingen, Auf der Morgenstelle 10, D-W-7400 Tübingen, Germany

In this paper we consider operator matrices closely related to linear Volterra equations of the type

$$\text{(VE)} \qquad \dot{u}(t) = Au(t) + \int_0^t B(t-s)u(s)ds + f(t), \qquad u(0) = x_0, \qquad t \geq 0$$

on a Banach space X for given linear operators A and $B(t)$.

A well known approach to study (VE) which was initiated by Miller [M] (see also [DS] and [CG]) is to transform (VE) into an abstract Cauchy problem

$$\text{(ACP)} \qquad \dot{v}(t) = \mathcal{A}v(t), \qquad v(0) = \binom{x_0}{f}.$$

Here $\mathcal{A}$ is the operator matrix

$$\mathcal{A} = \begin{pmatrix} A & \delta_0 \\ B & D_s \end{pmatrix} \text{ with domain } D(\mathcal{A}) = D(A) \times D(D_s)$$

on $X \times F(\mathbb{R}_+, X)$, where $F = F(\mathbb{R}_+, X)$ denotes a space of X-valued functions on $\mathbb{R}_+$, D_s the first derivative on F and δ_0 the dirac measure in 0. In this paper we choose $F = L^p(\mathbb{R}_+, X)$ for $1 \leq p < \infty$.

In fact, if $\mathcal{A}$ generates a strongly continuous semigroup $(\mathcal{T}(t))$, then the unique solution $u(.)$ of (VE) is given by the first component of $\mathcal{T}(.)\binom{x_0}{f}$.

This justifies a systematic analysis of operators $\mathcal{A}$ of the above form. In this paper we are mainly interested to transform $\mathcal{A}$ via a similarity transformation into a triangular or even a diagonal matrix. Note that these transformations neither change the spectral nor the generator property of $\mathcal{A}$. Since we want to demonstrate the technique we make the following quite restrictive assumtions in order to avoid technical details.

$$(H) \qquad \begin{aligned} &\text{(i) } A \in \mathcal{L}(X),\\ &\text{(ii) } B(.) \in C(\mathbb{R}_+, \mathcal{L}(X)),\\ &\text{(iii) } \|B(.)\| \in L^p(\mathbb{R}_+). \end{aligned}$$

Then it is an easy exercise to show the following.

Proposition 1. *If (H) is satisfied, then $\mathcal{A}$ generates a strongly continuous semigroup on $X \times L^p(\mathbb{R}_+, X)$.*

Next we define for arbitrary $Q \in \mathcal{L}(X, L^p(\mathbb{R}_+, X))$ and $L \in \mathcal{L}(L^p(\mathbb{R}_+, X), X)$ the operator matrices

$$\mathcal{S}_Q := \begin{pmatrix} Id & 0 \\ Q & Id \end{pmatrix} \qquad \text{and} \qquad \mathcal{S}^L := \begin{pmatrix} Id & L \\ 0 & Id \end{pmatrix}$$

on $X \times L^p(\mathbb{R}_+, X)$. Then $\mathcal{S}_Q$ and $\mathcal{S}^L$ are invertible and the inverses are given by $\mathcal{S}_{-Q}$ and $\mathcal{S}^{-L}$, respectively.

The problem is now to find Q, L such that

$$(\mathcal{S}_Q)^{-1} \mathcal{A} \mathcal{S}_Q = \begin{pmatrix} * & * \\ 0 & * \end{pmatrix} \qquad \text{or} \qquad (\mathcal{S}^L)^{-1} \mathcal{A} \mathcal{S}^L = \begin{pmatrix} * & 0 \\ * & * \end{pmatrix}.$$

As a first step we make the following definitions.

Definition 2. *For C, $G \in \mathcal{L}(X)$ and $x \in X$, $s \in \mathbb{R}_+$, $f \in \mathrm{L}^p(\mathbb{R}_+, X)$ define*

$$(Q_{C,G}x)(s) := C \cdot \mathrm{e}^{(A+C)s} x - \int_0^s B(\tau)(G + Id)\, \mathrm{e}^{(A+C)(s-\tau)} x \, d\tau$$

and

$$L_{C,G} f := -\int_0^\infty \mathrm{e}^{-(A+C)\tau} (G + Id) f(\tau) d\tau.$$

These are just formal definitions and we need additional assumptions to ensure that $Q_{C,G}$ and $L_{C,G}$ are bounded. In the next lemma we denote by $L^p_\omega(\mathbb{R}_+)$ the function space $L^p(\mathbb{R}_+, \mathrm{e}^{\omega s} ds)$, where it is assumed that $\omega \geq 0$.

Lemma 3.

(a) $Q_{C,G} \in \mathcal{L}(X, \mathrm{L}^p(\mathbb{R}_+, X))$ if one of the following conditions is satisfied.

(i) $\omega_0(A + C) < 0$.

(ii) $\|B(.)\| \in \mathrm{L}^p_\omega(\mathbb{R}_+)$,
$\omega_0(-(A + C)) < \omega$ and

$C = \int_0^\infty B(\tau)(G + Id)\, \mathrm{e}^{-(A+C)\tau} d\tau$.
In this case $(Q_{C,G}x)\,(s) = \int_s^\infty B(\tau)(G + Id)\, \mathrm{e}^{(A+C)(s-\tau)} d\tau$.

(b) $L_{C,G} \in \mathcal{L}(\mathrm{L}^p(\mathbb{R}_+, X), X)$ *if*
$-(A + C)$ *generates a bounded semigroup on* X *and* $p = 1$ *or*
$\omega_0(-(A + C)) < 0$ *and* $p > 1$.

Proof. Let $b(\tau) := \|B(\tau)\|$, $\epsilon_\mu(\tau) := \mathrm{e}^{\mu\tau}$.
(a.i): We have to show that $\|b * \epsilon_{\omega_0(A+C)}\|_p < \infty$, where $b * \epsilon_\mu(s) := \int_0^s b(\tau)\epsilon_\mu(s - \tau)d\tau$ denotes the convolution. Since $\epsilon_{\omega_0(A+C)} \in L^1(\mathbb{R}_+)$ this follows from the well known inequality $\|b * \epsilon_{\omega_0(A+C)}\|_p \leq \|b\|_p \cdot \|\epsilon_{\omega_0(A+C)}\|_1$ (see, e.g., [GLS, Chap.2, Thm.2.2]).
(a.ii): First observe that $(Q_{C,G}x)\,(s) = \int_s^\infty B(\tau)(G + Id)\, \mathrm{e}^{(A+C)(s-\tau)} d\tau$. Hence it suffices to show that for $\omega_0 < \omega$ we have $\int_0^\infty \left(\int_s^\infty b(\tau)\, \mathrm{e}^{\omega_0(\tau-s)} d\tau\right)^p ds < \infty$. Let $b_\omega := \epsilon_\omega \cdot b$. Then by Jensen's Inequality (see, e.g., [R, Chap.3, Thm.3.3]) we obtain

$$\begin{aligned}
\int_0^\infty \left(\int_s^\infty b(\tau)\, \mathrm{e}^{\omega_0(\tau-s)} d\tau\right)^p ds &= \int_0^\infty \mathrm{e}^{-\omega_0 ps} \left(\int_s^\infty b_\omega(\tau)\, \mathrm{e}^{(\omega_0-\omega)\tau} d\tau\right)^p ds \\
&\leq \int_0^\infty \mathrm{e}^{-\omega_0 ps} \int_s^\infty b_\omega(\tau)^p\, \mathrm{e}^{(p-1)(\omega_0-\omega)s} (\omega - \omega_0)^{(1-p)}\, \mathrm{e}^{(\omega_0-\omega)\tau} d\tau ds \\
&= \int_0^\infty b_\omega(\tau)^p (\omega - \omega_0)^{(1-p)} \left(\frac{\mathrm{e}^{((1-p)\omega-\omega_0)\tau} - 1}{(1-p)\omega - \omega_0}\right) \mathrm{e}^{(\omega_0-\omega)\tau} d\tau \\
&< \infty,
\end{aligned}$$

since $-p\omega \leq 0$ and $\omega_0 - \omega < 0$.
(b): Let $h \in L^p(\mathbb{R}_+)$, then we have to show that $\|\epsilon_{\omega_0(A+C)} \cdot h\|_1 < \infty$. Since $\epsilon_{\omega_0(A+C)} \in L^q(\mathbb{R}_+)$ for $\frac{1}{p} + \frac{1}{q} = 1$ it follows by Hölder's inequality that

$$\|\epsilon_{\omega_0(A+C)} \cdot h\|_1 \leq \|\epsilon_{\omega_0(A+C)}\|_q \cdot \|h\|_p.$$

□

The operators $Q_{C,G}$ and $L_{C,G}$ satisfy the following functional equations.

Lemma 4.
(a) If $Q_{C,G} \in \mathcal{L}(X, \mathrm{L}^p(\mathbb{R}_+, X))$, *then* $Q_{C,G}X \subset D(D_s)$ *and*

$$D_s Q_{C,G} = Q_{C,G}(A + C) - B(G + Id).$$

(b) If $L_{C,G} \in \mathcal{L}(\mathrm{L}^p(\mathbb{R}_+, X), X)$, *then*

$$L_{C,G} D_s = (G + Id)\delta_0 + (A + C)L_{C,G} \quad \text{on } D(D_s).$$

Proof. (a): Let $x \in X$, then $(Q_{C,G}x)(.)$ is differentiable and

$$\begin{aligned}
&(D_s(Q_{C,G}x))(s) = \\
&= C\, \mathrm{e}^{(A+C)s}(A + C)x - \int_0^s B(\tau)(G + Id)\, \mathrm{e}^{(A+C)(s-\tau)}(A + C)x d\tau - B(s)(G + Id)x \\
&= (Q_{C,G}(A + C)x)(s) - B(s)(G + Id)x.
\end{aligned}$$

(b): Let $f \in D(D_s)$, then

$$\begin{aligned} L_{C,G}D_sf &= -\int_0^\infty e^{-(A+C)\tau}(G+Id)f'(\tau)d\tau \\ &= (G+Id)f(0) - (A+C)\int_0^\infty e^{-(A+C)\tau}(G+Id)f(\tau)d\tau \\ &= (G+Id)f(0) + (A+C)L_{C,G}f. \end{aligned}$$

□

Using these equations we can now prove the main result.

Theorem 5.

(a) *If $Q_{C,0}$ is bounded, then* $\mathcal{A} \cong \begin{pmatrix} A+C & \delta_0 \\ 0 & D_s - Q_{C,0}\delta_0 \end{pmatrix}$.

(b) *If there exists $C \in \mathcal{L}(X)$ such that $L_{C,0}$ is bounded and $L_{C,0}B + C = 0$, then* $\mathcal{A} \cong \begin{pmatrix} A+C & 0 \\ B & D_s + BL_{C,0} \end{pmatrix}$.

(c) *If $Q_{C,0}$ is bounded and if there exists $G \in \mathcal{L}(X)$ such that $L_{C,G}$ is bounded and $L_{C,G}Q_{C,0} = G$, then* $\mathcal{A} \cong \begin{pmatrix} A+C & 0 \\ 0 & D_s - Q_{C,0}\delta_0 \end{pmatrix}$.

(d) *If $L_{C,0}$ is bounded and if there exists $G \in \mathcal{L}(X)$ such that $Q_{C,G}$ is bounded and $L_{C,0}Q_{C,G} = G$, then* $\mathcal{A} \cong \begin{pmatrix} A+C & 0 \\ 0 & D_s + BL_{C,0} \end{pmatrix}$.

Proof. Since the proofs of (a)–(d) are quite similar we only show (a). Define as above

$$\mathcal{S}_{Q_{C,0}} := \begin{pmatrix} Id & 0 \\ Q_{C,0} & Id \end{pmatrix}$$

on $X \times L^p(\mathbb{R}_+, X)$. Then

$$\left(\mathcal{S}_{Q_{C,0}}\right)^{-1}\mathcal{A}\mathcal{S}_{Q_{C,0}} = \begin{pmatrix} A + \delta_0 Q_{C,0} & \delta_0 \\ B + D_sQ_{C,0} - Q_{C,0}\delta_0Q_{C,0} - Q_{C,0}A & D_s - Q_{C,0}\delta_0 \end{pmatrix}$$

with domain $X \times D(D_s)$. Now $\delta_0 Q_{C,0} = C$, hence by Lemma 4, (a)

$$B + D_sQ_{C,0} - Q_{C,0}\delta_0Q_{C,0} - Q_{C,0}A = 0.$$

□

Since the semigroup generated by an operator matrix in triangular form can be computed explicitely (see Nagel [N, Prop.3.1]) we obtain the following corollary.

Corollary 6. *Let $(R(t))$ be the resolvent of the Volterra equation*

$$\dot{u}(t) = Au(t) + \int_0^t B(t-s)u(s)ds.$$

If (a), (b), (c) or (d) of the theorem is satisfied, then

$$\begin{aligned} R(t) \quad &\overset{(a)}{=} e^{(A+C)t} - \int_0^t e^{(A+C)(t-s)}\delta_0 \, e^{(D_s - Q_{C,0}\delta_0)s} ds \cdot Q_{C,0}, \\ &\overset{(b)}{=} e^{(A+C)t} + L_{C,0} \cdot \int_0^t e^{(D_s + BL_{C,0})(t-s)} B \, e^{(A+C)s} ds, \\ &\overset{(c)}{=} e^{(A+C)t}(Id + G) - L_{C,G} \, e^{(D_s - Q_{C,0}\delta_0)t} Q_{C,0}, \\ &\overset{(d)}{=} e^{(A+C)t}(Id + G) - L_{C,0} \, e^{(D_s + BL_{C,0})t} Q_{C,G}. \end{aligned}$$

In the one-dimensional case $X = \mathbb{C}$ the conditions of Lemma 3 and Theorem 5 can be reformulated. Here $\hat{B}$ denotes the Laplace transform of $B(.)$.

Corollary 7. *Let $X = \mathbb{C}$.*
(a) If $C \in \mathbb{C}$ such that

(i) $\mathrm{Re}(A + C) < 0$ *or*
(ii) $B(.) \in \mathrm{L}^p_\omega(\mathbb{R}_+)$, $\mathrm{Re}(A + C) > -\omega$ *and* $C = \hat{B}(A + C)$,

then

$$\mathcal{A} \simeq \begin{pmatrix} A + C & \delta_0 \\ 0 & D_s - Q_{C,0}\delta_0 \end{pmatrix}.$$

(b) If (a.ii) is satisfied, then

$$\mathcal{A} \simeq \begin{pmatrix} A + C & 0 \\ B & D_s + BL_{C,0} \end{pmatrix}.$$

(c,d) If (a.ii) is satisfied and C is a zero of order one of the characteristic equation $C - \hat{B}(A + C) = 0$, then

$$\mathcal{A} \simeq \begin{pmatrix} A + C & 0 \\ 0 & D_s - Q_{C,0}\delta_0 \end{pmatrix}$$

and

$$\mathcal{A} \simeq \begin{pmatrix} A + C & 0 \\ 0 & D_s + BL_{C,0} \end{pmatrix}.$$

Proof. We only have to show (c,d). First observe that $L_{C,0}Q_{C,G} = L_{C,G}Q_{C,0}$ for all C, $G \in \mathcal{L}(X) = \mathbb{C}$. Hence the assumptions of (c) and (d) of Theorem 5 are equivalent. We have to show that the equation

$$\begin{aligned} L_{C,G}Q_{C,0} &= -(G+1)\hat{Q}_{C,0}(A+C) \\ &= G \end{aligned}$$

has a solution $G \in \mathbb{C}$, i.e. that $\hat{Q}_{C,0}(A+C) \neq -1$. By Lemma 3 and Lemma 4

$$Q'_{C,0} - (A+C)Q_{C,0} + B = 0, \qquad Q_{C,0}(0) = \hat{B}(A+C),$$

hence

$$\lambda \cdot \hat{Q}_{C,0}(\lambda) - \hat{B}(A+C) - (A+C)\hat{Q}_{C,0}(\lambda) + \hat{B}(\lambda) = 0$$

for $\lambda \in \mathbb{C}$, $\mathrm{Re}\lambda > 0$. From this it follows that

$$\hat{Q}_{C,0}(\lambda) = -\frac{\hat{B}(\lambda) - \hat{B}(A+C)}{\lambda - (A+C)}.$$

Hence $\hat{Q}_{C,0}(A+C) = -1$ iff $\hat{B}'(A+C) = 1$, i.e. iff C is a zero of order greater or equal to 2 of the characteristic equation $C - \hat{B}(A+C) = 0$. □

Another immediate consequence of Theorem 5 is the final result.

Corollary 8. *If one of the conditions (a) or (b) of Theorem 5 is satisfied, then* $\sigma(A+C) \subset \sigma(\mathcal{A})$. *In particular* $\{\lambda \in \mathbb{C} : \mathrm{Re}\lambda < 0\} \subset \sigma(\mathcal{A})$.

References.

[CG] G. Chen, R. Grimmer, *Semigroups and integral equations*, J. Integral Eq. **2** (1980), 133–154.

[DS] W. Desch, W. Schappacher, *A semigroup approach to integrodifferential equations in Banach spaces*, J. Integral Eq. **10** (1985), 99–110.

[GLS] G. Gripenberg, S.-O. Londen, O. Staffans, *Volterra Integral and Functional Equations*, Cambridge University Press 1990.

[M] R.K. Miller, *Volterra integral equations in a Banach space*, Funkcial. Ekvac. **18** (1975), 163–194.

[N] R. Nagel, *Towards a "matrix theory" for unbounded operator matrices*, Math. Z. **201** (1989), 57–68.

[R] W. Rudin, *Real and Complex Analysis*, McGraw-Hill 1987.

Second Order Abstract Equations with Nonlinear Boundary Conditions: Applications to a von Karmàn System with Boundary Damping

A. FAVINI Dipartimento di Matematica, Università di Bologna, I-40127 Bologna, Italy

I. LASIECKA Department of Applied Mathematics, University of Virginia, Charlottesville, VA 22903, U.S.A.

1. INTRODUCTION

This paper is concerned with wellposedness of abstract nonlinear differential equations of the form

$$(1.1)\quad \begin{cases} Mu_{tt}(t) + Au(t) + A\,G\,G^{*}\,Au_t(t) + A\,G\,f(u)(t) = \mathcal{F}(u)(t); & t > 0 \\ u(t=0) = u_0\ ;\quad u_t(t=0) = u_1 \end{cases}$$

under the following assumptions:

(1.2) $\tilde{A}$ is a linear, closed, positive, selfadjoint operator on a Hilbert space H with $\mathcal{D}(\tilde{A}) \subset H$. The realization of $\tilde{A}$ from $\mathcal{D}(\tilde{A}^{1/2}) \to \mathcal{D}(\tilde{A}^{1/2})'$ will be called by A. We shall denote by $|\,-\,|$ and $\|\,-\,\|$ the norm of H and $\mathcal{D}(\tilde{A}^{1/2})$ respectively, and we shall use the same symbol $(\cdot\,,\cdot)$ to denote the scalar product on H and the duality pairing between $\mathcal{D}(\tilde{A}^{1/2})$ and $\mathcal{D}(\tilde{A}^{1/2})'$.

(1.3) Let V be another Hilbert space such that

$$\mathcal{D}(\tilde{A}^{1/2}) \subset V \subset H \subset V' \subset \mathcal{D}(\tilde{A}^{1/2})',$$

all injections being continuous and dense. We assume that $M \in \mathcal{L}(V, V')$ and $(\,,\,)$ is understood as a duality pairing between V and V'. Moreover $(M\,u, u) \geq \alpha\, |u|_V^2$ for some $\alpha > 0$. Hence $M^{-1} \in \mathcal{L}(V', V)$. Setting $\tilde{M} = M\,|_H$ with $\mathcal{D}(\tilde{M}) = \{u \in V;\ M\,u \in H\}$ we have $V = \mathcal{D}(\tilde{M}^{1/2})$.

(1.4) Let U be another Hilbert space with the scalar product denoted by $<\cdot\,,\cdot>$. We assume that the bounded linear operator G: $U \to H$ satisfies $\tilde{A}^{1/2}\,G \in \mathcal{L}(U; H)$. Hence $G^{*}A \in \mathcal{L}(\mathcal{D}(\tilde{A}^{1/2}); U)$.

(1.5) The nonlinear bounded operator $\mathcal{F}: \mathcal{D}(\tilde{A}^{1/2}) \to V'$ is assumed to be Frechet differentiable and its Frechet derivative, denoted by $D\,\mathcal{F}$, satisfies,

$$|D\,\mathcal{F}(u)\,h|_{V'} \leq C\,(\|u\|)\,\|h\|\,.$$

(1.6) The nonlinear bounded operator $f: \mathcal{D}(\tilde{A}^{1/2}) \to U$ is Frechet differentiable and

$$|Df(u)\,h|_U \le C(\|u\|)\,\|h\|\,.$$

Here $C(\|u\|)$ denotes a function which is bounded for bounded values of the argument $\|u\|$. Equations of the type (1.1) can be considered as abstract models of second order (in time) nonlinear problems with *nonlinear* boundary conditions (see [D-L-S] for the treatment of linear equations). In fact, the composition operator $A\,G$: $\mathcal{D}(\tilde{A}^{1/2}) \to \mathcal{D}(\tilde{A}^{1/2})'$ (whose domain in H typically contains only "zero" element) represents various boundary operators (see section 4). Examples motivating the above framework are equations of nonlinear elasticity with nonlinear boundary conditions. They include: nonlinear wave equations, Von Karman plate equations, nonlinear Euler Bernoulli and Kirchoff plates equations, etc. The interest in studying these equations has been spurred by recent developments in the area of control and stabilization hyperbolic and hyperbolic-like models (see [L.1], [L.4] [L-T.3] and references therein). In fact, the most recent advancements in *boundary* stabilization theory for elastic systems have brought to focus models with nonhomogeneous feedback boundary conditions (typically forces, shears, moments applied on the edge (or portion thereof) of a plate or a membrane). This, of course, brings up a question of wellposedness and of regularity properties of the solutions to such models. While there is a wealth of results dealing with wellposedness and regularity issues for *linear* equations (linear waves, plates) with linear boundary feedback (see [L.1], [L.4] and references therein), very few results are available in the nonlinear case (as considered in this paper). Indeed, the only results known to the authors are in the case of one dimensional Von Karman system (see [L-L]).

The goal of the present paper is to present a general "theory" of wellposedness and regularity valid for the abstract model (1.1). It should be noted that the main technical difficulties of the problem at the abstract level stem from: (i) the presence of unbounded operator A G in the model (1.1) which does not admit nontrivial realization on the basic space H, (ii) the lack of smoothing effects of the original dynamics unlike the "parabolic problems" (see for instance [B-F], [F-L-T]) where the smoothing character of the underlined evoluation "makes up" for the unboundedness of nonlinear terms. In fact, problem (1.1) with nonlinear term AGf is not monotone and it is not a (local) Lipschitz perturbation of a monotone problem. For these reasons the known results and methods for studying abstract nonlinear equations (see for instance [B.1], [B.2], [L.2], [C-S], [S.1], [T.2], [L-L.2], etc.) are not applicable here. It may be also worth noticing that the presence of the damping operator $A\,G\,G^*\,A\,u_t$ is critical to the theory.

The outline of the paper is as follows. In section 2 we formulate the results pertinent to the existence and regularity of local and global solutions to the abstract model (1.1). The proofs of these results are given in section 3. Section 4 deals with applications of abstract theory to the specific model of Von Karman plate equation. In Section 4 we prove the existence of local and global (under the additional structural hypothesis on the function $\tilde{f}$) weak solutions to Von Karman system. Moreover, under the additional assumptions on regularity and compatibility of initial conditions we prove that these weak solutions are, in fact, classical solutions. To our knowledge this result is original and new as all other results available in the literature deal either with homogeneous linear boundary conditions (see for instance [L.2], [B.3], [S.2] and references therein), or if boundary conditions are nonlinear, the problem is treated in the one dimension only (see [L-L]).

2. STATEMENT OF THE RESULTS.

We treat the equation

$$(2.1)\quad \begin{cases} Mu_{tt}(t) + Au(t) + \beta\, A\, G\, G^*\, Au_t(t) + A\, G\, f\,(u(t)) = \mathcal{F}(u(t)) \\ u(0) = u_0 \in \mathcal{D}(\tilde{A}^{1/2}); \quad u_t(0) = u_t \in V \end{cases}$$

under the assumptions (1.2) - (1.6) and a positive constant β.

Definition 2.1 We say that the function $\tilde{u}(t) = (u(t), u_t(t))$ is a *strong* solution to (2.1) on [0,T] iff $\tilde{u} \in C\,[0T;\ \mathcal{D}(\tilde{A}^{1/2}) \times \mathcal{D}(\tilde{A}^{1/2})]$; $u_{tt} \in C\,[0T;\, V]$; $\tilde{u}\,(0) = (u_0, u_1)$; and relation (2.1) holds for all $t \in [0, T]$ in the sense of $\mathcal{D}(\tilde{A}^{1/2})'$ topology.

In order to define weak solutions to the problem (2.1), we first define weak solution to the nonhomogeneous problem

$$(2.2)\quad \begin{cases} Mu_{tt}(t) + Au_t(t) + \beta\, A\, G\, G^*\, Au_t(t) = -A\, G\, f + \mathcal{F}; \\ u(0) = u_0,\ u_t(0) = u_1\ . \end{cases}$$

where f (resp. $\mathcal{F}$) are given elements in $L_1(0T;\, U)$ (resp. $L_1(0T;\, V')$).

Definition 2.2. We say that the function $\tilde{u} \in C\,[0T;\ \mathcal{D}(\tilde{A}^{1/2}) \times V]$ is a *weak* solution to (2.2) *iff* there exists a sequence of functions $f_n \in L_1(0T;\ U)$; $\mathcal{F}_n \in L_1\,(0T;\ V')$ and of the corresponding strong solutions $\tilde{u}_n(t)$ of (2.2) such that $f_n \to f$ in $L_1\,(0T;\ U)$; $\mathcal{F}_n \to \mathcal{F}$ in $L_1\,(0T;\ V')$ and $\tilde{u}_n \to \tilde{u}$ in $C\,[0T;\ \mathcal{D}(\tilde{A}^{1/2}) \times V]$.

Definition 2.3 We say that the function $\tilde{u} \in C\,[0T;\, \mathcal{D}(\tilde{A}^{1/2}) \times V]$ is a *weak* solution to (2.1) *iff* $\tilde{u}$ is a weak solution to the nonhomogeneous problem (2.2) with $f = f\,(u)$ and $\mathcal{F} = \mathcal{F}\,(u)$.

Theorem 2.1. (local existence). For each initial data $(u_0, u_1) \in \mathcal{D}(\tilde{A}^{1/2}) \times V$, there exists $T_0 > 0$ such that there exists a unique *weak* solution $(u(t), u_t(t))$ to the problem (2.1) on $(0, T_0)$. Moreover,

$$(2.3)\quad \int_0^{T_0} |G^*\, Au_t(t)|_U^2\, dt \le C_{T_0,\beta}\,(\|u_0\|\,;\ |u_1|_V)\,,$$

and the weak solution $\tilde{u}\,(t)$ satisfies

$$(2.4)\quad \frac{d}{dt}\,(Mu_t(t), \phi) + (Au(t), \phi) + \beta \langle G^*\, Au_t(t), G^* A\, \phi\rangle + \langle f\,(u(t)), G^*\, A\, \phi\rangle = (\mathcal{F}(u(t)), \phi)$$

for all $\phi \in \mathcal{D}(\tilde{A}^{1/2})$, where the above equality holds in $H^{-1}\,(0, T_0)$.

Theorem 2.2 (regularity)

Assume that the initial data (u_0, u_1) satisfy

$$(2.5)\quad u_1 \in \mathcal{D}(\tilde{A}^{1/2});$$

$$(2.6)\quad A\,(u_0 + \beta\, G\, G^*\, Au_1 + Gf(\,u_0)) \in V'\,.$$

Moreover, assume that

(2.7) $|\tilde{A}^{-1/2} D\mathcal{F}(u)h|_H \le C(\|u\|)\,|h|_V$; $\quad |\tilde{A}^{1/2} GDf(u)h|_H \le C(\|u\|)\,|h|_U$.

Then the solution to (2.1) is *strong* on $[0, T_0]$. Moreover,

(2.8) $A(u(t) + \beta G G^* Au_t(t) + G f(u(t))) \subset C[0T_0; V']$
and (2.4) holds for *all* $t \in [0, T]$ and $\phi \in \mathcal{D}(\tilde{A}^{1/2})$.

Remark 2.1

In the linear case (when $f \equiv 0$ and $\mathcal{F} \equiv 0$) the result of Theorem 2.1 can be obtained by using variational techniques as in, for example, [S.1]. Also, if $\mathcal{F} \neq 0$ but *still* $f = 0$, a combination of the variational approach with a contraction argument would lead to the result. What makes this problem more interesting is the presence of the nonlinear term represented by the function f. In fact, in this case the result depends critically on the strict positivity of the constant β. The reason for this is that, in general, the regularity of the "undamped" linear model is not sufficient to control the "boundary" terms A G f(u).

In order to obtain *more* regular solutions, additional hypotheses on the nonlinear term need to be added.

Theorem 2.3 (regularity revisited)

In addition to the assumptions of Theorem 2.2, we assume that $f = 0$ and $\mathcal{F}$ is *twice* differentiable $\mathcal{D}(\tilde{A}^{1/2}) \to V'$. Moreover, we assume that

(2.9) $\tilde{M}^{-1} \in \mathcal{L}(H; \mathcal{D}(\tilde{A}^{1/2}))$;

(2.10) $\mathcal{F}(u_0) \in H$;

(2.11a) $u_0 + \beta G G^* Au_1 \in \mathcal{D}(\tilde{A})$;

(2.11b) $A(u_1 + \beta G G^* A M^{-1} [\tilde{A}(u_0 + \beta G G^* Au_1) + \mathcal{F}(u_0)]) \in V'$.

Then,

(2.12) $u_{tt} \in C[0, T_0; \mathcal{D}(\tilde{A}^{1/2})]$;

(2.13) $u_{ttt} \in C[0T_0; V]$;

(2.14) $\begin{cases} \tilde{A}(u + \beta G G^* Au_t) - \mathcal{F}(u) \in C[0T_0; H]; \\ \tilde{A}(u_t + \beta G G^* Au_{tt}) - D\mathcal{F}(u)u_t \in C[0T_0; V']; \end{cases}$

$Mu_{tt}(t) + A(u(t) + \beta G G^* Au_t(t)) - \mathcal{F}(u(t)) = 0$
for all $t \ge 0$ where the above equation holds on H.

To obtain global solutions, we need to impose some structional conditions on the functions f and $\mathcal{F}$.

Theorem 2.4. (global existence)

In addition to the assumptions of Theorem 2.1 we assume that for all $\tilde{u} \equiv (u, u_t) \in C[0T_0; \mathcal{D}(\tilde{A}^{1/2}) \times V]$ and such that $G^* Au_t \in L_2[0T_0; U]$ the following inequalities hold for all $t \in T_0$

(2.15) $\int_0^t (\mathcal{F}(u(\tau)), u_t(\tau))\, d\tau \le C_1 \int_0^t \|u(\tau)\|^2 + |u_t(\tau)|_V^2\, d\tau + C_2 \quad (\|u_0\|\,;\ |u_1|_V) \equiv C_0\,;$

(2.16) $-\int_0^t \langle f(\, u(\tau)),\ G^*\, Au_t(\tau) \rangle\, d\tau \le C_0\,.$

Then, the weak solution (u(t) u_t(t)) of Theorem 2.1 is global on [0, T] for any T > 0.

3. PROOFS OF THEOREMS 2.1 - 2.4.

3.1. Preliminary Lemmas

We define a linear operator

$$\mathcal{A}:\ \mathcal{H} \to \mathcal{H};\quad \mathcal{H} \equiv \mathcal{D}(\tilde{A}^{1/2}) \times V;$$

(3.1) $$\mathcal{A}\begin{bmatrix} u \\ v \end{bmatrix} \equiv \begin{bmatrix} -v \\ M^{-1}\, A\,(u + \beta\, G\, G^*\, Av) \end{bmatrix};$$

$$\mathcal{D}(\mathcal{A}) = \{(u, v) \in \mathcal{D}(\tilde{A}^{1/2}) \times \mathcal{D}(\tilde{A}^{1/2})\,;\ A\,(u + \beta\, G\, G^*\, Av) \in V'\,\}.$$

Proposition 3.1

The operator $\mathcal{A}$ generates a C_0 semigroup of contractions on $\mathcal{H}$ which we denote by $e^{-\mathcal{A}t}$.

proof: is rather standard and based on application of Lumer Phillips Theorem (see [P.1]). It suffices to show that $\mathcal{A}$ is maximal monotone.

Step 1. $\mathcal{A}$ is monotone. Indeed, with $\bar{u} = (u, v) \in \mathcal{D}(\mathcal{A})$ we have

$$(\mathcal{A}(u, v)\,,\ (u, v))_{\mathcal{H}} = -(\tilde{A}^{1/2}\, u,\ \tilde{A}^{1/2}\, v) + (M^{-1}\,(A\,(u + \beta\, G\, G^*\, Au)),\ v)_V$$
$$= -(A\, u,\ v) + (A\,(u + \beta\, G\, G^*\, Av)\,,\ v) = \beta\ |G^*\, Av\,|_V^2 \ge 0\,.$$

Step 2. $\mathcal{A}$ is maximal monotone. By Minty's Theorem (see [B.1]) it suffices to prove that there exists a solution $(u, v) \in \mathcal{D}(\mathcal{A})$ to the following equations

(3.2) $$\begin{cases} \lambda\, u - v = g\,, \\ \lambda\, v + M^{-1}\, A\,[u + G\, G^* Av] = h\,, \end{cases}$$

with $\lambda > 0$ and $g \in \mathcal{D}(\tilde{A}^{1/2})$; $h \in V$. System (3.2) reduces to

(3.3) $$A\, u + \lambda^2\, Mu + \beta\lambda\, A\, G\, G^*\, Au = \lambda\, Mg + Mh + \beta\, A\, G\, G^*\, A\, g \in \mathcal{D}(\tilde{A}^{1/2})'.$$

The operator A is maximal monotone and coercive $\mathcal{D}(\tilde{A}^{1/2}) \to \mathcal{D}(\tilde{A}^{1/2})'$. The sum of two operators $\lambda^2\, M + \beta\, A\, G\, G^*\, A$ is continuous and monotone $\mathcal{D}(\tilde{A}^{1/2}) \to \mathcal{D}(\tilde{A}^{1/2})'$. Hence (see [B.1])

$$A + \lambda^2\, M + \beta\, A\, G\, G^*\, A:\ \mathcal{D}(\tilde{A}^{1/2}) \to \mathcal{D}(\tilde{A}^{1/2})'$$

is maximal monotone and coercive, hence boundedly invertible. This implies that there exists $u \in \mathcal{D}(\tilde{A}^{1/2})$ - solution to (3.3), and from (3.2) we obtain that $v = \lambda\, u - g \in \mathcal{D}(\tilde{A}^{1/2})$. Going back to the second equation in (3.2) we infer that

$$M^{-1}[A(u+\beta G G^* Av)] = h - \lambda v \in V$$

hence $A(u+\beta G G^* Av) \in V'$ as desired. ■

We consider the linear part of equation (2.1)

$$(3.4)\quad \begin{cases} Mu_{tt} + Au + \beta A G G^* A u_t = 0 \\ u(0) = u_0 \in \mathcal{D}(\tilde{A}^{1/2})\,; \quad u_t(0) = u_1 \in V. \end{cases}$$

Corollary 3.1.

(i) For each $(u_0, u_1) \in \mathcal{D}(\mathcal{A})$ there exists a unique *strong* solution to (3.4).

(ii) For each $(u_0, u_1) \in \mathcal{D}(\tilde{\mathcal{A}}^{1/2}) \times V$ there exists a unique weak solution to (3.4). Moreover, the weak solution $\tilde{u} = (u, u_t)$ satisfies the estimate

$$(3.5)\quad \int_0^T |G^* A u_t(t)|_U^2 \, dt \le \frac{2}{\beta} [\|u_0\|^2 + |u_1|_V^2].$$

proof: All the statements except (3.5) follow from Proposition 3.1 combined with standard results in linear semigroup theory (see [B.1]). To prove (3.5) we consider first strong solutions $\tilde{u}_n(t)$ corresponding to the initial data $(u_{0n}, u_{1n}) \in \mathcal{D}(\mathcal{A})$, such that $u_{0n} \to u_0$ in $\mathcal{D}(\tilde{A}^{1/2})$ and $u_{1n} \to u_1$ in V. Since $\tilde{u}_n(t)$ is a strong solution, each term in equation (3.4) is a continuous function on [0, T] will the values in $\mathcal{D}(\tilde{A}^{1/2})'$. Hence for all $t \ge 0$ and $\phi \in \mathcal{D}(\tilde{A}^{1/2})$,

$$(Mu_{ntt}(t)\,,\ \phi) + (Au_n(t)\,, \phi) + \beta\,(G^* A\, u_{nt}(t), G^* A\,\phi) = 0$$

Setting $\phi \equiv u_{nt}(t) \in \mathcal{D}(\tilde{A}^{1/2})$ yields

$$(3.6)\quad |u_{nt}(t)|_V^2 + \|u_n(t)\|^2 + 2\beta \int_0^t |G^* A\, u_{nt}(\tau)|_U^2 \, d\tau = \|u_{n0}\|^2 + |u_{n1}|_V^2 .$$

Similarly, we obtain

$$(3.7)\quad \lim_{n, m \to \infty} |(u_{nt} - u_{mt})(t)|_V^2 + \|(u_n - u_m)(t)\|^2 + 2\beta \int_0^t |G^* A\,(u_{nt} - u_{mt})|_U^2 \, d\tau = 0 .$$

Hence

$$(3.8)\quad \tilde{u}_n \to \tilde{u} \ \text{ in } C\,[0T;\ \mathcal{D}(\tilde{A}^{1/2}) \times V]\,,$$

$$(3.9)\quad G^* Au_{nt} \to g \ \text{ in } L_2\,(0T;\ U)\,.$$

From (3.8) and the regularity $G^* A^{1/2} \in \mathcal{L}(H; U)$ we infer that

$$(3.10)\quad G^*Au_n \to G^*Au \ \text{ in } C\,[0T;\ U] \ \text{ and } \ \frac{d}{dt} G^*Au_n \to \frac{d}{dt} G^*Au \ \text{ in } H^{-1}\,(0T;\ U).$$

By the uniqueness of the strong limit we must have that $g = \frac{d}{dt} G^*Au \in L_2(0T;\ U)$ and

$$(3.11)\quad G^*Au_{nt} \to \frac{d}{dt} G^* Au \ \text{ in } L_2[0T;\ U].$$

On the other hand we also have from (3.8)

$$u_{nt} \to u_t \text{ in } H^{-1}(0T;\ \mathcal{D}(\tilde{A}^{1/2})),$$

and since $G^* A^{1/2} \in \mathcal{L}(H; U)$

(3.12) $G^* A u_{nt} \to G^* A u_t$ in $H^{-1}(0T;\ \mathcal{D}(\tilde{A}^{1/2}))$.

Comparing (3.11) with (3.12) yields $g = G^* A u_t$ and

(3.13) $G^* A u_{nt} \to G^* A u_t$ on $L_2[0T;\ U]$.

Passage with the limit on (3.6) after taking into account (3.8) and (3.13) yields (3.5). ■

We introduce the following operators:

$$\mathcal{B}:\ U \to \mathcal{D}(\mathcal{A}^*)'$$

where $\mathcal{D}(\mathcal{A}^*) \subset \mathcal{H} \subset \mathcal{D}(\mathcal{A}^*)'$, and

(3.14) $\mathcal{B}g \equiv \begin{bmatrix} 0 \\ M^{-1} A G g \end{bmatrix}$. Notice that

(3.15) $\mathcal{A}^{-1} \mathcal{B} g = \begin{bmatrix} A^{-1} M^{-1} A G g \\ 0 \end{bmatrix} \in \mathcal{H}$.

$\mathcal{L}:\ L_2(0T;\ U) \to C[0T;\ \mathcal{D}(\mathcal{A}^*)']$ defined by

(3.16) $(\mathcal{L} g)(t) = \int_0^t e^{-\mathcal{A}(t-s)} \mathcal{B} g(s)\, ds$.

The following regularity result plays a crucial role in the proof of Theorem 2.1

Lemma 3.1

The operator $\mathcal{L}$ defined by (3.16) admits a bounded extension from $L_2(0T;\ U) \to C[0T;\ \mathcal{H}]$.

proof: From (3.15) and (3.16) it follows that $\mathcal{A}^{-1} \mathcal{L} \in \mathcal{L}(L_2(0T; U) \to C[0T; \mathcal{H}])$. Hence (see [K.1]) $\mathcal{L}$ is closable. It is straightforward to verify (see [L-T.1]) that $H_0^1(0T; U) \subset \mathcal{D}(\mathcal{L})$, which implies that $\mathcal{L}$ is densely defined. Thus, by using the duality argument of [L-T.1], it suffices to prove that

(3.17) $\int_0^T |\mathcal{B}^* e^{-\mathcal{A}^* t} \tilde{u}|_U^2\, dt \le C_T\, |\tilde{u}|_{\mathcal{H}}^2$ for $\tilde{u} = (u, v) \in \mathcal{D}(\mathcal{A}^*) \subset \mathcal{D}(\tilde{A}^{1/2}) \times \mathcal{D}(\tilde{A}^{1/2})$.

Here $<\mathcal{B}^* v, g>_U \equiv (v, \mathcal{B}g)_{\mathcal{H}}$ for $g \in U$, $v \in \mathcal{D}(\mathcal{A}^*)$ and $(\cdot, \cdot)_{\mathcal{H}}$ denotes the duality pairing in $\mathcal{D}(\mathcal{A}^*) \times \mathcal{D}(\mathcal{A}^*)'$.

Straightforward computations show that with $(u, v) \in \mathcal{D}(\mathcal{A}^*)$ $\tilde{z}(t) \equiv (z(t), -z_t(t)) \equiv e^{-\mathcal{A}^* t}(u, v)$ is characterized as a strong solution to

(3.18) $\begin{cases} M z_{tt} + A z + \beta A G G^* A z_t = 0 \\ z(0) = u,\ \ z_t(0) = -v\,. \end{cases}$

Notice that $(u, v) \in \mathcal{D}(\mathcal{A}^*)$ is equivalent to $(u, -v) \in \mathcal{D}(\mathcal{A})$. Thus $z \in C[0T; \mathcal{D}(\tilde{A}^{1/2})]$; $z_t \in C[0T; \mathcal{D}(\tilde{A}^{1/2})]$ and $z_{tt} \in C[0T; V]$.
Applying the inequality (3.5) to (3.18) yields

$$(3.19) \quad \int_0^T |G^* A\, z_t(t)|_U^2\, dt \le \frac{2}{\beta} [\|u\|^2 + |v|_V^2].$$

On the other hand with $(u, v) \in \mathcal{D}(\mathcal{A}^*) \subset \mathcal{D}(\tilde{A}^{1/2}) \times \mathcal{D}(\tilde{A}^{1/2})$, we have

$$<g, \mathcal{B}^* \begin{bmatrix} u \\ v \end{bmatrix}> = (\mathcal{B}g, \begin{bmatrix} u \\ v \end{bmatrix})_H = (M^{-1} A\, Gg, v)_V = (A\, Gg, v) = (Gg, Av) = <g, G^* Av>.$$

Hence with $(u, v) \in \mathcal{D}(\mathcal{A}^*) \subset \mathcal{D}(\tilde{A}^{1/2}) \times \mathcal{D}(\tilde{A}^{1/2})$

$$(3.20) \quad \mathcal{B}^* \begin{bmatrix} u \\ v \end{bmatrix} = G^* A\, v.$$

Combining (3.18) - (3.20) yields the desired inequality in (3.17). ■

Remark 3.1

Notice that the inequality (3.17) or—equivalently—the result of Lemma 3.1 *does not* follow from the regularity properties of the solutions provided by the semigroup theory. (3.17) is an independent regularity result which critically relies on the assumption that $\beta > 0$. In fact, it can be shown for a number of pde examples that "trace regularity" property (3.17) is not valid if $\beta = 0$ (see [L-T.2]).

Our next step is to obtain regularity properties of the solution to the *nonhomogeneous* problem (2.2).

Lemma 3.2

(i) For every $(f, \mathcal{F}) \in H_0^1(0T; U \times V')$ and $\tilde{u}(0) \in \mathcal{D}(\mathcal{A})$ there exists a unique strong solution to the problem (2.2).

(ii) For each $(f, \mathcal{F}) \in L_1(0T; U \times V')$, $\tilde{u}(0) \in \mathcal{H}$, there exists a unique weak solution to the problem (2.2). Moreover, this weak solution $\tilde{u}(t)$ is represented by the following formula

$$(3.21) \quad \tilde{u}(t) = (\mathcal{L}f)(t) + (\hat{\mathcal{L}}\mathcal{F})(t) + e^{-\mathcal{A}t}\, \tilde{u}(0)$$

where $\mathcal{L}$ is defined by (3.16) and

$$(\hat{\mathcal{L}}\mathcal{F})(t) \equiv \int_0^t e^{-\mathcal{A}(t-s)} \begin{bmatrix} 0 \\ M^{-1}\, \mathcal{F}(s) \end{bmatrix} ds.$$

(iii) Weak solutions to the problem (2.2) satisfy the following inequalities:

$$(3.22) \quad \int_0^T |G^* A\, u_t(t)|_U^2\, dt \le C_{T,\beta} [\|u_0\|^2 + |u_1|_V^2 + |f|^2_{L_2(0T;U)} + |\mathcal{F}|^2_{L_1(0T;V')}],$$

$$
\begin{aligned}
&|u_t(t)|_V^2 + \|u(t)\|^2 + 2\beta \int_0^t |G^* Au_t(s)|_U^2\, ds + 2\int_0^t <f(s), G^* Au_t(s)> ds \\
(3.23)\quad & -2\int_0^t ((\mathcal{F}(s), u_t(s))\, ds \le \|u_0\|^2 + |u_1|_V^2
\end{aligned}
$$

proof: Notice that with $f \in H_0^1(0T; U)$, $\frac{d}{dt}(\mathcal{L} f)(t) = \mathcal{L}\left[\frac{d}{dt} f\right] + \mathcal{A}^{-1}\mathcal{B} f(t)$. Hence, by the result of Lemma 3.1 and (3.15)

$$(3.24)\quad \frac{d}{dt}\mathcal{L} \in \mathcal{L}(H_0^1(0T; U);\ C[0T; \mathcal{H}]).$$

Assuming also that $\mathcal{F} \in H^1[0T; U]$, we obtain

$$\frac{d}{dt}(\hat{\mathcal{L}}\mathcal{F})(t) = \mathcal{A}^{-1}\begin{bmatrix} 0 \\ M^{-1}\mathcal{F}(t) \end{bmatrix} + \hat{\mathcal{L}}\left[\frac{d}{dt}\mathcal{F}\right]$$

and

$$(3.25)\quad \frac{d}{dt}\hat{\mathcal{L}} \in \mathcal{L}(H_0^1(0T; V');\ C[0T; \mathcal{H}]).$$

By using (3.24), (3.25) and Proposition 3.1 along with the standard semigroup arguments one easily shows that strong solutions to the problem (2.2) arc given by the formula (3.21). This proves part (i) of Lemma 3.2. To obtain weak solutions of part (ii) it is just enough to recall the boundedness of the operator $\mathcal{L}: L_2[0T; U] \to C[0T; \mathcal{H}]$ (Lemma 3.1) and of the operator $\hat{\mathcal{L}}: L_1[0T; V'] \to C[0T; \mathcal{H}]$. As for part (iii) of the Lemma, it suffices to establish inequalities (3.22) and (3.23) for strong solutions and then, by the same arguments as these in Corollory 3.1 to pass to the limit. Let $\tilde{u}(t) = (u(t), u_t(t))$ be a strong solution to (2.2). Then $u_{tt} \in C[0T; V]$, $u_t \in C[0T; \mathcal{D}(\tilde{A}^{1/2})]$, and $G^* A u_t \in C[0T; U]$. Thus, $\tilde{u}(t)$ satisfies

$$(3.26)\quad (Mu_{tt}(t), \phi) + (Au_t(t), \phi) + \beta <G^* Au_t(t), G^* A\phi> = -<f(t), G^* Au_t(t)> + (\mathcal{F}(t), \phi)$$

for *all* $\phi \in \mathcal{D}(\tilde{A}^{1/2})$ and $t \ge 0$. Thus, setting $\phi \equiv u_t(t)$ and integrating (3.26) from 0 to t (as in Corollory 3.1) yields inequality (3.23). From (3.23) we obtain

$$
\begin{aligned}
&|u_t(t)|_V^2 + \|u(t)\|^2 + 2\beta\int_0^t |G^* u_t(\tau)|_U^2\, d\tau \le 2\int_0^t |f(\tau)|_U\, |G^* Au_t(\tau)|_U\, d\tau \\
&\quad + 2\int_0^t |\mathcal{F}(\tau)|_{V'}\, |u_t(\tau)|_V\, d\tau.
\end{aligned}
$$

Hence

$$(3.27)\quad |u_t(t)|_V^2 + \|u(t)\|^2 + \int_0^t |G^* Au_t(\tau)|_U^2\, d\tau \le C_\beta \int_0^t |f(\tau)|_U^2\, d\tau + \int_0^t |\mathcal{F}(\tau)|_{V'}\, |u_t(\tau)|_V\, d\tau.$$

By using Lemma A-5 in [B.2] we obtain

$$|u_t(t)|_V \le C_\beta |f|_{L_2[0T;\,U]} + \int_0^t |\mathcal{F}(\tau)|_{V'}\, d\tau$$

which inequality together with (3.27) leads to the desired result in (3.22) for strong solutions. Passage to the limit along the same arguments as in Corollary 3.1 proves these estimates for weak solutions (here, careful attention must be paid — as in Corollary 3.1 — in passing to the limit on the term $G^* Au_t$, since this term is not bounded for $\tilde{u} \in C\,[0T;\, \mathcal{H}]$ and $G^* A$ is typically unclosable). ■

3.2. Proof of Theorem 2.1

To prove the Theorem, we shall construct a fixed point for the map $\tilde{u} = \Lambda(\tilde{u})$ where

$$(3.28) \qquad (\Lambda \tilde{u})(t) \equiv e^{-\mathcal{A}t}\, \tilde{u}_0 + \mathcal{L} f(u)(t) + \hat{\mathcal{L}} \mathcal{F}(u)(t)\,.$$

Let B_R denote a ball in $\mathcal{H}$ with a radius R. We shall show that Λ admits the unique fixed point in the closed subspace $C\,[0T_0;\, B_R]$ for sufficiently large R and sufficiently small T_0. To accomplish this we need to prove that Λ is a contraction and that

$$(3.29) \qquad \Lambda\,(C\,[0T_0;\, B_R]) \subset C\,[0T_0;\, B_R]\,.$$

The contraction property of Λ follows now from Lemma 3.1 and the following computations.

$$|(\mathcal{L} f(u_1) - \mathcal{L} f(u_2))(t)|^2_{\mathcal{H}} \le C_\beta \int_0^t |f\,(u_1(s)) - f(\ u_2(s))|^2\, ds$$

by assumption (1.6)

$$(3.30) \qquad \le C_\beta(R) \int_0^t \|u_1(s) - u_2(s)\|^2\, ds \le C_\beta(R)\, t\, |\tilde{u}_1 - \tilde{u}_2|^2_{C\,[0T;\, \mathcal{H}]}$$

Similarly, using hypothesis (1.5) and the bound $\hat{\mathcal{L}} \in \mathcal{L}(L_1\,(0T;\, V') \to C\,[0T;\, \mathcal{H}])$ we obtain

$$(3.31) \qquad \begin{aligned} |\hat{\mathcal{L}} \mathcal{F}(u_1)(t) - \hat{\mathcal{L}} \mathcal{F}(u_2)(t)|_{\mathcal{H}} &\le C \int_0^t |\mathcal{F}(u_1(s)) - \mathcal{F}(u_2(s))|_{V'}\, ds \\ &\le C\,(R) \int_0^t \|u_1(s) - u_2(s)\|\, ds \le C\,(R)\, t\, |\tilde{u}_1 - \tilde{u}_2|_{C\,[0t;\, \mathcal{H}]}\,. \end{aligned}$$

Thus for a given R we select a sufficiently small $T_0(R)$ so that Λ is a contraction. To prove (3.29) it is enough to take R large enough (depending on the initial data $\tilde{u}_0$) and to perform computations similar to these in (3.30), (3.31). Application of Fixed Point Theorem yields the existence of weak solutions. To complete the proof of Theorem 2.1 we need to justify the validity of inequality (2.3). To do this we shall use the result of Lemma 3.2. Indeed, since for all weak solutions by assumption (1.5), (1.6) we have

$$|f\,(u)|_{L_2\,(0T_0;\, U)} \le C_{T_0,\beta}\, (\|u_0\|,\ |u_1|_V);$$

$$|\mathcal{F}(u)|_{L_1(0T_0,\, V')} \le C_{T_0}\, (\|u_0\|,\ |u_1|_V)\,,$$

we are in a position to apply inequality (3.22) of Lemma 3.2. This yields the result in (2.3). Derivation of (2.4) is now straightforward, via the usual semigroup argument (see [P.1] or [B.1]). ■

3.3. Proof of Theorem 2.2

By using regularity properties (2.7) and (2.3) one easily shows that $\tilde{z} \equiv \tilde{u}_t$ (derivative in the sense of distribution)) with $\tilde{u}$ – weak solution guaranteed by Theorem 2.1 satisfies the equation

$$(3.32) \quad \tilde{z}(t) = e^{-\mathcal{A}t}\,\tilde{z}(0) + \int_0^t e^{-\mathcal{A}(t-s)} \begin{bmatrix} 0 \\ M^{-1}\,\mathcal{D}\mathcal{F}(u(s))\,u_t(s) \end{bmatrix} ds + \mathcal{L}(Df(u)\,G^*\,Au_t)(t)\,; \text{ in } \mathcal{D}(\mathcal{A}^*)'$$

$\tilde{z}(0) = -\mathcal{A}\tilde{u}_0 + \begin{bmatrix} 0 \\ M^{-1}\,\mathcal{F}(u_0) \end{bmatrix}$. Properties (2.7) and (2.3) are used to assert that weak solutions $\tilde{u}$ satisfy

$$(3.33) \quad \mathcal{A}^{-1} \begin{bmatrix} 0 \\ M^{-1}\,D\,\mathcal{F}(u)u_t \end{bmatrix} \in C\,[0T_0;\,\mathcal{H}]\,;$$

$$(3.34) \quad \mathcal{A}^{-1}\,\mathcal{B}\,\mathcal{D}f(u)\,G^*\,Au_t \in L_2(0T_0,\,U)\,.$$

These regularity properties, allow us to compute $\tilde{z} = \tilde{u}_t$ in $\mathcal{D}(\mathcal{A}^*)'$ as in (3.32). In view of (3.32), to prove Theorem 2.2 if suffices to show that the following integral equation in the variable $\tilde{z} = (z, z_t)$

$$(3.35) \quad \tilde{z}\,(t) = e^{-\mathcal{A}t}\,\tilde{z}(0) + \int_0^t e^{-\mathcal{A}(t-s)} \begin{bmatrix} 0 \\ M^{-1}\,\mathcal{D}\mathcal{F}(u(s))\,z(s) \end{bmatrix} ds + \mathcal{L}(Df((u(s))\quad z(\cdot))\,(t)$$

admits *unique* solution in $C\,[0T_0,\,\mathcal{H}]$ for any $\tilde{z}\,(0) \in \mathcal{H}$ and fixed $\tilde{u}$ - weak solution to (2.1). Indeed, assuming for a moment solvability of (3.35) we easily check that the unique solution $\tilde{z}\,(t)$ of (3.35) with

$$(3.36) \quad \tilde{z}\,(0) \equiv -\mathcal{A}\,\tilde{u}_0 + \begin{bmatrix} 0 \\ M^{-1}\,\mathcal{F}(u_0) \end{bmatrix} = \begin{bmatrix} u_1 \\ -M^{-1}\,[A(u_0 + \beta\,G\,G^*\,Au_1 + Gf\,(u_0)) - \mathcal{F}(u_0)] \end{bmatrix}$$

is precisely the solution of (3.32), hence it coincides with $\tilde{u}_t = (u_t, u_{tt}) \in C\,[0T_0;\,\mathcal{H}]$. To claim this we use hypothesis (2.5) and (2.6) which give

$$M^{-1}\,[A\,(u_0 + \beta\,G\,G^*\,Au_1 + Gf(\ u_0)] \in V\,,$$

$$M^{-1}\,\mathcal{F}(u_0) \in V\,,$$

$$u_1 \in \mathcal{D}(\tilde{A}^{1/2})\,.$$

Hence, by (3.36), $\tilde{z}\,(0) \in \mathcal{H}$. Thus, to complete the proof of the Theorem we need to prove solvability of (3.35) in $C\,[0T_0;\,\mathcal{H}]$ with $z\,(0) \in \mathcal{H}$. To this end notice first that for a fixed $u \in C\,[0T_0;\,\mathcal{H}]$, equation (3.35) is *linear* in $\tilde{z}$. Thus, provided that the appropriate Lipschitz continuity of the terms in (3.35) (3.3) (in the variable $\tilde{z}$) holds, we are in a position to use the Contraction Mapping Principle. This is first done locally on $C\,[0T_1,\,\mathcal{H}]$ where $T_1 \ll T_0$ and then, by linearity, globally for all $t \in [0,T_0]$. The aforementioned Lipschitz continuity follows from the following estimates

$$(3.37)\quad |M^{-1} D\,\mathcal{F}(u)\,[z_1 - z_2]|_V \le C(\|u\|)\,\|z_1 - z_2\|,$$

$$(3.38)\quad |D f(u)\;(z_1 - z_2)|_U \le C(\|u\|)\,\|z_1 - z_2\|.$$

By Lemma 3.1 and (3.38)

$$(3.39)\quad \begin{aligned} |\mathcal{L}(Df(u), (z_1 - z_2))|^2_{C[0T, \mathcal{H}]} &\le C_T \int_0^T |Df(u(t))\;(z_1 - z_2)(t)|^2_U\, dt \\ &\le C_T(\|u\|_{C[0T; \mathcal{D}(\tilde{A}^{1/2})]})\, T\, \|z_1 - z_2\|^2_{C[0T; \mathcal{D}(\tilde{A}^{1/2})]}. \end{aligned}$$

Similarly the operator

$$(\mathcal{L}_1 z)(t) \equiv \int_0^t e^{-\mathcal{A}(t-s)} \begin{bmatrix} 0 \\ M^{-1} D\,\mathcal{F}(u(s))\, z(s) \end{bmatrix} ds$$

satisfies the Lipschitz condition

$$(3.40)\quad |\mathcal{L}_1 (z_1 - z_2)|_{C[0,T; \mathcal{H}]} \le C_T[\|u\|_{C[0T; \mathcal{D}(\tilde{A}^{1/2})]}]\, T\, \|z_1 - z_2\|_{C[0T; \mathcal{D}(\tilde{A}^{1/2})]}.$$

Bounds in (3.39), (3.40) allow for the application of the Contraction Mapping Principle on $C[0T_1; \mathcal{D}(\tilde{A}^{1/2})]$, where T_1 is sufficiently small and depends on the norms of initial data and $C[0T; \mathcal{H}]$ norm of the weak solution $\tilde{u}(t)$. This completes the proof of the existence of a strong solution. Relation (2.8) in Theorem 2.2 can be directly read off from the equation. ■

3.4. Proof of Theorem 2.3

The idea of the proof is similar to that of Theorem 2.2. Indeed, having already solutions u(t) with regularity as in Theorem 2.2 we differentiate once more formula (3.32) (with respect to time). This leads us to the following equation in the variable $\tilde{z} \equiv (u_{tt}, u_{ttt})$.

$$(3.41)\quad z(t) = e^{-\mathcal{A}t}\,\tilde{z}(0) + \int_0^t e^{-\mathcal{A}(t-s)} \begin{bmatrix} 0 \\ M^{-1} D\,\mathcal{F}(u)\, u_{tt}(s) \end{bmatrix} ds + \int_0^t e^{-\mathcal{A}(t-s)} \begin{bmatrix} 0 \\ M^{-1} D^2\,\mathcal{F}(u)\,(u_t, u_t) \end{bmatrix} ds.$$

where

$$\tilde{z}(0) \equiv \mathcal{A}\left[\mathcal{A}\tilde{u}_0 + \begin{matrix} 0 \\ M^{-1}\,\mathcal{F}(u_0) \end{matrix}\right] + \begin{bmatrix} 0 \\ M^{-1} D\,\mathcal{F}(u_0)\, u_1 \end{bmatrix}.$$

Notice that since $u_{tt} \in C[0T_0; V]$, $u_t \in C[0T_0, \mathcal{D}(\tilde{A}^{1/2})]$ and for a fixed $u \in \mathcal{D}(\tilde{A}^{1/2})$, $D^2\,\mathcal{F}(u)$ is a bilinear continuous transformation

$$(3.42)\quad \mathcal{D}(\tilde{A}^{1/2}) \times \mathcal{D}(\tilde{A}^{1/2}) \to V'$$

(see [A-1] p. 21 Theorem 4.3), all the terms on RHS of (3.41) are continuous elements in $C[0T_0; \mathcal{D}(\mathcal{A}^*)']$. What we need to prove is that

$$(3.43)\quad \tilde{z} \in C[0T_0; \mathcal{H}].$$

Rewriting (3.41) as an integral equation yields

(3.44) $$\tilde{z}(t) = e^{-\mathcal{A}t}\tilde{z}(0) + \int_0^t e^{-\mathcal{A}(t-s)}\begin{bmatrix}0\\ M^{-1} D\mathcal{F}(u(s))\, z(s)\end{bmatrix} ds + \int_0^t e^{-\mathcal{A}(t-s)}\begin{bmatrix}0\\ M^{-1}\mathcal{D}^2\mathcal{F}(u)\,(u_t, u_t)(s)\end{bmatrix} ds$$

Define

$$a(t) \equiv \int_0^t e^{-\mathcal{A}(t-s)}\begin{bmatrix}0\\ M^{-1} D^2\mathcal{F}(u)(u_t(s), u_t(s))\end{bmatrix} ds$$

From (3.42) and regularity of u_t, i.e. $u_t \in C[0T_0; \mathcal{D}(\tilde{A}^{½})]$, we obtain

$$D^2\mathcal{F}(u)\,(u_t, u_t) \in C[0T_0; V'],$$

hence

(3.45) $a \in C[0T_0; \mathcal{H}]$.

Regularity of the initial data u_0, u_1 postulated by (2.9) - (2.11) implies that

(3.46) $\tilde{z}(0) \in \mathcal{H}$.

Returning to (3.44) we obtain

(3.47) $\tilde{z}(t) = e^{-\mathcal{A}t}\tilde{z}(0) + \hat{\mathcal{L}}_1(z(\cdot))(t) + a(t)$

where, we recall, $\hat{\mathcal{L}}_1$ is defined in the formula below (3.39). By using the estimate (3.40), regularity in (3.45), and (3.46) we easily show (as in Theorem 2.2) that the linear equation (3.47) has a unique global solution on $C[0T_0; \mathcal{H}]$. This completes the proof of regularity in (2.12), (2.13). The remaining statement of the Theorem follows directly from the equation and the regularity of u_{tt}. ■

3.5. Proof of Theorem 2.4

To prove the Theorem it suffices to establish the following apriori bound.

Lemma 3.3

Let $\tilde{u} = (u, u_t)$ be a weak solution to (2.1). Assume hypotheses of Theorem 2.4. Then

$$\|u(t)\| + |u_t(t)|_V \le C(\|u_0\|, |u_1|_V)$$

for $t \le T_0$.

proof:

We shall use the result of Lemma 3.2 with $f(t) \equiv f(u(t))$, $\mathcal{F}(t) \equiv \mathcal{F}(u(t))$.

Since $u \in C[0T_0; \mathcal{D}(\tilde{A}^{½})]$, by the assumptions imposed on f and $\mathcal{F}$ (see (1.5) and (1.6)) we obtain

$$|f|_{C[0T_0;\, U]} \le C_{T_0}(\|u_0\|; |u_1|_V),$$

$$|\mathcal{F}|_{C[0T_0;\, V']} \le C_{T_0}(\|u_0\|; |u_1|_V).$$

Thus, we are in a position to apply inequality (3.23) of Lemma 3.2. This yields

$$|u_t(t)|_V^2 + \|u(t)\|^2 + 2\beta \int_0^t |G^* Au_t(s)|_U^2 \, ds + 2\int_0^t (f(u(s)), G^* Au_t(s) \, ds \tag{3.48}$$
$$- 2\int_0^t (\mathcal{F}(u(s)), u_t(s) \, ds \le \|u_0\|^2 + |u_1|_V^2 .$$

From inequality (2.3) in Theorem 2.1 we conclude that hypotheses (2.15), (2.16) of Theorem 2.4 are applicable to weak solutions (u, u_t). Hence, from (3.48) and (2.15), (2.16) we infer that

$$|u_t(t)|_V^2 + \|u(t)\|^2 \le \|u_0\|^2 + |u_1|_V^2 + C\int_0^t |u_t(\tau)|_V^2 + \|u(\tau)\|^2 \, d\tau \quad t \le T_0 . \tag{3.49}$$

Application of Gronwell's inequality to (3.49) completes the proof of Lemma 3.3. ■

4. VON KARMAN SYSTEM

Let Ω be an open bounded domain in R^2 with sufficiently smooth boundary Γ. We consider the following model of a dynamic Von Karman plate in the variable u(t, x)

$$u_{tt} - \gamma\Delta u_{tt} + \Delta^2 u = [F(u), u] \text{ in } \Omega \times (0, T) \tag{4.1}$$

with initial conditions

$$u(t=0) = u_0 \in H^2(\Omega) , \; u_t(t=0) = u_1 \in H_0^1(\Omega) \text{ in } \Omega , \tag{4.2}$$

and boundary conditions

$$\begin{cases} u|_\Gamma = 0 \\ \Delta u|_\Gamma = -\beta \dfrac{\partial}{\partial \nu} u_t + \tilde{f}(\dfrac{\partial}{\partial \nu} u(t, x)) \text{ on } \Gamma \times (0, T) . \end{cases} \tag{4.3}$$

The nonlinear operator F: $H^2(\Omega) \to H^2(\Omega)$ is defined by

$$\begin{cases} \Delta^2 F(u) = -[u, u] \quad \text{in } \Omega , \\ F = \dfrac{\partial F}{\partial \nu} = 0 \text{ on } \Gamma ; \end{cases} \tag{4.4}$$

$$\text{with } [\psi, \phi] \equiv \psi_{xx}\phi_{yy} + \psi_{yy}\phi_{xx} - 2\psi_{xy}\phi_{xy}.$$

Here $\tilde{f} \in C^1(R)$ and it is assumed to be polynomially bounded i.e.

$$|\tilde{f}(s)| \le C[1 + s^P] \text{ for } 0 \le P < \infty : \quad s \in R .$$

The constants β and γ are strictly positive.

Remark 4.1

One could also consider Von Karman plate equation with different boundary conditions than in (4.3), (for instance clamped boundary conditions or hinged). Since the technicalities are similar to those in (4.3) we shall concentrate only on the latter. Also one may consider more general structure of the operator $\tilde{f}(u, \nabla u, u_t)$. Since this level of generality does not introduce new (conceptual) difficulties for simplicity of exposition we take $\tilde{f}(\frac{\partial}{\partial \nu} u)$.

Von Karman plate equations have attracted a considerable attention in the past. However, to the author's best knowledge the results on wellposedness available in the literature for two dimensional problem deal with the case when the boundary conditions are *homogeneous*, i.e. right hand side of (4.3) equal to zero (see [B.3]], [L.2], [S.1]. In fact, in [L.2], the existence and uniqueness result for the *homogeneous* (on the boundary) problem was established by using Faedo-Gelerkin method. The problem becomes more difficult when the boundary conditions are nonhomogeneous and nonlinear (as they often arise in boundary stabilization problems, see [L.1]). In this case the existing techniques (see [L.2]) are not applicable. The reason for this is that in order to "handle" the nonlinear term on the boundary $f(\frac{\partial}{\partial \nu} u)$, the regularity of the solutions fo the homogeneous Von Karman plate is not sufficient (this precludes the use of standard perturbation or approximation technique). On the other hand, as we shall see below, the results on wellposedness (and regularity) will follow from the abstract theory presented in section 2. To accomplish this we need to put the problem (4.1) in to the abstract framework. We introduce the following spaces and operators

$H = L_2(\Omega);\ V = H_0^1(\Omega);\ U = L_2(\Gamma)$.

$A_D: L_2(\Omega) \to L_2(\Omega)$ defined by

$A_D u = -\Delta u;\ u \in \mathcal{D}(A_D) \equiv H^2(\Omega) \cap H_0^1(\Omega)$ $\quad$ $D: L_2(\Gamma) \to L_2(\Omega)$ given by

$Dg = v$ iff $\Delta v = 0$ in Ω and $v|_\Gamma = g$.

We set

$\tilde{A} \equiv A_D^2$ hence $\tilde{A}^{1/2} = A_D$; $\mathcal{D}(\tilde{A}^{1/2}) = H^2(\Omega) \cap H_0^1(\Omega)$ and

$\|u\| = |A_D u| = |\Delta u|_{L_2(\Omega)}$; $u \in \mathcal{D}(A_D)$, $\quad \tilde{M} \equiv (I + \gamma A_D)$, hence

$M \in \mathcal{L}(H_0^1(\Omega); H^{-1}(\Omega))$ and $|u|_V^2 = |\tilde{M}^{1/2} u|^2 = ((I + \gamma\Delta)u, u) = |u|^2_{L_2(\Omega)} + \gamma |\nabla u|^2_{L_2(\Omega)}$.

$G \equiv A_D^{-1} D$, hence $\tilde{A}^{1/2} G = A_D A_D^{-1} D \in \mathcal{L}(U; H)$.

From [L-T.3] we also have

(4.5) $\quad G^* Au = D^* A_D u = -\frac{\partial}{\partial \nu} u|_\Gamma$ for $u \in \mathcal{D}(A_D)$.

(4.6) $\quad \mathcal{F}(u) \equiv [F(u), u]$ where $\Delta^2 F(u) = [-u, u]$ in Ω

$$F = 0 \text{ in } \Gamma;\quad \frac{\partial F}{\partial \nu} = 0 \text{ on } \Gamma.$$

(4.7) $f(u)(t,x) \equiv \tilde{f}(\frac{\partial u}{\partial \nu}(t,x))$.

Notice that by (4.5), (4.7)

(4.8) $\frac{\partial}{\partial \nu} u_t = -G^* A u_t$ and

(4.9) $\tilde{f}(\frac{\partial}{\partial \nu} u) = \tilde{f}(-G^* Au) \equiv f(u)$.

With the above notation it is known (see [B-L]) that the abstract form of equation (4.1) becomes precisely equation (2.1). Thus, in order to apply the results of Section 2 we need to verify hypotheses (1.2) - (1.6). Notice that the hypotheses (1.2) - (1.4) follow directly from the definitions of the operators. As for hypothesis (1.5), this follows from the inequality

$$|D\mathcal{F}(u)h|_{H^{-\varepsilon}(\Omega)} \le C|u|^2_{H^2(\Omega)} |h|_{H^2(\Omega)} : 0 \le \varepsilon \le 1$$

which inequality, in turn, has been proved in [B-L] (see also [L.3]). By using Sobolev's Imbeddings Theorems and differentiability properties of Nemytskii Operators on L_p spaces (see [A-P]) one can easily show that (1.6) is satisfied. Thus we are in a position to apply the result of Theorem 2.1, which specialized to our situation gives:

Theorem 4.1. (local existence)

For any $u_0 \in H^2(\Omega) \cap H_0^1(\Omega)$; $u_1 \in H_0^1(\Omega)$ there exists unique solution (u, u_t) to (4.1) - (4.4) such that

(4.10) $u \in C[0T_0; H^2(\Omega) \cap H_0^1(\Omega)]$,

(4.11) $u_t \in C[0T_0; H_0^1(\Omega)]$,

(4.12) $\frac{\partial}{\partial \nu} u_t \in L_2(0T_0; L_2(\Gamma))$,

for some $T_0 > 0$.

Remark 4.2.

Notice that boundary regularity in (4.12) does not follow from interior regularity in (4.10) - (4.11). It is an additional regularity result.

Now we shall turn to the question of global existence of the solutions to (4.1)-(4.4). At this point we need to assume some structural condition on the function $\tilde{f}$. We shall make the following hypothesis

(4.13) $\tilde{f}(s)\, s \le 0$ for $s \in R$.

Theorem 4.2 (global existence)

Under the additional hypothesis (4.13) the solutions to (4.1)-(4.3) are global.

proof: It suffices to verify hypotheses (2.15), (2.16) and to apply the result of Theorem 2.4. To accomplish this we first note that (see [K.2])

(4.14) $([\psi, \phi], f)_{L_2(\Omega)} = ([f, \phi], \psi)_{L_2(\Omega)} = ([\psi, f], \phi)_{L_2(\Omega)}$; $\psi, f, \Phi \in H^2(\Omega) \cap H_0^1(\Omega)$.

By (4.12), (4.14), Green's Formula and integration by parts procedure we obtain

$$\int_0^t (\mathcal{F}(y), y_t)_{L_2(\Omega)} \, d\Omega = -\tfrac{1}{4} |\Delta F(y(t))|^2_{L_2(\Omega)} + \tfrac{1}{4} |\Delta F(y(0))|^2_{L_2(\Omega)} \le \tfrac{1}{4} |\Delta F(y(0))|^2_{L_2(\Omega)}$$

(4.15) $\le C |y(0)|^4_{H^2(\Omega)}$.

As for (2.16) we write with $\tilde{f}_1(s) \equiv \int_0^s \tilde{f}(\tau)\, d\tau \ge 0$

$$-\int_0^t < f(y(z)), \frac{\partial}{\partial \nu} y_t >_{L_2(\Gamma)} d\Omega \equiv -\int_0^t < \tilde{f}(-\frac{\partial}{\partial \nu} y), \frac{\partial}{\partial \nu} y_t >_{L_2(\Gamma)} d\tau$$

$$= + \int_0^t \frac{d}{d\tau} \int_\Gamma \tilde{f}_1 (-\frac{\partial}{\partial \nu} y)\, d\Gamma\, d\tau = \int_\Gamma \tilde{f}_1 (\frac{-\partial}{\partial \nu} y(t))\, d\Gamma + \int_\Gamma \tilde{f}_1 (-\frac{\partial}{\partial \nu} y(0))\, d\Gamma$$

by Sobolev Imbeddings and Trace Theorem

$$\le C \left| \frac{\partial}{\partial \nu} y(0) \right|^{\frac{1}{p+1}}_{L_{p+1}(\Gamma)} \le C \left| \frac{\partial}{\partial \nu} y(0) \right|^{\frac{1}{p+1}}_{H^{1/2}(\Gamma)} \le C (|y(0)|_{H^2(\Omega)})$$

which together with (4.15) proves the desired inequality in (2.16) and in (2.15). ■

We finally turn to the question of regularity of solutions to (4.1)-(4.3). To simplify the exposition we shall assume $\tilde{f} = 0$ (this restriction is, of course, not essential at the regularity level). Our final result states that assuming more smoothness on initial data, the solutions to (4.1) are classical.

Theorem 4.3. (regularity). Assume

(4.16) $u_0 \in H^4(\Omega), u_1 \in H^3(\Omega) \in H_0^1(\Omega)$,

$$(4.17) \quad \begin{cases} \Delta u_1 |_\Gamma = -\beta \dfrac{\partial}{\partial \nu} A_D^{-1} [\Delta^2 u_0 + \mathcal{F}(u_0)], \text{ on } \Gamma, \\ \Delta u_0 |_\Gamma = -\beta \dfrac{\partial}{\partial \nu} u_1, \text{ on } \Gamma. \end{cases}$$

Then,

(4.18) $u_{tt} \in C[0T; H^2(\Omega)]$,

(4.19) $u_{ttt} \in C[0T; H_0^1(\Omega)]$,

(4.20) $u \in C[0T; H^4(\Omega)]$.

The result of Theorem 4.4 follows from Theorem 2.3 once we show that the hypotheses of Theorem 2.3 hold true. Verification of the hypotheses is rather straightforward though tedious and lengthy. Because of space limitations we omit the details.

Acknowledgments. Part of this work was performed while the second author was visiting the Dipartimento di Matematica of Universitá di Bologna, whose support and hospitality is greatfully acknowledged.

References

[A.1] R. Adams. Sobolev Spaces, Academic Press, New York, (1975).

[A-P] A. Ambrosetti, G. Prodi. Analisi Non Lineare, Pisa, (1973).

[B.1] V. Barbu. Nonlinear Semigroups and Differential Equations in Banach Spaces, Noordhoff International Publishing, Leyden, The Netherlands, (1976).

[B.2] H. Brezis. Operateurs Maximaux Monotones et Semigroups de Contractions dans les espaces de Hilbert, North Holland, Mathematics Studies, Amsterdam, (1973).

[B-F] V. Barbu, A. Favini. Existence for implicit differential equations in Banach spaces, to appear Rendiconti Acad. Nat. Lincei.

[B-L] M. E. Bradley, I. Lasiecka. Local exponential stabilization of a nonlinearly perturbed von Karman plate, Journal of Nonlinear Analysis, Theory, Methods and Applications, Vol. 18, No. 4, pp. 333-343, (1992).

[B.3] W. V. Bolotin. Nonconservative Problems of the Theory of Elastic Stability, Fizmatgiz, Moscov 1691. English Translation, Pergamon Press, Oxford and Macmillan, New York, (1963).

[C-S] R. D. Carroll, R.E. Showalter. Singular and Degenerate Cauchy Problems, Academic Press, New York, (1976).

[D-L-S] W. Desch, I. Lasiecka, W. Shappacher. Feedback boundary control problems for linear semigroups, Israel Journal of Mathematics, Vol. 51, No. 3, 177-207, (1985).

[F.1] W. F. Fitzgibbon. Strongly damped quasilinear evolution equations, J. Math. Anal. Appl., 42, 536-550, (1981).

[F-L-T] A. Favini, I. Lasiecka, W. Tanabe. Abstract differential equations and nonlinear dispersive systems.

[K.1] T. Kato. Perturbation Theory for Linear Operators, Springer Verlag, (1966).

[K.2] S. Kesavan. Topics in Functional Analysis and Applications, J. Wiley

[G.1] P. Grisvard. Elliptic Problems in Nonsmooth Domains, Pitman, London, (1985).

[L.1] J. Lagnese. Boundary Stabilization of Thin Plates, SIAM, Philadelphia, (1983).

[L.2] J. L. Lions. Quelques méthodes de résolution des problèmes aux limites nonlinéaires, Dunod, Paris, (1969).

[L.3] J. Lagnese. Local controllability of dynamic von Karman plates Control and Cybernetics 19, 155-168, (1990).

[L.4] J. L. Lions. Contrôllabilité exact des systèmes distribués, Masson, Paris, (1988).

[L-L] J. Lagnese, G. Leugering. Uniform stabilization of a nonlinear beam by nonlinear feedback, Journal Differential Equations 91: 355-388, (1991).

[L-L.2] V. Lakshmikantham and S. Leela. Nonlinear Differential Equations in Abstract Spaces, Pergamon Press, Oxford, (1981).

[L-M] J. L. Lions, E. Magenes. Non Homogeneous Boundary Value Problems and Applications, Vol. I, Springer-Verlag, New York, (1972).

[L-T.1] I. Lasiecka, R. Triggiani. Lifting Theorem for the time regularity of solutions to abstract equations. Proceedings of the American Mathematical Society, Vol. 104, No. 3, 745-755, (1988).

[L-T.2] I. Lasiecka, R. Triggiani. Trace regularity of the solutions of the wave equation with homogeneous Neumann boundary conditions and data supported away from the boundary, *J. of Math. Anal. Appl.* Vol. 141, No. 1, pp. 49-71, (1989).

[L-T.3] I. Lasiecka, R. Triggiani. Differential and Algebraic Riccati Equations with Applications to Boundary/Point Control Problems: Continuous Theory and Approximation Theory, Vol. 164, Springer Verlag, LNCIS, (1991).

[P.1] A. Pazy. Semigroups of Operators and Applications to Partial Differential Equations, Springer-Verlag, New York, (1983).

[S.1] R. E. Showalter. Hilbert Space Methods for Partial Differential Equations, Pitman, London, (1977).

[S.2] A. Stahel. A remark on the equation of a vibrating plate. Proc. Royal Soc. Edinburgh, 136A, 307-314, (1987).

[T.2] H. Tanabe. Equation of Evolution, Pitman London, (1979).

Linear Parabolic Differential Equations of Higher Order in Time

ANGELO FAVINI Dipartimento di Matematica, Università di Bologna, I-40127 Bologna, Italy

HIROKI TANABE Department of Mathematics, Osaka University, Toyonaka 560, Osaka, Japan

ABSTRACT. In this paper we are interested in extending the results of A.Favini and E.Obrecht [3] on the solvability of parabolic evolution equations of second order in time to equations of third order. We explaine by an example the utility of the idea of [3] as well as the extension of S.G.Krein's method of transforming the original equation to a first order system ([6]). We also solve equations with coefficients depending on the time variable applying the classical result of T.Kato and H.Tanabe ([5]).

The object of this paper is to describe a method of solving the initial value problem of linear parabolic differential equations of higher order in time by the following example:

(1) $$u''' + A_2 u'' + A_1 u' + A_0 u = f.$$

Let

$$\mathcal{B} = -\sum_{i,j=1}^{n} a_{ij} \frac{\partial^2}{\partial x_i \partial x_j} + \sum_{i=1}^{n} b_i \frac{\partial}{\partial x_i} + c$$

be a strongly elliptic linear partial differential operator of second order with real coefficients in a bounded domain Ω in R^n with smooth boundary $\partial\Omega$. Let B_0 be the realization of $\mathcal{B}$ under the Dirichlet boundary condition in $L^p(\Omega)$, $1 \le p < \infty$:

$$D(B_0) = \begin{cases} W^{2,p}(\Omega) \cap W_0^{1,p}(\Omega) \quad \text{if} \quad 1 < p < \infty, \\ \{u \in W_0^{1,q}(\Omega) \quad \text{for any} \quad 1 \le q < n/(n-1) \quad \text{and} \quad \mathcal{B}u \in L^1(\Omega) \\ \quad \text{in the sense of distributions}\} \quad \text{if} \quad p = 1, \end{cases}$$

$$B_0 u = \mathcal{B}u \quad \text{for} \quad u \in D(B_0).$$

Similarly let B_1 be the realization of $\mathcal{B}^2$ under the Dirichlet boundary condition in $L^p(\Omega), 1 \le p < \infty$:

$$D(B_1) = \begin{cases} W^{4,p}(\Omega) \cap W_0^{2,p}(\Omega) \quad \text{if} \quad 1 < p < \infty, \\ \{u \in W^{3,q}(\Omega) \cap W_0^{2,q}(\Omega) \quad \text{for any} \quad 1 \le q < n/(n-1) \quad \text{and} \\ \quad \mathcal{B}^2 u \in L^1(\Omega) \quad \text{in the sense of distributions}\} \quad \text{if} \quad p = 1, \end{cases}$$

$$B_1 u = \mathcal{B}^2 u \quad \text{for} \quad u \in D(B_1).$$

The operators A_0, A_1, A_2 are defined by

$$A_2 = B_0 B_1, \quad A_1 = B_1 A_2, \quad A_0 = A_2^2 = B_0 B_1 A_2.$$

Let

$$P(\lambda) = \lambda^3 + \lambda^2 A_2 + \lambda A_1 + A_0$$

be the operator pencil associated with (1). We first show that $P(\lambda)$ is parabolic in the sense of A. Favini and E. Obrecht [3] and E. Obrecht [7], [8], [9]:

$$\text{(2)} \qquad \begin{aligned} &\|P(\lambda)^{-1}\| \le C|\lambda|^{-3}, \quad \|A_2 P(\lambda)^{-1}\| \le C|\lambda|^{-2}, \\ &\|A_1 P(\lambda)^{-1}\| \le C|\lambda|^{-1}, \quad \|A_0 P(\lambda)^{-1}\| \le C \end{aligned}$$

if λ belongs to some sector $\Sigma = \{\lambda : \arg\lambda \in (-\theta_0, \theta_0)\}, \pi/2 < \theta_0 < \pi$, and $|\lambda|$ is sufficiently large. Based on the estimates (2) we solve the initial value problem for (1) transforming it to a system of first order in t following the idea of S. G. Kreĭn [6]. We can show that the solution constructed by this method satisfies the initial conditions in a suitable sense without relying upon the Brézis-Fraenkel condition (cf. [2]) which was essentially used in [3], [8], [9].

Following A. Favini and E. Obrecht [3] we consider another operator

$$\text{(3)} \quad P_0(\lambda) = (\lambda + B_0)(\lambda + B_1)(\lambda + A_2) = P(\lambda) + \lambda^2(B_0 + B_1) + \lambda(B_0 B_1 + B_0 A_2).$$

As is well known $-B_0$ and $-B_1$ generate analytic semigroups. As for A_2 the corresponding fact does not seem evident since it is not defined variationally. We show that this is true by verifying Agmon's condition in order that any ray other than the positive real axis is a direction of minimal growth of the resolvent of A_2 (cf. [1]). By definition A_2 is the realization of $\mathcal{B}^3$ under the boundary conditions $u = \partial u/\partial \nu = \mathcal{B}^2 u = 0$ on $\partial\Omega$. The characteristic polynomial of $\mathcal{B}^3$ is

$$(\sum_{i,j=1}^{n} a_{ij}(x)\xi_i\xi_j)^3 = (A(x)\xi, \xi)^3,$$

where $A(x)$ is the matrix $(a_{ij}(x))$.

PROPOSITION 1. *$-A_2$ generates an analytic semigroup in $L^p(\Omega)$ for $1 \leq p < \infty$. More precisely, A_2 is of type (0,M) for some M in the sense of T. Kato* [4].

PROOF Fix $x \in \partial\Omega$ and $\xi \in R^n$ be tangential to $\partial\Omega$ at x. Write A instead of $A(x)$ for simplicity. For $r > 0$ and $0 < \theta < 2\pi$ the algebraic equation for s

$$(A(\xi + s\nu), \xi + s\nu)^3 = re^{i\theta}$$

has no real roots. Let s_1, s_2, s_3 be the roots with positive imaginary parts. We are going to show that the polynomials 1, s, $(A(\xi + s\nu), \xi + s\nu)^2$ are linearly independent modulo $(s - s_1)(s - s_2)(s - s_3)$. This is equivalent to saying that the matrix

$$\begin{pmatrix} 1 & s_1 & (A(\xi + s_1\nu), \xi + s_1\nu)^2 \\ 1 & s_2 & (A(\xi + s_2\nu), \xi + s_2\nu)^2 \\ 1 & s_3 & (A(\xi + s_3\nu), \xi + s_3\nu)^2 \end{pmatrix}$$

is nonsingular since s_1, s_2, s_3 are distinct. By an elementary calculus and noting

$$\begin{aligned} &(A(\xi + s_1\nu), \xi + s_1\nu) + (A(\xi + s_2\nu), \xi + s_2\nu) + (A(\xi + s_3\nu), \xi + s_3\nu) \\ &\quad = r^{1/3}e^{\theta i/3}(1 + e^{2\pi i/3} + e^{4\pi i/3}) = 0 \end{aligned}$$

we see that what is to be shown reduces to

$$\begin{aligned} &4(A\xi, \nu)^2 - (A\xi, \xi)(A\nu, \nu) + 2(s_1 + s_2 + s_3)(A\xi, \nu)(A\nu, \nu) \\ &\quad + (s_2s_3 + s_3s_1 + s_1s_2)(A\nu, \nu)^2 \neq 0. \end{aligned} \tag{4}$$

Here we use the following classical result by Hermit and Biehler on polynomials:

Let $f(z)$ be a polynomial whose roots have imaginary parts of the same sign. Decomposing the coefficients of $f(z)$ into real and imaginary parts write $f(z) = u(z) + iv(z)$. Then, the roots of $u(z)$ and $v(z)$ are real distinct and separate each other.

Suppose (4) is false. Then

$$\begin{aligned} &4(A\xi, \nu)^2 - (A\xi, \xi)(A\nu, \nu) + 2(s_1 + s_2 + s_3)(A\xi, \nu)(A\nu, \nu) \\ &\quad + (s_2s_3 + s_3s_1 + s_1s_2)(A\nu, \nu)^2 = 0. \end{aligned} \tag{5}$$

Let

$$\begin{aligned} f(z) &= (z - s_1)(z - s_2)(z - s_3) \\ &= z^3 - (s_1 + s_2 + s_3)z^2 + (s_2s_3 + s_3s_1 + s_1s_2)z - s_1s_2s_3, \end{aligned}$$

$$u(z) = z^3 - \mathrm{Re}(s_1 + s_2 + s_3) \cdot z^2 + \mathrm{Re}(s_2s_3 + s_3s_1 + s_1s_2) \cdot z - \mathrm{Re} s_1s_2s_3, \tag{6}$$

$$v(z) = -\mathrm{Im}(s_1 + s_2 + s_3) \cdot z^2 + \mathrm{Im}(s_2s_3 + s_3s_1 + s_1s_2) \cdot z - \mathrm{Im} s_1s_2s_3. \tag{7}$$

According to the theorem of Hermit and Biehler we have

$$u(z) = (z - \alpha_1)(z - \alpha_2)(z - \alpha_3),$$

$$v(z) = -\mathrm{Im}(s_1 + s_2 + s_3) \cdot (z - \beta_1)(z - \beta_2),$$

where $\alpha_1 < \beta_1 < \alpha_2 < \beta_2 < \alpha_3$. Noting

$$u'\left(\frac{\alpha_1 + \alpha_2}{2}\right) = -\left(\frac{\alpha_1 - \alpha_2}{2}\right)^2 < 0, \quad u'\left(\frac{\alpha_2 + \alpha_3}{2}\right) = -\left(\frac{\alpha_2 - \alpha_3}{2}\right)^2 < 0$$

we see that $u'(z) < 0$ for $\frac{\alpha_1 + \alpha_2}{2} < z < \frac{\alpha_2 + \alpha_3}{2}$. This and $\frac{\alpha_1 + \alpha_2}{2} < \frac{\beta_1 + \beta_2}{2} < \frac{\alpha_2 + \alpha_3}{2}$ implies $u'\left(\frac{\beta_1 + \beta_2}{2}\right) < 0$. From (5) and (7) it follows that $\frac{\beta_1 + \beta_2}{2} = -\frac{(A\xi, \nu)}{(A\nu, \nu)}$ and hence

$$u'\left(-\frac{(A\xi, \nu)}{(A\nu, \nu)}\right) < 0. \tag{8}$$

On the other hand differentiating both sides of (6), substituting $z = -\frac{(A\xi, \nu)}{(A\nu, \nu)}$ in the resulting equality and using the hypothesis (5) we get

$$u'\left(-\frac{(A\xi, \nu)}{(A\nu, \nu)}\right) = \frac{(A\xi, \xi)(A\nu, \nu) - (A\xi, \nu)^2}{(A\nu, \nu)^2} \geq 0$$

in view of the Schwarz inequality. This contradicts (8) and the proof of the proposition is complete.

Noting

$$\begin{aligned}
A_2 P_0(\lambda)^{-1} &= A_2(\lambda + A_2)^{-1}(\lambda + B_1)^{-1}(\lambda + B_0)^{-1}, \\
A_1 P_0(\lambda)^{-1} &= B_1 A_2(\lambda + A_2)^{-1}(\lambda + B_1)^{-1}(\lambda + B_0)^{-1} \\
&= B_1\{I - \lambda(\lambda + A_2)^{-1}\}(\lambda + B_1)^{-1}(\lambda + B_0)^{-1} \\
&= B_1(\lambda + B_1)^{-1}(\lambda + B_0)^{-1} - \lambda B_1(\lambda + A_2)^{-1}(\lambda + B_1)^{-1}(\lambda + B_0)^{-1}, \\
A_0 P_0(\lambda)^{-1} &= A_2^2(\lambda + A_2)^{-1}(\lambda + B_1)^{-1}(\lambda + B_0)^{-1} \\
&= A_2\{I - \lambda(\lambda + A_2)^{-1}\}(\lambda + B_1)^{-1}(\lambda + B_0)^{-1} \\
&= B_0 B_1(\lambda + B_1)^{-1}(\lambda + B_0)^{-1} - \lambda A_2(\lambda + A_2)^{-1}(\lambda + B_1)^{-1}(\lambda + B_0)^{-1} \\
&= B_0(\lambda + B_0)^{-1} - \lambda B_0(\lambda + B_1)^{-1}(\lambda + B_0)^{-1} \\
&- \lambda A_2(\lambda + A_2)^{-1}(\lambda + B_1)^{-1}(\lambda + B_0)^{-1},
\end{aligned}$$

we easily see that the inequalities (2) hold with $P(\lambda)$ replaced by $P_0(\lambda)$. It is also easy to show that

$$\|\lambda^2(B_0 + B_1)P_0(\lambda)^{-1}\| = O(|\lambda|^{-1/3})$$
$$\|\lambda(B_0 B_1 + B_0 A_2)P_0(\lambda)^{-1}\| = O(|\lambda|^{-1/2}).$$

Combining this and (3) we conclude that (2) hold and $P(\lambda)$ is parabolic in the sense of A. Favini and E. Obrecht [3] and E. Obrecht [7], [8], [9].

THEOREM 1. *Suppose* $1 \le p < \infty$. *For any initial values* $u(0), u'(0), u''(0)$ *satisfying*

$$u(0) \in D(A_1), \quad u'(0) \in D(A_2), \quad u''(0) \in L^p(\Omega)$$

and a Hölder continuous function f *with values in* $L^p(\Omega)$ *the solution of* (1) *exists and is unique. The initial conditions are satisfied in the following sense:*

$$u(t) \to u(0) \quad in \quad D(A_1), \quad u'(t) \to u'(0) \quad in \quad D(A_2), \quad u''(t) \to u''(0) \quad in \quad L^p(\Omega).$$

PROOF We transform the equation (1) to a first order system as follows (cf. S. G. Kreĭn [6: Chap. III, §3]). Let

$$A_1 u = u_1 - u_2, \quad A_2 u' = u_2 - u_3, \quad u_3 = u''. \tag{9}$$

Then,

$$\frac{d}{dt}\begin{pmatrix} u_1 \\ u_2 \\ u_3 \end{pmatrix} = -(\mathfrak{A} + \mathfrak{B})\begin{pmatrix} u_1 \\ u_2 \\ u_3 \end{pmatrix} + \begin{pmatrix} f \\ f \\ f \end{pmatrix}, \tag{10}$$

where

$$\mathfrak{A} = \begin{pmatrix} B_0 & 0 & 0 \\ B_0 & B_1 & 0 \\ B_0 & B_1 & A_2 \end{pmatrix}, \quad \mathfrak{B} = -\begin{pmatrix} 0 & B_0 & 0 \\ 0 & B_0 & B_1 \\ 0 & B_0 & B_1 \end{pmatrix}. \tag{11}$$

It is easy to show that $-\mathfrak{A}$ generates an analytic semigroup in $(L^p(\Omega))^3$ and

$$\|\mathfrak{B}(\lambda - \mathfrak{A})^{-1}\| \le C|\lambda|^{-1/3},$$

if $\lambda \in \Sigma$ and $|\lambda|$ is sufficiently large. We complete the proof of the theorem returning to the original equation.

Next we consider the case where the coefficients of the operator are dependent on t:

$$\mathcal{B}(t) = -\sum_{i,j=1}^{n} a_{ij}(x,t)\frac{\partial^2}{\partial x_i \partial x_j} + \sum_{i=1}^{n} b_i(x,t)\frac{\partial}{\partial x_i} + c(x,t).$$

We again assume that the coefficients are sufficiently smooth.

For the sake of clarity we consider the equation of the form

$$A(t, D_t)u \equiv u''' + (A_2 u)'' + (A_1 u)' + A_0 u = f. \tag{12}$$

If $1 < p < \infty$, the domains of $B_0(t)$ and $B_1(t)$ do not depend on t, but that of $A_2(t)$ does.

THEOREM 2. *Suppose* $1 < p < \infty$. *For any initial values* $u(0), u'(0), u''(0)$ *satisfying*

$$u(0) \in D(A_1(0)), \quad u'(0) \in D(B_1(0)),$$
$$B_1(0)u'(0) + B_1'(0)u(0) \in D(B_0(0)), \quad u''(0) \in L^p(\Omega)$$

and a Hölder continuous function f *with values in* $L^p(\Omega)$ *the solution of* (12) *exists and is unique. The initial conditions are satisfied in the following sense:*

$$u(t) \to u(0) \quad in \quad W^{10,p}(\Omega), \quad u'(t) \to u'(0) \quad in \quad W^{6,p}(\Omega),$$
$$u''(t) \to u''(0) \quad in \quad L^p(\Omega).$$

PROOF Slightly modifying (9) we put

$$A_1 u = u_1 - u_2, \quad (A_2 u)' = u_2 - u_3, \quad u_3 = B_1^{-1}(B_1 u)''.$$

Then, ${}^t(u_1, u_2, u_3)$ satisfies (10) where $\mathfrak{A}$ is the same as that of (11) and

$$\mathfrak{B} = -\begin{pmatrix} 0 & B_0 & -2(B_1^{-1})'B_1 \\ 0 & B_0 & B_1 - 2(B_1^{-1})'B_1 \\ 0 & B_0 & B_1 - 2(B_1^{-1})'B_1 \end{pmatrix} + \text{bounded operators.}$$

The classical result by T. Kato and H. Tanabe [5] on evolution equations of parabolic type

$$du(t)/dt + A(t)u(t) + B(t)u(t) = f(t),$$

where

$$\|(\partial/\partial t)(\lambda - A(t))^{-1}\| \le C|\lambda|^{-\rho}, \quad 0 < \rho \le 1, \tag{13}$$

$$\|B(t)(\lambda - A(t))^{-1}\| \le C|\lambda|^{-\gamma}, \quad 0 < \gamma \le 1 \tag{14}$$

is applicable to the present situation. In fact, it is easily verified that (13) and (14) hold with $\rho = 1$ and $\gamma = 1/3$ respectively in our case. Thus we conclude the proof of the theorem returning to the original equation (12).

The method of the proof of Theorem 2 cannot be applied if $p = 1$, since B_0, B_1 have variable domains in this case. Hence, we solve the problem by a different method constructing the fundamental solution of the original equation (12) without transforming it to a first order system.

Suppose $1 \le p < \infty$. The operator valued function $U(t,s)$ is called a fundamental solution of (12) if it satisfies

$$A(t, D_t)U(t,s) = 0 \quad \text{in} \quad 0 \le s < t \le T,$$
$$U(s,s) = (\partial/\partial t)U(t,s)|_{t=s} = 0, \quad (\partial/\partial t)^2 U(t,s)|_{t=s} = I.$$

Let

$$P(t,\lambda)=\lambda^3+\lambda^2A_2(t)+\lambda A_1(t)+A_0(t),$$
$$U_0(t,s)=\frac{1}{2\pi i}\int_\Gamma e^{\lambda(t-s)}P(t,\lambda)^{-1}d\lambda,$$

where Γ is a smooth curve connecting $\infty e^{-i\theta_0}$ and $\infty e^{i\theta_0}$ in Σ. The fundamental solution $U(t,s)$ is constructed by

$$U(t,s)=U_0(t,s)+\int_s^t U_0(t,\tau)R(\tau,s)d\tau,$$
$$R(t,s)+R_1(t,s)+\int_s^t R_1(t,\tau)R(\tau,s)d\tau=0,$$
$$R_1(t,s)=A(t,D_t)U(t,s).$$

It is possible to express the solution of the initial value problem satisfying the initial conditions

(15) $$u(0)=u_0,\quad u'(0)=u_1,\quad u''(0)=u_2,$$

if the following rather restrictive assumptions are satisfied:

(16) $$u_0\in D(A_0(0)),\quad v_1\equiv u_1-D_tA_0(t)^{-1}A_0(0)u_0|_{t=0}\in D(A_0(0)).$$

Put

(17) $$v_2\equiv u_2-D_t^2A_0(t)^{-1}A_0(0)u_0|_{t=0}-2D_tA_0(t)^{-1}A_0(0)v_1|_{t=0}.$$

Then

(18) $$u(t)=\sum_{i=0}^{2}u_i(t)+\int_0^t U(t,s)f(s)ds,$$

where

(19) $$u_0(t)=A_0(t)^{-1}A_0(0)u_0-\int_0^t U(t,s)A(s,D_s)(A_0(s)^{-1}A_0(0)u_0)ds,$$

(20) $$u_1(t)=tA_0(t)^{-1}A_0(0)v_1-\int_0^t U(t,s)A(s,D_s)(sA_0(s)^{-1}A_0(0)v_1)ds,$$

(21) $$u_2(t)=U(t,0)v_2,$$

is the solution of the initial value problem.

Next, we consider the uniqueness in case $p=1$. If $1<p<\infty$ the following relation holds for any $0\le s<t\le T$ and $\phi\in L^p(\Omega)$:

(22) $$U(t,s)A_0(s)^{-1}\phi$$
$$=\frac{1}{2}(t-s)^2A_0(t)^{-1}\phi-\int_s^t U(t,\sigma)A(\sigma,D_\sigma)\left(\frac{1}{2}(\sigma-s)^2A_0(\sigma)^{-1}\phi\right)d\sigma,$$

since both sides are solutions of (12) with $f=0$ in $(s,T]$ satsifying the initial conditions $u(s)=u'(s)=0, u''(s)=A_0(s)^{-1}\phi$, and the uniqueness is already known in this case. Let ϕ be an arbitrary element of $L^1(\Omega)$. Approximating ϕ by a sequence of elements of $L^p(\Omega), 1<p<\infty$, we see that (22) holds also for $\phi\in L^1(\Omega)$. Thus following the proof of Theorem 3 of [10] we see that the uniqueness also holds in the space $L^1(\Omega)$, and we have established the following theorem.

THEOREM 3. *Suppose $p = 1$. If (16) is satisfied and f is a Hölder continuous function with values in $L^1(\Omega)$, then the function u defined by (17), (18), (19), (20), (21) is a solution of the initial value problem (12), (15). The initial conditions are satisfied in the following sense:*

$$u(t) \to u(0) \quad \text{in} \quad W^{9,q}(\Omega), \quad u'(t) \to u'(0) \quad \text{in} \quad W^{5,q}(\Omega)$$

$$\text{for any} \quad q \in [1, n/(n-1)), \quad \text{and} \quad u''(t) \to u''(0) \quad \text{in} \quad L^1(\Omega).$$

Furthermore both $\lim_{t\downarrow 0} A_1(t)u(t)$ and $\lim_{t\downarrow 0}(A_2(t)u(t))'$ exist in $L^1(\Omega)$. The solution is unique in this class of functions.

REFERENCES

[1] S. Agmon, *On the eigenfunctions and on the eigenvalues of general elliptic boundary value problems*, Comm. Pure Appl. Math. **15** (1962), 119-147.

[2] H. Brézis and L. E. Fraenkel, *A function with prescribed initial derivatives in different Banach spaces*, J. Funct. Anal. 29 (1978), 328-335.

[3] A. Favini and E. Obrecht, *Conditions for parabolicity of second order abstract differential equations*, preprint.

[4] T. Kato, *Fractional powers of dissipative operators*, J. Math. Soc. Japan **13** (1961), 246–274.

[5] T. Kato ane H. Tanabe, *On the abstract evolution equation*, Osaka Math. J. **14** (1962), 107–133.

[6] S. G. Kreĭn, "Linear Differential Equations in Banach space," Nauka, Moscow, 1967 (in Russian)(English translation: Translations of Mathematical Monographs 29, American Mathematical Society, 1972).

[7] E. Obrecht, *Evolution operators for higher order abstract parabolic equations*, Czechoslovak Math. J. **36(111)** (1986), 210-222.

[8] E. Obrecht, *The Cauchy problem for time-dependent abstract parabolic equations of higher order*, J. Math. Anal. Appl. **125** (1987), 508-530.

[9] E. Obrecht, *Second order abstract parabolic equations*, Semesterbericht Funktionalanalysis Tübingen, Sommersemester 1986, 191-206.

[10] H. Tanabe, *On fundamental solutions of linear parabolic equations of higher order in time and associated Volterra equations*, J. Differential Equations **73** (1988), 288-308.

Analytic and Gevrey Class Semigroups Generated by $-A + iB$, and Applications

A. FAVINI Dipartimento di Matematica, Università di Bologna, I-40127 Bologna, Italy

R. TRIGGIANI Department of Applied Mathematics, University of Virginia, Charlottesville, VA 22903, U.S.A.

1. INTRODUCTION; STATEMENT OF MAIN RESULTS

Throughout this note, unless otherwise stated, X is a complex Hilbert space with inner product (,) and norm $\| \ \|$. We consider the (linear) dissipative operator

$$\mathcal{G} = -A + iB; \quad X \supset \mathcal{D}(\mathcal{G}) = \mathcal{D}(A) \cap \mathcal{D}(B) \to X, \tag{1.1}$$

$\mathcal{D}(A) \cap \mathcal{D}(B)$ being dense in X, under the following standing assumptions:
(H.1) A: $X \supset \mathcal{D}(A) \to X$ is a (strictly) positive self-adjoint operator on X;
(H.2) B: $X \supset \mathcal{D}(B) \to X$ is a self-adjoint operator on X.

It readily follows that $\mathcal{G}^{-1} \in \mathcal{L}(X)$: indeed, for $x \in \mathcal{D}(\mathcal{G})$:

$$\|\mathcal{G}x\| \ \|x\| \geq |(\mathcal{G}x, x)| \geq \|A^{1/2} x\|^2 \geq c\|x\|^2,$$

and thus $\mathcal{G}$ has a continuous inverse on its range, range ($\mathcal{G}$), which is dense in X, since the null space of $\mathcal{G}^*$ is the trivial subspace. In particular, $\mathcal{G}$ is closed on X. In addition, we shall assume a condition of comparison between A and B, which in particular includes the case where "B is comparable to A^α, $\alpha \geq 1$," in the technical sense explained below. We let B^+ be the (unique) non-negative self-adjoint square root of the non-negative operator B^2: $B^+ = (B^2)^{1/2}$, so that the fractional powers $(B^+)^s$, $0 \leq s \leq 1$, are well-defined by the self-adjoint calculus. We shall distinguish two cases: $\alpha = 1$ and $\alpha > 1$. The case $\alpha > 1$ is our main result. One readily sees that, under assumptions (H.1) and (H.2), the operator $\mathcal{G}$ in (1.1) with domain $\mathcal{D}(\mathcal{G})$ is the generator of a strongly continuous (s.c.) contraction semigroup $e^{\mathcal{G}t}$ on X (see section 2, Lemma 2.1).

1.1. Case where $e^{\mathcal{G}t}$ is an analytic semi-group: $\alpha = 1$.

In this subsection we assume further that

(H.3A) $$\mathcal{D}(A^{1/2}) \subset \mathcal{D}((B^+)^{1/2}); \tag{1.2a}$$

equivalently rewritten as

$$(B^+)^{1/2} A^{-1/2} \in \mathcal{L}(X) \iff \|(B^+)^{1/2} x\| \le \sqrt{\rho_2}\, \|A^{1/2} x\| \qquad x \in \mathcal{D}(A^{1/2}); \tag{1.2b}$$

(by the closed graph theorem) $\Updownarrow$

$$(B^+x, x) \le \rho_2 (Ax, x) \tag{1.2c}$$

for a constant $0 < \sqrt{\rho_2} = \|(B^+)^{1/2} A^{1/2}\| < \infty$. The main distinctive feature of this case is that the s.c. contraction semigroup $e^{\mathcal{G}t}$ is also analytic.

Theorem 1.1. Under assumptions (H.1), (H.2), (H.3A) = (1.2), the operator $-\mathcal{G}$ is m-accretive (or $\mathcal{G}$ is m-dissipative) and sectorial with vertex 0 and semi-angle $< \pi/2$. Thus, the s.c. contraction semigroup $e^{\mathcal{G}t}$ is also analytic (holomorphic) on X □

Remark 1.1. If $\mathcal{D}(A) \subset \mathcal{D}(B) \iff B^2 \le \rho_2^2 A^2 \iff \|Bx\| \le \rho_2 \|Ax\|, \quad x \in \mathcal{D}(A)$,

Lowner's Theorem, or interpolation, yields a fortiori (H.3A) = (1.2). Notice that, in this case, B is A-bounded with an A-bound which is allowed to be arbitrarily large. Thus, standard perturbation theory for analytic semigroup [K.1], [P.1] does not cover this case. □

1.2. Case where $e^{\mathcal{G}t}$ is a Gevrey class semigroup: $\alpha > 1$

In this subsection, in place of hypothesis (H.3A) = (1.2), we shall assume the following condition

(H.3G) $$\mathcal{D}(A^{1/2}) = \mathcal{D}((B^+)^{1/2\alpha}) = \mathcal{D}((B^+)^{\theta/2}) \quad \alpha > 1; \; \theta = 1/\alpha < 1, \tag{1.3a}$$

which can be *equivalently* rewritten [when, without loss of generality for the problem here considered, $B^{-1} \in \mathcal{L}(X)$, for otherwise we replace $B^+ \ge 0$ with $(B^+ + I) > 0$] as

$$(B^+)^{\theta/2} A^{-1/2} \in \mathcal{L}(X) \text{ and } A^{1/2} ((B^+)^{-\theta/2}) \in \mathcal{L}(X); \tag{1.3b}$$

equivalently

$$\sqrt{\rho_1}\, \|(B^+)^{\theta/2} x\| \le \|A^{1/2} x\| \le \sqrt{\rho_2}\, \|(B^+)^{\theta/2} x\|, \quad 0 < \rho_1 < \rho_2 < \infty, \; x \in \mathcal{D}((B^+)^{\theta/2}) \tag{1.3c}$$

equivalently

$$\rho_1 ((B^+)^\theta x, x) \le (Ax, x) \le \rho_2 ((B^+)^\theta x, x), \quad x \in \mathcal{D}((B^+)^{\theta/2}); \tag{1.3d}$$

with B^+ positive, this is *equivalently* rewritten as (setting $x = (B^+)^{-\theta/2} y$)

$$\rho_1 \|y\|^2 \le (S_\theta y, y) \le \rho_2 \|y\|^2, \quad y \in X, \tag{1.3e}$$

where S_θ is the bounded, boundedly invertible, self-adjoint operator

$$S_\theta \equiv (B^+)^{-\theta/2} A (B^+)^{-\theta/2}. \tag{1.4}$$

The main distinctive feature of this case is that the s.c. contraction semigroup $e^{\mathcal{G}t}$ is of Gevrey class $> \alpha$ for all $t > 0$. To state the theorem properly, we introduce (with B^+ positive), the spaces $[\mathcal{D}((B^+)^{\theta/2})]'$ and $\mathcal{D}((B^+)^{1-\theta/2})$, $\theta = 1/\alpha < 1$, the first being the dual of $\mathcal{D}((B^+)^{\theta/2})$ with respect to the X-topology, with norms

$$\|y\|_{[\mathcal{D}((B^+)^{\theta/2})]'} = \|(B^+)^{-\theta/2}\, y\|_X \; ; \quad \|y\|_{\mathcal{D}((B^+)^{1-\theta/2})} = \|(B^+)^{1-\theta/2}\, y\|_X \; ; \tag{1.5}$$

as well as the intermediate spaces in between

$$X_{\theta,s} \equiv [\mathcal{D}((B^+)^{1-\theta/2}),\ [\mathcal{D}((B^+)^{\theta/2})]']_s = \mathcal{D}((B^+)^{1-s-\theta/2}), \quad 0 < s < 1, \tag{1.6a}$$

where we have set conventionally $\mathcal{D}((B^+)^{-p}) = [\mathcal{D}((B^+)^p)]'$ for $p > 0$, with norms for all $0 \le s \le 1$ which generalize (1.5):

$$\|y\|_{X_{\theta,s}} = \|(B^+)^{1-s-\theta/2}\|_X \tag{1.6b}$$

We next recall that a s.c. semigroup T(t) on X is of Gevrey class $\delta > 1$ for $t > t_0$ if T(t) is infinitely differentiable for $t \in (t_0, \infty)$ and for any compact $\mathcal{K} \subset (t_0, \infty)$ and each $k > 0$, there exists a constant
$c = c(k, \mathcal{K})$ such that $\|T^{(n)}(t)\| \le c\, k^n\, (n!)^\delta$ for all $t \in \mathcal{K}$ and $n = 0, 1, \ldots$. The case $\delta = 1$ corresponds to analyticity. See [Ta.1] where a theory of Gevrey class semigroups is developed that parallels and extends the theory of differentiable semigroups in [P.1].

Theorem 1.2. Under assumptions (H.1), (H.2), (H.3G) = (1.3), the resolvent operator $R(\lambda, \mathcal{G}) = (\lambda I - \mathcal{G})^{-1}$ of $\mathcal{G}$ satisfies the following uniform estimates on the imaginary axis $\lambda = i\tau$, $\tau \in \Re$

$$|\tau|^{1/\alpha}\, [\|R(i\tau, \mathcal{G})\|_{\mathcal{L}([\mathcal{D}((B^+)^{\theta/2})]')} + \|R(i\tau, \mathcal{G})\|_{\mathcal{L}(\mathcal{D}((B^+)^{1-\theta/2}))}] \le \mathrm{Const}_\theta\,, \quad \theta = 1/\alpha \tag{1.7}$$

and by interpolation a similar estimate holds true on each of the spaces $X_{\theta,s}$ defined in (1.6). Thus, [Ta.1] $e^{\mathcal{G}t}$ extends/restricts as a s.c. semigroup of Gevrey class $> \alpha$ on each such space, in particular on $X = X_{\theta,\, s = 1-\theta}$, for all $t > 0$ □

Remark 1.2. Since $\mathcal{G}$ is a s.c. semigroup generator, estimate (1.7) can be extended by Taylor expansion of the resolvent operator $R(\lambda, \mathcal{G})$ to hold true with $i\tau$ in (1.7) replaced by any complex number λ, with $\mathrm{Re}\,\lambda \ge 0$, $|\lambda|$ sufficiently large, and—more interestingly—on the region

$$\Sigma \equiv \{\lambda \in \mathbb{C} : \mathrm{Re}\,\lambda \ge -c\,(1 + |\mathrm{Im}\,\lambda|)^{1/\alpha}\}, \quad c > 0 \tag{1.8}$$

whereby the spectrum of $\mathcal{G}$ is then contained in the complement Σ^c of Σ in $\mathbb{C}$; see [F-Y.1]. □

Remark 1.3. If, with $\alpha > 1$, $\theta = 1/\alpha < 1$:

$$\mathcal{D}(A^\alpha) = \mathcal{D}(B) \quad \Leftrightarrow \quad \rho_1^{2\alpha} B^2 \le A^{2\alpha} \le \rho_2^{2\alpha} B^2$$

or else

$$\mathcal{D}(A) = \mathcal{D}((B^+)^\theta) \quad \Leftrightarrow \quad \rho_1^2\, (B^+)^{2\theta} \le A^2 \le \rho_2^2\, (B^+)^{2\theta}\,,$$

where $(B^+)^2 = B^2$, then Lowner's theorem, or interpolation, implies (1.3) □

Remark 1.4. If A and B commute, than the right hand side inequality $A \le \rho_2 (B^+)^\theta$ in (1.3d) may be dropped from the assumptions of Theorem 1.2. Thus, in this case, it suffices to assume

$$\mathcal{D}(A^{1/2}) \subset \mathcal{D}((B^+)^{1/2\alpha}) \quad \Leftrightarrow \quad (B^+)^{\theta/2}\, A^{-1/2} \in \mathcal{L}(X) \tag{1.9}$$

to obtain the full conclusion of Theorem 1.2. See Remark 3.2.1 and Remark 4.1 below □

2. PROOF OF THEOREM 1.1 (ANALYTIC CASE)

We shall verify that $-\mathcal{G}$, with $\mathcal{D}(\mathcal{G}) = \mathcal{D}(A) \cap \mathcal{D}(B)$ is an m-sectorial operator [K.1, p. 280]; i.e. explicitly that:

(i) $\mathcal{G}$ is m-dissipative, [K.1, p. 279]; indeed, only under assumptions (H.1) and (H.2).

(ii) $-\mathcal{G}$ is sectorial with vertex 0 and semi-angle $\omega < \pi/2$ [K.1, p. 280].

Once (i) and (ii) are established, one then invokes a known result [K.1, Theorem 1.24, p. 490] to obtain the desired conclusion that $e^{\mathcal{G}t}$ is a s.c. analytic semigroup on X. We state the property in (i) for the general case.

Lemma 2.1. Assume (H.1), (H.2). Then, the operator $\mathcal{G}$ in (1.1) with dense domain $\mathcal{D}(A) \cap \mathcal{D}(B)$ is m-dissipative and thus generates a s.c. contraction semigroup on X.

Proof of Lemma 2.1. Let $x \in \mathcal{D}(\mathcal{G}) = \mathcal{D}(A) \cap \mathcal{D}(B)$ and let $\lambda > 0$. Then

$$\left.\begin{array}{l} \|(\lambda - \mathcal{G})\, x\| \\ \|(\lambda - \mathcal{G}^*)\, x\| \end{array}\right\} \geq \lambda \|x\| \tag{2.1}$$

as it plainly follows, since B is self-adjoint and A is positive self-adjoint via [W.1, Thm. 5.27, p. 111], from

$$\|(\lambda + A \pm i\,B)\,x\|\,\|x\| \geq |((\lambda + A \pm i\,B)\,x, x)| \geq ((\lambda + A)\,x, x) \geq \lambda \|x\|\,\|x\| \tag{2.2}$$

Thus, (2.1) says that $(\lambda - \mathcal{G})$ is boundedly invertible on its range $\mathcal{R}(\lambda - \mathcal{G})$. But then, if $\mathcal{N}$ stands for "null space",

$$\overline{\mathcal{R}(\lambda - \mathcal{G})} = X \Leftrightarrow \mathcal{N}(\lambda - \mathcal{G}^*) = \{0\} \tag{2.3}$$

where triviality of $\mathcal{N}$ follows from (2.1). Thus, (2.1) and (2.3) together say that $(\lambda - \mathcal{G})^{-1} \in \mathcal{L}(X)$ and that

$$\|(\lambda - \mathcal{G})^{-1}\| = \|(\lambda + A - i\,B)^{-1}\| \leq \frac{1}{\lambda} \quad \forall\, \lambda > 0 \tag{2.4}$$

so that $\mathcal{G}$ is m-dissipative and thus $\mathcal{G}$ generates a s.c. contraction semi-group $e^{\mathcal{G}t}$ on X, $t \geq 0$ (Hille-Yosida, or Lumar-Phillips). □

Proof of (ii). We must show that the numerical range $\{z \in \mathbb{C}: z = (-\mathcal{G}x, x),\ \forall\, x \in \mathcal{D}(A),\ \|x\| = 1\}$ is contained in the triangular sector centered at the origin (containing the positive semi-axis) of semi-angle $\omega < \pi/2$. Let B^+ be the non-negative, self-adjoint square root of B^2 defined below Eq. (1.1). Spectral resolution shows (see (A. 10) in Appendix at the end)

$$(Bx, x) = (B^+ x, x) - 2\,((\Pi^- B^+)\,x, x) \quad \forall\, x \in \mathcal{D}((B^+)^{1/2}) \tag{2.5}$$

where the operator Π^- is an orthogonal projection on X defined by (A.5) below; $\Pi^- B^+$ is a non-negative self-adjoint operator defined by (A.8) below and where

$$\mathcal{D}(A^{1/2}) \subset \mathcal{D}((B^+)^{1/2}) \subset \mathcal{D}((\Pi^- B)^{1/2}) \tag{2.6}$$

as the first inclusion in (2.6) is hypothesis (H.3A) = (1.2a), while the second inclusion in (2.6) is noted in (A.13) below. Analogously to the equivalences in (1.2), one obtains from (2.6) the inequality

$$((\Pi^- B^+)x, x) \leq c_2 (Ax, x), \quad \forall x \in \mathcal{D}(A^{1/2}) \tag{2.7}$$

Then (2.5) yields, by (1.1) and (1.2c), (2.7)

$$\begin{aligned} |\text{Im}(-\mathcal{G}x, x)| = |(Bx, x)| &\leq (B^+ x, x) + 2((\Pi^- B^+)x, x) \\ &\leq \rho_2 (Ax, x) + 2c_2 (Ax, x), \quad \forall x \in \mathcal{D}(A^{1/2}) \end{aligned} \tag{2.8}$$

Thus (2.8) is rewritten by (1.1) as

$$|\text{Im}(-\mathcal{G}x, x)| \leq (\rho_2 + 2c_2)\,\text{Re}(-\mathcal{G}x, x), \quad \forall x \in \mathcal{D}(A^{1/2}) \tag{2.9}$$

and $-\mathcal{G}$ is sectorial as desired □

Remark 2.1. A direct proof of the resolvent estimate for analyticity is as follows under assumption (H.3A) = (1.2). Taking the inner product of

$$(\lambda + A - iB)x = f; \quad x = (\lambda - \mathcal{G})^{-1} f \tag{2.10}$$

with $x \in \mathcal{D}(A^{1/2})$, and restricting to real parts yields

$$(\text{Re}\,\lambda)\|x\|^2 + \|A^{1/2} x\|^2 \leq \|f\|\,\|x\|, \quad x \in \mathcal{D}(A^{1/2}) \tag{2.11}$$

hence

$$(\text{Re}\,\lambda)\|x\| \leq \|f\|; \quad \|A^{1/2} x\|^2 \leq \|f\|\,\|x\|; \tag{2.12}$$

while restricting to imaginary parts yields by use of (2.8)

$$\begin{aligned} |\text{Im}\,\lambda|\,\|x\|^2 \leq \|f\|\,\|x\| + |(Bx, x)| &\leq \|f\|\,\|x\| + c\|A^{1/2} x\|^2 \\ &\leq (1+c)\|f\|\,\|x\| \end{aligned} \tag{2.13}$$

where in the last step we have used (2.12) (right). Thus (2.13) and (2.12) (left) yield

$$|\lambda|\,\|x\| \leq K\|f\| = K\|(\lambda + A - iB)x\| = K\|(\lambda - \mathcal{G})x\| \tag{2.14}$$

Eq. (2.4) written for the adjoint $(\bar{\lambda} - \mathcal{G}^*)$ yields $\mathcal{N}(\bar{\lambda} - \mathcal{G}^*) = \{0\}$ and then from (2.14) we have $(\lambda - \mathcal{G})^{-1} \in \mathcal{L}(X)$ and, as desired,

$$\|(\lambda - \mathcal{G})^{-1}\| \leq \frac{c}{|\lambda|} \quad \text{Re}\,\lambda > 0 \;\square \tag{2.15}$$

3. PROOF OF THEOREM 1.2 (GEVREY CLASS, $\theta = 1/\alpha < 1$)

3.1. General outline of the proof

The idea of the proof is as follows (see [C-T.1] for a similar idea applied to a different, second order problem in time), as well as [L-T.3].

a) First, one considers the special case $A = \rho(B^+)^\theta$, $\theta = 1/\alpha < 1$, $\rho > 0$, whereby $\mathcal{G}$ in (1.1) specializes then to

$$\mathcal{B}_{\rho\theta} = -\rho\,(B^+)^\theta + iB\,,\quad \mathcal{D}(\mathcal{B}_{\rho\theta}) = \mathcal{D}(B) = \mathcal{D}(B^+)\,,\quad 0<\theta<1\,,\quad \rho>0\,. \tag{3.1.1}$$

In this special case, the effort to prove the resolvent estimate on $X_{\theta,s}$ in (1.6)

$$|\tau|^\theta \,\|R(i\tau,\ \mathcal{B}_{\rho\theta})\|_{\mathcal{L}(X_{\theta,s})} \le \mathrm{Const}_{\theta,s,\rho} \qquad \forall\, \tau\in\mathcal{R} \tag{3.1.2}$$

either for $s = 0$ ($X_{\theta,0} = \mathcal{D}((B^+)^{1-\theta/2})$; or for $s = 1$ ($X_{\theta,1} = [\mathcal{D}((B^+)^{\theta/2})]'$); or for $s = 1 - \frac{\theta}{2}$ ($X_{\theta,1-\theta/2} = X$); or any s, is the same. This special case exploits the property that $\mathcal{B}_{\rho\theta}$ is a *normal* operator on X, see section 3.2.

b) Next, one considers $\mathcal{G}$ in (1.1) as a perturbation of $\mathcal{B}_{\rho\theta}$ in (3.1.1) ($\mathcal{G}$ and $\mathcal{B}_{\rho\theta}$ have the same dominant term iB and differ at the lower order term $\rho\,(B^+)^\theta$ versus A). Thus

$$\mathcal{G} = \mathcal{B}_{\rho\theta} + \mathcal{P}_{\rho\theta} \tag{3.1.3}$$

$$\mathcal{P}_{\rho\theta} = -\,(A - \rho\,(B^+)^\theta) = -\,(B^+)^{\theta/2}\,(S_\theta - \rho\, I)\,(B^+)^{\theta/2} \tag{3.1.4}$$

where S_θ is the operator defined in (1.4). With reference to the constants $0<\rho_1<\rho_2<\infty$ in (1.3c), henceforth we select, unless otherwise stated, a fixed ρ which satisfies $0<\rho<\rho_1$, so that by assumption (1.3c) we have

$$0 < (\rho_1 - \rho)\, I \le S_\theta - \rho\, I \le (\rho_2 - \rho)\, I \tag{3.1.5}$$

and hence

$$0 < \frac{1}{\rho_2 - \rho}\, I \le (S_\theta - \rho\, I)^{-1} \le \frac{1}{\rho_1 - \rho}\, I\,. \tag{3.1.6}$$

With reference to (3.1.3), the (usual) perturbation formula for the resolvents is

$$R(i\tau,\ \mathcal{G}) = R(i\tau,\ \mathcal{B}_{\rho\theta})\,[I - \mathcal{P}_{\rho\theta}\, R(i\tau,\ \mathcal{B}_{\rho\theta})]^{-1}\,. \tag{3.1.7}$$

Having the required resolvent estimates (3.1.2) for $\mathcal{B}_{\rho\theta}$, one then uses (3.1.7) to find the corresponding resolvent estimates for $\mathcal{G}$,

b_1) first on $[\mathcal{D}((B^+)^{\theta/2})]' = X_{\theta,1}$ with weaker norm:

$$|\tau|^\theta\,\|R(i\tau,\ \mathcal{G})\|_{\mathcal{L}([\mathcal{D}((B^+)^{\theta/2})]')} \le \mathrm{Const}_\theta\,,\quad \forall\,\tau\in\mathcal{R}; \tag{3.1.8}$$

b_2) next on $\mathcal{D}((B^+)^{1-\theta/2}) = X_{\theta,0}$ with stronger norm:

$$|\tau|^\theta\,\|R(i\tau,\ \mathcal{G})\|_{\mathcal{L}(\mathcal{D}((B^+)^{1-\theta/2}))} \le \mathrm{const}_\theta\,,\quad \forall\,\tau\in\Re \tag{3.1.9}$$

and finally by interpolation [L-M.1, Thm 5.1, p. 27] on each space $X_{\theta,s}$, $0<s<1$, in (1.6). The *key step* is to show estimate (3.1.8) in b_1). This is done in section 3.3. As a consequence of (3.1.8), one then obtains (3.1.9) (in section 3.4). Once the resolvent estimates are proved, the claim that the s.c. semigroup of contraction $e^{\mathcal{G}t}$ is also of Gevrey class $>\alpha$ for all $t>0$ follows from [Ta.1}.

3.2. The special case $A = \rho(B^+)^\theta$, $\theta = 1/\alpha < 1$.

Here we study $\mathcal{B}_{\rho\theta}$ defined in (3.1.1).

Theorem 3.2.1. The normal operator $\mathcal{B}_{\rho\theta}$ in (3.1.1) generates a s.c. contraction semigroup on X

$$e^{\mathcal{B}_{\rho\theta} t} x = \int_{-\infty}^{\infty} e^{(-\rho |\mu|^\theta + i\mu)t} \, dE_\mu \, x, \quad x \in X \tag{3.2.1}$$

E_μ being the resolution of the identity associated with B, μ real, see Appendix, with resolvent on the imaginary axis $\lambda = i\tau$ given by

$$R(i\tau, \mathcal{B}_{\rho\theta})\, x = \int_{-\infty}^{\infty} \frac{dE_\mu \, x}{\rho |\mu|^\theta + i(\tau - \mu)}, \quad x \in X. \tag{3.2.2}$$

Finally, the resolvent estimate (3.1.2) holds true. Thus, $e^{\mathcal{B}_{\rho\theta} t}$ is of Gevrey class $> \alpha$ on X for all $t > 0$, and extends/restricts as a s.c. contraction semigroup of Gevrey class $> \alpha$ for all $t > 0$ on each space $X_{\theta, s}$ in (1.6). (However, plainly, $e^{\mathcal{B}_{\rho\theta} t}$ is *not* analytic for $\theta < 1$) □

Proof of Theorem 3.2.1. One readily verifies that $\mathcal{B}_{\rho\theta}$ is normal and hence that formulas (3.2.1) and (3.2.2) hold true with respect to the resolution of the identity E_μ, μ real, of the self-adjoint operator B, as in the Appendix.

To complete the proof, we shall now show the resolvent estimate (3.1.2) as a consequence of formula (3.2.2). Consider the curve $z(\mu) = -\rho |\mu|^\theta + i\mu$ in the complex plane, so that

$$|i\tau - z(\mu)|^2 = \rho^2 |\mu|^{2\theta} + |\tau - \mu|^2 \tag{3.2.3}$$

There is a constant $c_\theta > 0$ such that

$$|i\tau - z(\mu)| \geq c_\theta |\tau|^\theta. \tag{3.2.4}$$

In fact given, say, $\tau > 0$, it suffices to consider the branch with $\mu > 0$. Let $0 < \varepsilon < 1$ be given. We distinguish two cases. If $\mu \geq \varepsilon\tau$, then (3.2.3) yields dropping its second term

$$|i\tau - z(\mu)| \geq \rho |\mu|^\theta \geq \rho \varepsilon^\theta \tau^\theta \tag{3.2.5}$$

as desired. If, instead, $0 < \mu \leq \varepsilon\tau$, for $\tau \geq 1$, then (3.2.3) yields dropping its first term

$$|i\tau - z(\mu)| \geq (\tau - \mu) \geq (1 - \varepsilon)\tau \geq (1 - \varepsilon)\tau^\theta \tag{3.2.6}$$

since $\theta < 1$. Thus (3.2.5), (3.2.6) yield (3.2.4). Then returning to (3.2.2) and using (3.2.3), (3.2.4) we obtain with $x \in X$

$$\|R(i\tau, \mathcal{B}_{\rho\theta})\, x\|^2 = \int_{-\infty}^{\infty} \frac{\|E_\mu x\|^2}{\rho^2 |\mu|^{2\theta} + |\tau - \mu|^2} \tag{3.2.7}$$

$$\leq \frac{1}{c_\theta^2 |\tau|^{2\theta}} \|x\|^2 \tag{3.2.8}$$

uniformly in $\tau \in \mathfrak{R}$. Thus (3.2.8) proves the resolvent estimate (3.1.2) at least on $X = X_{\theta, 1-\theta/2}$.

From here one immediately finds the same resolvent estimate also on $X_{\theta,s}$ with norms as in (1.6b) (e.g. $\|(B^+)^\theta R(i\tau; \mathcal{B}_{p\theta})x\| \le \frac{1}{\rho}\|x\|$) □

Remark 3.2.1. Suppose that A and B commute and thus they possess a common resolution of the identity E_μ, μ real. Assume, moreover, that

$$\rho_1 (B^+)^\theta \le A \tag{3.2.9}$$

(the lower bound inequality in (1.3d)). Then, $\mathcal{G} = -A + iB$, $\mathcal{D}(\mathcal{G}) = \mathcal{D}(A) \cap \mathcal{D}(B)$ is still a normal operator and (3.2.1), (3.2.7) are replaced now by

$$e^{\mathcal{G}t} x = \int_{-\infty}^{\infty} e^{(-|a(\mu)| + i\mu)t} dE_\mu x, \quad x \in X \tag{3.2.10}$$

$$\|R(i\tau, \mathcal{G})x\|^2 = \int_{-\infty}^{\infty} \frac{\|dE_\mu x\|^2}{|a(\mu)|^2 + |\tau - \mu|^2}, \quad x \in X \tag{3.2.11}$$

where $\rho_1 |\mu|^\theta \le |a(\mu)|$ by (3.2.9). Setting now $w(\mu) = -|a(\mu)| + i\mu$, one has in place of (3.2.3), (3.2.4)

$$|i\tau - w(\mu)|^2 = |a(\mu)|^2 + |\tau - \mu|^2 \ge \rho_1^2 |\mu|^{2\theta} + |\tau - \mu|^2 \ge c_{1\theta} |\tau|^\theta \tag{3.2.12}$$

and thus (3.2.8) is replaced now via (3.2.11) and (3.2.12) by

$$c_{1\theta} |\tau|^\theta \|R(i\tau, \mathcal{G})\| \le 1, \quad \forall \tau \in \mathfrak{R} \tag{3.2.13}$$

which proves, in the commutative case, the desired estimate for $\mathcal{G} = -A + iB$, under the sole lower bound inequality (3.2.9).

3.3. The general case for $\mathcal{G}$ on the larger space $[\mathcal{D}((B^+)^{\theta/2})]'$.

We return to the operator $\mathcal{G} = -A + iB$, $\mathcal{G}^{-1} \in \mathcal{L}(X)$, which can be extended as

$$\mathcal{G}: \mathcal{D}(\mathcal{G}) = \mathcal{D}((B^+)^{1-\theta/2}) \text{ onto } [\mathcal{D}((B^+)^{\theta/2})]' \tag{3.3.1}$$

as it follows from (A.14): B: $\mathcal{D}((B^+)^{1-\theta/2})$ onto $[\mathcal{D}((B^+)^{\theta/2})]'$, as well as from the following relation obtained via (1.4), where $x \in \mathcal{D}((B^+)^{1-\theta/2}) \subset \mathcal{D}((B^+)^{\theta/2})$ (since $1 - \theta/2 \ge \theta/2$ for $\theta \le 1$):

$$Ax = (B^+)^{\theta/2} (B^+)^{-\theta/2} A (B^+)^{-\theta/2} (B^+)^{\theta/2} x = (B^+)^{\theta/2} S_\theta (B^+)^{\theta/2} x \in [\mathcal{D}((B^+)^{\theta/2})]'$$

Theorem 3.3.1. (i) The operator $\mathcal{G}$ generates a s.c. contraction semigroup $e^{\mathcal{G}t}$ on X, or else on the space $[\mathcal{D}((B^+)^{\theta/2})]'$ with norm (1.5). (ii) The resolved estimate (3.1.8) holds true, so that $e^{\mathcal{G}t}$ is of Gevrey class $> \alpha$ on $[\mathcal{D}((B^+)^{\theta/2})]'$ for all $t > 0$.

Proof of Theorem 3.3.1. Step 1. Generation of a s.c. contraction semigroup $e^{\mathcal{G}t}$ on X was shown in Lemma 2.1.

Step 2. We return to the perturbation formula (3.1.7). In view of Theorem 3.2.1 that guarantees estimate (3.1.2) for s = 1, i.e. on $[\mathcal{D}((B^+)^{\theta/2})]'$, it suffices to show that the operator (see (3.1.4))

$$[I - \mathcal{P}_{\rho\,\theta}\, R(i\tau, \mathcal{B}_{\rho\theta})]^{-1} = [I + (B^+)^{\theta/2} (S_\theta - \rho\, I)\, (B^+)^{\theta/2}\, R(i\tau, \mathcal{B}_{\rho\theta})]^{-1} \tag{3.3.2}$$

satisfies the following uniform bound for $\theta < 1$ and ρ fixed in $0 < \rho < \rho_1$ (see above (3.1.5)). Henceforth, we drop the subindeces ρ, θ from $\mathcal{B}$.

Proposition 3.3.2. Let $\theta < 1$, $0 < \rho < \rho_1$, be fixed. We have

a) $$\| [I + (B^+)^{\theta/2} (S_\theta - \rho\, I)\, (B^+)^{\theta/2}\, R\,(i\tau, \mathcal{B})]^{-1} \|_{\mathcal{L}([\mathcal{D}((B^+)^{\theta/2})]')} \le c_{\rho,\theta}\,, \quad \forall\, \tau \in \Re \tag{3.3.3}$$

b) equivalently

$$\| I + (S_\theta - \rho\, I)\, R\,(i\,\tau, \mathcal{B})\, (B^+)^{\theta}]^{-1} \|_{\mathcal{L}(X)} \le c_{\rho,\,\theta}\,, \quad \forall\, \tau \in \Re \tag{3.3.4}$$

c) under the condition on the right of (3.1.6) i.e.

$$(S_\theta - \rho\, I)^{-1} \le \frac{1}{\rho_1 - \rho}\, I\,, \;\; 0 < \rho < \rho_1 \;\Leftrightarrow\; \rho_1\, (B^+)^{\theta} \le A \;\Leftrightarrow\; \mathcal{D}(A^{1/2}) \subset \mathcal{D}((B^+)^{\theta/2})\,, \tag{3.3.5}$$

estimate (3.3.4) holds true provided that

$$\| [S_\theta - \rho\, I)^{-1} + R\,(i\tau, \mathcal{B})\, (B^+)^{\theta}]^{-1} \|_{\mathcal{L}(X)} \le c_{\rho\,\theta}\,, \quad \forall\, \tau \in \Re \tag{3.3.6}$$

Proof of Proposition 3.3.2. First, if we recall the norms in (1.5), equivalence between (3.3.3) and (3.3.4) is seen to follow from

$$\begin{aligned}
&(B^+)^{-\theta/2}\, [I + (B^+)^{\theta/2} (S_\theta - \rho\, I)\, (B^+)^{\theta/2}\, R\,(i\tau, \mathcal{B})]^{-1}\, (B^+)^{\theta/2}\, (B^+)^{-\theta/2} \\
&\quad = \{(B^+)^{-\theta/2}\, [I + (B^+)^{\theta/2} (S_\theta - \rho\, I)\, (B^+)^{\theta/2}\, R\,(i\tau, \mathcal{B})]\, (B^+)^{\theta/2}\}^{-1}\, (B^+)^{-\theta/2} \\
&\quad = \{I + (S_\theta - \rho\, I)\, R(i\tau, \mathcal{B})\, (B^+)^{\theta}\,\}^{-1}\, (B^+)^{-\theta/2}
\end{aligned} \tag{3.3.7}$$

by commutativity between B^+ and $R\,(i\tau, \mathcal{B})$. Next, to prove part c), we compute

$$\begin{aligned}
[I + (S_\theta - \rho\, I)\, R\,(i\tau, \mathcal{B})\, (B^+)^{\theta}]^{-1} &= \{(S_\theta - \rho\, I)\, [(S_\theta - \rho\, I)^{-1} + R\,(i\tau, \mathcal{B})\, (B^+)^{\theta}\,\}^{-1} \\
&= [(S_\theta - \rho\, I)^{-1} + R\,(i\tau, \mathcal{B})\, (B^+)^{\theta}]^{-1}\, (S_\theta - \rho\, I)^{-1}
\end{aligned} \tag{3.3.8}$$

and thus, under condition (3.3.5), we see from (3.3.8) that (3.3.6) implies (3.3.4) □

Thus, it remains to show (3.3.4). To this end we preliminarily prove

Step 3. Lemma 3.3.3. The following inequality holds true

$$\mathrm{Re}\,(R\,(i\tau, \mathcal{B})\, (B^+)^{\theta}\, x, x) \ge 0 \quad \forall\, \tau \in \Re,\; x \in X \tag{3.3.9}$$

Proof of Lemma 3.3.3. From (3.2.2) we have

$$\begin{aligned}
(R\,(i\tau, \mathcal{B})\, (B^+)^{\theta}\, x, x) &= \int_{-\infty}^{\infty} \frac{|\mu|^{\theta}\, (dE_\mu\, x, x)}{\rho\, |\mu|^{\theta} + i\,(\tau - \mu)} \\
&= \int_{-\infty}^{\infty} \frac{[\rho\, |\mu|^{\theta} - i(\tau - \mu)]\, |\mu|^{\theta}\, (dE_\mu\, x, x)}{\rho^2\, |\mu|^{2\theta} + (\tau - \mu)^2}
\end{aligned}$$

so that, as desired

$$\mathrm{Re}\,(R\,(i\,\tau,\,\mathcal{B})\,(B^+)^\theta\,x,\,x) = \int_{-\infty}^{\infty} \frac{\rho\,|\mu|^{2\theta}\,(dE_\mu\,x,\,x)}{\rho^2\,|\mu|^{2\theta} + (\tau-\mu)^2} \geq 0 \;\square$$

Remark 3.3.1 Notice that Lemma 3.3.3 holds true for any power $(B^+)^p$, not only for $p = \theta$ □

Step 4.

Proposition 3.3.4. Under the condition on the left of (3.1.6), i.e. with $0 < \rho < \rho_2$

$$0 < \frac{1}{\rho_2 - \rho}\, I \leq (S_\theta - \rho\, I)^{-1}\,, \quad \Leftrightarrow \; A \leq \rho_2\,(B^+)^\theta \Leftrightarrow \mathcal{D}((B^+)^{\theta/2}) \subset \mathcal{D}(A^{1/2}) \tag{3.3.10}$$

estimate (3.3.6) holds true.

Proof of Proposition 3.3.4. We first show that with $C > 0$

$$\|[(S_\theta - \rho\, I)^{-1} + R\,(i\,\tau,\,\mathcal{B})\,(B^+)^\theta\,]\,x\| \;\geq\; C\|x\|\,, \quad \forall\,\tau \in \Re, \quad x \in X. \tag{3.3.11}$$

In fact

$$\|[(S_\theta - \rho\, I)^{-1} + R\,(i\tau,\,\mathcal{B})\,(B^+)^\theta]\,x\|\;\|x\| \geq |([(S_\theta - \rho\, I)^{-1} + R\,(i\tau,\,\mathcal{B})\,(B^+)^\theta]\,x,\,x)|$$

$$= \{[((S_\theta - \rho\, I)^{-1}\,x,\,x) + \mathrm{Re}\,(R\,(i\,\tau,\,\mathcal{B})\,(B^+)^\theta\,x,\,x)]^2 \tag{3.3.12}$$

$$+ [\mathrm{Im}\,(R\,(i\tau,\,\mathcal{B})\,(B^+)^\theta\,x,\,x)]^2\}^{1/2}$$

$$\geq ((S_\theta - \rho\, I)^{-1}\,x,\,x) \geq \frac{1}{\rho_2 - \rho}\,\|x\|^2 \tag{3.3.13}$$

where in going from (3.3.12) to (3.3.13) we have dropped the Re - term by (3.3.9), as well as the Im - term, and we have used (3.3.10). Thus (3.3.13) proves (3.3.11) with $c = 1/(\rho_2 - \rho)$. Next, we see that
$\forall\,\tau \in \Re$:

$$\overline{\mathrm{range}}\,\{(S_\theta - \rho\, I)^{-1} + R\,(i\,\tau,\,\mathcal{B})\,(B^+)^\theta\} = X$$

since equivalently

$$\mathcal{N}\{[(S_\theta - \rho\, I)^{-1} + R\,(i\,\tau,\,\mathcal{B})\,(B^+)^\theta]^*\} = \mathcal{N}\{(S_\theta - \rho\, I)^{-1} + R\,(-i\,\tau,\,\mathcal{B})\,(B^+)^\theta\}$$
$$= \{0\}\,, \quad \forall\,\tau \in \Re \tag{3.3.14}$$

where the statement that the null space $\mathcal{N}$ is trivial follows from (3.3.11) with τ replaced by $-\tau$ (we note that B^+ and $R\,(\cdot\,,\,\mathcal{B})$ commute). Thus, (3.3.11) and (3.1.14) together say that estimate (3.3.6) holds true □

Thus, choosing $0 < \rho < \rho_1$, we see that by (3.1.6), we have proved estimate (3.3.6), hence ultimately estimate (3.3.3). By (3.3.2) and Theorem 3.2.1, we have thus proved Theorem 3.3.1.

3.4. The general case for $\mathcal{G}$ on the smaller space $\mathcal{D}((B^+)^{1-\theta/2})$

We now return to the operator

$$\mathcal{G} = -A + iB, \quad \mathcal{D}(\mathcal{G}) = \mathcal{D}((B^+)^{1-\theta/2}) \text{ onto } X_{\theta,1} = [\mathcal{D}((B^+)^{\theta/2})]' \tag{3.4.1}$$

of section 3.3, $\mathcal{G}^{-1} \in \mathcal{L}(X_{\theta,1})$ which we now restrict on the space $\mathcal{D}(\mathcal{G})$ this time equipped, however, with its natural norm derived from the underlying space $X_{\theta,1}$ (with norm as in (1.5)), i.e.

$$\|y\|_{\mathcal{D}(\mathcal{G})} \equiv \|\mathcal{G}y\|_{[\mathcal{D}((B^+)^{\theta/2})]'} \quad y \in \mathcal{D}((B^+)^{1-\theta/2}) = \mathcal{D}(\mathcal{G}) \tag{3.4.2}$$

On the other hand, on the subspace $\mathcal{D}(\mathcal{G}) = \mathcal{D}((B^+)^{1-\theta/2})$ we have the original norm in (1.5) (right) i.e.

$$\|y\|_{\mathcal{D}((B^+)^{1-\theta/2})} = \|(B^+)^{1-\theta/2}y\|_X, \; y \in \mathcal{D}((B^+)^{1-\theta/2}) \tag{3.4.3}$$

Lemma 3.4.1. The two norms (3.4.1) and (3.4.2) are equivalent on $\mathcal{D}(\mathcal{G}) = \mathcal{D}((B^+)^{1-\theta/2})$.

Proof of Lemma 3.4.1. Since

$$B^+ : \text{ isomorphism } \mathcal{D}((B^+)^{1-\theta/2}) \text{ onto } [\mathcal{D}((B^+)^{\theta/2})]' \tag{3.4.4}$$

we deduce from (3.4.1) and (3.4.4) via the closed graph theorem that with $X_{\theta,1} = [\mathcal{D}((B^+)^{\theta/2})]'$:

$$(B^+)\mathcal{G}^{-1} \text{ and } \mathcal{G}(B^+)^{-1} \text{ are both in } \mathcal{L}(X_{\theta,1}). \tag{3.4.5}$$

Let now $y \in \mathcal{D}((B^+)^{1-\theta/2})$. Then, one direction of the norm equivalence follows from

$$\begin{aligned} \|\mathcal{G}y\|_{[\mathcal{D}((B^+)^{\theta/2})]'} &= \|\mathcal{G}(B^+)^{-1}(B^+)y\|_{X_{\theta,1}} \\ \text{(by (3.4.5) (right))} \quad &\leq \|\mathcal{G}(B^+)^{-1}\|_{\mathcal{L}(X_{\theta,1})} \|(B^+)y\|_{[\mathcal{D}((B^+)^{\theta/2})]'} \\ \text{(by (1.5))} \quad &= \|\mathcal{G}(B^+)^{-1}\|_{\mathcal{L}(X_{\theta,1})} \|y\|_{\mathcal{D}((B^+)^{1-\theta/2})} \end{aligned} \tag{3.4.6}$$

The other direction of the norm equivalence follows now from

$$\begin{aligned} \|y\|_{\mathcal{D}((B^+)^{1-\theta/2})} &= \|(B^+)y\|_{[\mathcal{D}((B^+)^{\theta/2})]'} = \|(B^+)\mathcal{G}^{-1}\mathcal{G}y\|_{X_{\theta,1}} \\ \text{(by (3.4.5))} \quad &\leq \|(B^+)\mathcal{G}^{-1}\|_{\mathcal{L}(X_{\theta,1})} \|\mathcal{G}y\|_{[\mathcal{D}((B^+)^{\theta/2})]'} \quad \square \end{aligned} \tag{3.4.7}$$

Theorem 3.4.2. (i) The operator $\mathcal{G}$ in (3.4.1) restricts to the space $\mathcal{D}((B^+)^{1-\theta/2})$ with norm (3.4.3) to generate here a s.c. semigroup $e^{\mathcal{G}t}$, $t \geq 0$;
(ii) its resolvent satisfies the estimate

$$|\tau|^\theta \|R(i\tau, \mathcal{G})\|_{\mathcal{D}((B^+)^{1-\theta/2})} \leq C_\theta, \quad \forall \tau \in \Re \tag{3.4.8}$$

so that $e^{\mathcal{G}t}$ is of Gevrey class $> \alpha$ for all $t > 0$ on $\mathcal{D}((B^+)^{1-\theta/2})$.

Remark 3.4.1 Notice that $e^{\mathcal{G}t}$ is not claimed to be a contraction in the norm of (3.4.3), only in the norm of (3.4.2). □

Proof of Theorem 3.4.2. (i) The operator $\mathcal{G}$ in (3.4.1) generates a s.c. contraction semigroup $e^{\mathcal{G}t}$ on $[\mathcal{D}((B^+)^{\theta/2})]'$ with norm (1.5) by Theorem 3.3.1. Such $e^{\mathcal{G}t}$ restricts on the space $\mathcal{D}(\mathcal{G})$ with norm (3.4.2) as a s.c. contraction semigroup, while contraction need not be guaranteed in the norm (3.4.3).
(ii) The resolvent estimate (3.1.8) is equivalent to the resolvent estimate

$$|\tau|^\theta \|R(i\tau, \mathcal{G})\|_{\mathcal{D}(\mathcal{G})} \leq c_\theta, \quad \forall \tau \in \Re \tag{3.4.9}$$

since

$$\|R(i\tau, \mathcal{G})\, y\|_{[\mathcal{D}((B^+)^{\theta/2})]'} = \|\mathcal{G} R(i\tau, \mathcal{G})\, \mathcal{G}^{-1} y\|_{[\mathcal{D}((B^+)^{\theta/2})]'} \tag{3.4.10}$$

(by (3.4.2))
$$= \| R(i\tau, \mathcal{G})\, \mathcal{G}^{-1} y\|_{\mathcal{D}(\mathcal{G})}$$

while again by (3.4.2)

$$\|y\|_{[\mathcal{D}((B^+)^{\theta/2})]'} = \|\mathcal{G}\mathcal{G}^{-1} y\|_{[\mathcal{D}((B^+)^{\theta/2})]'} = \|\mathcal{G}^{-1} y\|_{\mathcal{D}(\mathcal{G})} \tag{3.4.11}$$

Finally, the resolvent estimate (3.4.9) is equivalent to the desired resolvent estimate (3.4.8) by Lemma 3.4.1 □

4. ADDITIONAL RESULTS

In this section we shall briefly pursue a different technique, which is more direct, and which yields a Gevrey semigroup for $\alpha > 1$, the more interesting case, of class $> 2\alpha$ on X, (rather than of class $> \alpha$ and on a scale of spaces containing X as in section 3), however under the sole assumption

$$\mathcal{D}(A^{\alpha}) \subset \mathcal{D}(B) \Leftrightarrow \|B A^{-\alpha} x\| \le c\|x\| \Leftrightarrow \|By\| \le c\|A^{\alpha} y\|, \quad y \in \mathcal{D}(A^{\alpha}) \tag{4.1}$$

(which then implies $\mathcal{D}(A^{½}) \subset \mathcal{D}((B^+)^{½\alpha})$ as in (1.9); compare with (1.3) in the general case, and with the same containment (1.9) in the commutative case).

Proposition 4.1. Assume, in addition to (H.1) and (H.2), that there exists $\gamma \in (0, 1]$ such that

$$\|A^{\gamma} R(\lambda, \mathcal{G})\| \le C \qquad \forall \operatorname{Re}\lambda > 0 \tag{4.2}$$

Then, the following estimate holds true

$$|\lambda|^{\gamma/\alpha} \|R(\lambda, \mathcal{G})\| \le M_{\gamma\alpha}, \quad \operatorname{Re}\lambda > 0, \quad |\lambda| > 1 \tag{4.3}$$

Thus, the s.c. contraction semigroup, $e^{\mathcal{G}t}$, guaranteed by assumptions (H.1) and (H.2) by Lemma 2.1, is moreover [Ta.1] of Gevrey class $> \alpha/\gamma$ for all $t > 0$ □

Proof of Proposition 4.1.

Step 1.

Lemma 4.2. Assume (H.1), (H.2) and (4.1). Then (the bar denotes closure)

$$\overline{A^{-(\alpha-\gamma)} B A^{-\gamma}} \in \mathcal{L}(X) \qquad 0 < \gamma \le \alpha \tag{4.4}$$

Proof of Lemma 4.2. We first assume that B is non-negative and use the Heinz-Kato inequality, or interpolation, on inequality (4.1) (right). We obtain

$$\|B^{\sigma} y\| \le c_{\sigma} \| A^{\alpha\sigma} y\| \Leftrightarrow \|B^{\sigma} A^{-\alpha\sigma} x\| \le c_{\sigma} \|x\|, \quad 0 \le \sigma \le 1 \tag{4.5}$$

We then write for $0 \le \gamma \le \alpha$:

$$A^{\gamma-\alpha} B A^{-\gamma} = A^{-\alpha(1-\gamma/\alpha)} B^{1-\gamma/\alpha} B^{\gamma/\alpha} A^{-\gamma} \tag{4.6}$$

where

$$B^{\gamma/\alpha} A^{-\gamma} \in \mathcal{L}(X) \quad \text{by (4.5) with } 0 < \sigma = \gamma/\alpha \le 1 \tag{4.7}$$

$$A^{-\alpha(1-\gamma/\alpha)}\, B^{1-\gamma/\alpha} \in \mathcal{L}(X) \quad \text{by (4.5) with } 0 \le \sigma = 1-\gamma/\alpha \le 1 \tag{4.8}$$

by taking the adjoint, the closure of $A^{-\alpha\sigma} B^{\sigma}$. Thus (4.7) and (4.8) used in (4.6) prove (4.4), as desired, at least for $B \ge 0$. The general case of B uses the decomposition $B = B^+ - 2(\Pi^- B^+)$, where $B^+ \ge 0$ and $\Pi^- B^+ \ge 0$ as in (A.12), so that the preceding argument applies to both B^+ and $(\Pi^- B^+)$ □

Step 2. We compute

$$\begin{aligned}(\lambda + A - iB)^{-1} - (\lambda + A^{\alpha})^{-1} &= (\lambda + A^{\alpha})^{-1} (A^{\alpha} - A + iB)(\lambda + A - iB)^{-1}\\ &= (\lambda + A^{\alpha})^{-1} A^{\alpha} (\lambda + A - iB)^{-1} - (\lambda + A^{\alpha})^{-1} A (\lambda + A - iB)^{-1}\\ &+ i(\lambda + A^{\alpha})^{-1} B (\lambda + A - iB)^{-1} = (1) + (2) + (3)\end{aligned} \tag{4.9}$$

where of course (since $-A^{\alpha}$ generates an analytic semigroup)

$$\|(\lambda + A^{\alpha})^{-1}\| \le \frac{c}{|\lambda|} \qquad \forall \operatorname{Re}\lambda > 0 \tag{4.10}$$

and hence by interpolation via the moment inequality one obtains, as is well known

$$\|(A^{\alpha})^{r} (\lambda + A^{\alpha})^{-1}\| \le \frac{c_{\alpha r}}{|\lambda|^{1-r}} \qquad 0 \le r \le 1, \ \operatorname{Re}\lambda > 0 \tag{4.11}$$

Step 3. We estimate $(1) = A^{\alpha-\gamma} (\lambda + A^{\alpha})^{-1} [A^{\gamma} (\lambda + A - iB)^{-1}]$ by use of (4.11) with $r = 1 - \gamma/\alpha$, $0 < \gamma \le \alpha$, as well as of assumption (4.2). We obtain with $\alpha r = \alpha - \gamma$:

$$\begin{aligned}\|(1)\| &= \|A^{\alpha-\gamma} (\lambda + A^{\alpha})^{-1} A^{\gamma} R(\lambda, \mathcal{G})\|\\ &\le \|A^{\alpha-\gamma} (\lambda + A^{\alpha})^{-1}\| \, \|A^{\gamma} R(\lambda, \mathcal{G})\| \le \frac{c_{\alpha\gamma}}{|\lambda|^{\gamma/\alpha}}, \quad \operatorname{Re}\lambda > 0, \ 0 < \gamma \le \alpha\end{aligned} \tag{4.12}$$

Step 4. Similarly for $(2) = A^{1-\gamma} (\lambda + A^{\alpha})^{-1} A^{\gamma} (\lambda + A - iB)^{-1}$ we obtain this time by (4.11) with $\alpha r = 1 - \gamma$, $0 < \gamma \le 1$, and by (4.2)

$$\begin{aligned}\|(2)\| &= \|A^{1-\gamma} (\lambda + A^{\alpha})^{-1} A^{\gamma} R(\lambda, \mathcal{G})\|\\ &\le \|A^{r\alpha} (\lambda + A^{\alpha})^{-1}\| \, \|A^{\gamma} R(\lambda, \mathcal{G})\| \le \frac{c_{\alpha\gamma}}{|\lambda|^{1-\frac{1}{\alpha}+\frac{\gamma}{\alpha}}} \qquad 0 < \gamma \le 1, \ \operatorname{Re}\lambda > 0\end{aligned} \tag{4.13}$$

Step 5. We rewrite (3) as

$$(3) = i(\lambda + A^{\alpha})^{-1} A^{\alpha-\gamma} [A^{-(\alpha-\gamma)} B A^{-\gamma}] [A^{\gamma} (\lambda + A - iB)^{-1}] \tag{4.14}$$

so that we estimate (3) this time by (4.11) with $r = 1 - \gamma/\alpha$ as before, by (4.4) of Lemma 4.2, and by assumption (4.2). We obtain from (4.14) and $\alpha - \gamma = \alpha r$:

$$\|(3)\| \le \|A^{\alpha-\gamma} (\lambda + A^{\alpha})^{-1}\| \, \|\overline{A^{-(\alpha-\gamma)} B A^{-\gamma}}\| \, \|A^{\gamma} R(\lambda, \mathcal{G})\| \le \frac{c_{\alpha\gamma}}{|\lambda|^{\gamma/\alpha}} \ \ 0 < \gamma \le \alpha, \tag{4.15}$$

$\operatorname{Re}\lambda > 0$

Step 6. We put together in (4.9) estimates (4.12), (4.13), (4.15) for $\operatorname{Re}\lambda > 0$, $|\lambda| > 1$, whereby the smaller exponent γ/α dominates, and we obtain

$$\|(\lambda + A - iB)^{-1}\| - \|(\lambda + A^{\alpha})^{-1}\| \le \frac{c_{\alpha\gamma}}{|\lambda|^{\gamma/\alpha}}, \quad \operatorname{Re}\lambda > 0, \ |\lambda| > 1 \tag{4.16}$$

Hence the desired estimate (4.3) follows from (4.16) via (4.10) □

We now show that assumption (4.2) holds true with $\gamma = \frac{1}{2}$.

Theorem 4.3. Assume (H.1) and (H.2).

(i) Condition (4.2) holds true for $\gamma = \frac{1}{2}$.
(ii) Thus, Proposition 4.1 gives that estimate (4.3) holds with $\gamma = \frac{1}{2}$ and hence [Ta.1] the operator $\mathcal{G} = -A + iB$, with $\mathcal{D}(\mathcal{G}) = \mathcal{D}(B)$ generates a s.c. contraction semigroup on X by Lemma 2.1, which moreover is of Gevrey class $> 2\alpha$ for all $t > 0$ □

Proof. It suffices to show (i). Taking the inner product with $u \in \mathcal{D}(B^{1/2}) \subset \mathcal{D}(A^{1/2})$ on the identity

$$(\lambda + A - iB)\, u = f; \quad u = (\lambda + A - iB)^{-1} f, \ \operatorname{Re}\lambda \ge 0 \tag{4.17}$$

we obtain by restricting to real parts

$$c_0^2 \|u\|^2 \le \|A^{1/2} u\|^2 \le (\operatorname{Re}\lambda)\|u\|^2 + \|A^{1/2} u\|^2 \le \|f\|\,\|u\| \tag{4.18}$$

$c_0 > 0$, so that $\|u\| \le \|f\|/(c_0^2)$, which inserted into the right hand side of (4.18) yields

$$\|A^{1/2} u\| \le \frac{1}{c_0}\|f\|; \ \text{ or } \|A^{1/2}(\lambda + A - iB)^{-1}\| \le \frac{1}{c_0}, \ \operatorname{Re}\lambda > 0 \tag{4.19}$$

and condition (4.2) holds true with $\gamma = \frac{1}{2}$ □

Theorem 4.4. (i) In addition to the hypotheses of Theorem 4.3, assume that

$$(Au, Bu) = \text{real} \in \mathfrak{R} \quad \forall\, u \in \mathcal{D}(B) \tag{4.20}$$

Then condition (4.2) holds with $\gamma = 1$. (ii) Hence the s.c. contraction semigroup $e^{\mathcal{G}t}$ on X is of Gevrey class $> \alpha$ for all $t > 0$ □

Proof. It suffices to show (i). This time we take the inner product of identity (4.17) (left) with Au, for $u \in \mathcal{D}(B)$ and obtain after restricting to real parts and using (4.20)

$$(\operatorname{Re}\lambda)\|A^{1/2}u\|^2 + \|Au\|^2 \le \|f\|\,\|Au\|, \quad \operatorname{Re}\lambda \ge 0. \tag{4.21}$$

Hence, (4.21) implies $\|Au\| \le \|f\|$ i.e. explicitly (4.2) with $\gamma = 1$ and $c = 1$ by (4.17) (right) □

Remark 4.1. If A and B commute then (4.20) holds true □

Remark 4.2. The results of this section can be extended to reflexive Banach spaces.

(i) First, let $-A_0$ be the infinitesimal generator of a s.c. semigroup of contractions on the Banach space X. Let -A be a dissipative operator such that

$$\mathcal{D}(A_0) \subset \mathcal{D}(A^{\alpha}) \quad \alpha > 1 \tag{4.22}$$

with A positive in the sense of [Tr. 1 p. 91] and with $A^{it} \in \mathcal{L}(X)$, t real, $|t| < \varepsilon$ [Tr.1 p. 103]. Then, the operator $-A - A_0$, with domain equal to $\mathcal{D}(A_0)$, generates a s.c. contraction semigroup on X.

(ii) Next, let the Banach space X be reflexive, and assume with $\alpha > 1$

$$\mathcal{D}(A_0^{*\alpha}) \subset \mathcal{D}(A_0^*) \quad \text{and} \quad \mathcal{D}(A^\alpha) \subset \mathcal{D}(A_0) \tag{4.23}$$

Then, the conclusion as in Lemma 4.1.4 holds true in the reflexive Banach space X: the closure of $(A^{-(\alpha-\gamma)} A_0 A^{-\gamma}) \in \mathcal{L}(X)$, $0 < \gamma \leq \alpha$. The proof relies on a Banach space counterpart of the Hilbert space interpolation proof of Lemma 4.1.4, i.e. on the three lines theorem [D-S.1, p. 520] as applied to the function

$$f(z)\, x \equiv A^{-(\alpha+z)} A_0 A^z x \qquad x \in \mathcal{D}(A^\alpha)$$

analytic in the strip $-\alpha < \operatorname{Re} z < 0$. Assumptions (4.23) are then invoked to check the required hypotheses of the three lines theorem at the line $z = -\alpha + i t$, $\alpha > 1$, $t \in \mathfrak{R}$, and at the line $z = it$, $t \in \mathfrak{R}$, to conclude at $z = -\gamma$. The remaining steps of the proof of Proposition 4.1 work also in a Banach space setting. One then concludes that the counterpart of (4.3) holds true; i.e.

$$|\lambda|^{\gamma/\alpha} \, \|(\lambda + A_0 + A)^{-1}\| \leq M_{\gamma\alpha} \qquad \operatorname{Re} \lambda > 0, \ |\lambda| > 1$$

so that the s.c. semigroup generated on X by $-A - A_0$, as guaranteed by part (i), is moreover of Gevrey class $> \alpha/\gamma$ for all $t > 0$. Details are omitted □

5. APPLICATIONS TO OPTIMAL CONTROL PROBLEMS WITH UNBOUNDED INPUT → SOLUTION MAP.

In this section we present an application of Theorem 1.1 and 1.2 to feedback dynamics such as they arise in the theory of optimal control problems with quadratic cost over an infinite time horizon and related algebraic (operator) Riccati equations, when the map from the control function in $L_2(0, T; U)$ to the solution at time T in X is *unbounded.* Here U and X are Hilbert spaces. Consider the abstract equation

$$\dot{x}(t) = Lx(t) + Mu(t)\,; \qquad x(0) = x_0 \in X \tag{5.1}$$

where L is the generator of a s.c. semigroup on X and the (control) operator M is continuous from U to $[\mathcal{D}(L^*)]'$, the dual space of $\mathcal{D}(L^*)$ with respect to the X-topology, where L^* is the adjoint of L in X. It is shown in [L-T.1] that the unique solution $\{u^0(t; x_0),\ x^0(t; x_0)\}$ of the optimal control problem associated with (5.1) which minimizes over all $u \in L_2(0, \infty; U)$ the quadratic cost

$$J(u, x) = \int_0^\infty \{\|Rx(t)\|_X^2 + \|u(t)\|_U^2\}\, dt\,, \tag{5.2}$$

$R \in \mathcal{L}(X)$, $R^* R \geq c I$, $c > 0$, is given by

$$u^0(t; x_0) = -M M^* P x^0(t; x_0) \in L_2(0, \infty; U)\,, \tag{5.3}$$

where P is the (unique) solution of the corresponding Algebraic Riccati Equation (A.R.E.)

$$L^* P + P L + R^* R = P M M^* P \tag{5.4}$$

(in the technical sense explained in [L-T.1]), provided that mild assumptions hold true. These, in particular, include the cases R = I, and L either self-adjoint or skew adjoint. The optimal dynamics

$$x^0(t; x_0) = \Phi(t) x_0 \tag{5.5}$$

defines, in such generality, a one-time integrated semigroup $\Phi(t)$ on X, $t \geq 0$, with (feedback) generator

$$L_F = L - M M^* P. \tag{5.6}$$

In special cases, the one-time integrated semigroup $\Phi(t)$ may well be a bona-fid, s.c. semigroup on X with additional properties such as analyticity or Gevrey class. This is illustrated by the canonical example below.

Example 5.1. Consider the "abstract hyperbolic" case where

$$L = i S; \quad M = iS^{\theta}, \quad 0 \leq \theta \leq 1; \quad R = I, \tag{5.7}$$

and S is a positive self-adjoint operator on the Hilbert space X = U. One then verifies that the positive, self-adjoint operator $P = S^{-\theta}$ satisfies the A.R.E. (5.4) (Notice that $P^{-1} = S^{\theta}$ is an *unbounded* operator, a *feature* of the general theory in [L-T.1]. By contrast, unboundedness of P^{-1} cannot occur within the more regular theory of [L-T.2], [F-L-T] where the input → solution map: $u \to x$ is continuous from $L_2(0, T; U) \to C([0, T]; X)$). Thus, the feedback dynamical operator L_F in (5.6) is

$$L_F = - S^{\theta} + iS. \tag{5.8}$$

Hence, L_F is of the general form as in (3.1.1) with $\rho = 1$, S = B. According to Theorems 1.1 and 1.2 (or Theorem 3.2.1), we conclude that

$$x^0(t; x_0) = \Phi(t) x_0 = e^{L_F t} x_0 \tag{5.9}$$

where $e^{L_F t}$ is a s.c. analytic semigroup on X if $\theta = 1$; of Gevrey class $> 1/\theta$ for all $t > 0$ if $0 < \theta < 1$; and, finally, a group if $\theta = 0$. Thus, the original free dynamics described by the unitary group $e^{Lt} = e^{iSt}$ is transformed by virtue of the Riccati operator into a feedback optimal dynamics $x^0(t; x_0)$ which is described by the semigroup $\Phi(t) = e^{L_F t}$, with analytic ($\theta = 1$), or Gevrey class properties ($0 < \theta < 1$); this shows an interesting smoothing or regularizing effect due to P.

6. APPLICATIONS TO INHOMOGENEOUS PROBLEMS FOR $\alpha > 1$

6.1. First order equations

In this final section we apply some results of the first section to the Cauchy problem

$$\begin{cases} u'(t) = \mathcal{G}u(t) + f(t), & 0 < t \leq T, \\ u(0) = u_0, \end{cases} \tag{6.0}$$

where the operator $\mathcal{G}$ satisfies assumption (H.3G) = (1.3). Moreover, f is a given continuous function from [0,T] into X and $u_0 \in X$. According to [F-Y.1], [K.2 p. 54], a *classical* or *weakened* solution of problem (P) is a function u = u(t) which is continuous from [0,T] into X, is strongly continuously differentiable on (0, T], satisfies $u(t) \in \mathcal{D}(\mathcal{G})$ for $0 < t \leq T$, and is such that (6.0) holds.

If one assumes (H.3G) = (1.3) with $\alpha > 1$, and since $\mathcal{G}$ generates a strongly continuous semigroup, it is not difficult to recognize, by Taylor's expansion of the resolvent, starting from (1.7), that the estimate

$$|\lambda|^{1/\alpha} \|R(\lambda, \mathcal{G})\|_{\mathcal{L}(X)} \leq \mathrm{Const}_\alpha \tag{6.1}$$

holds true for any complex number λ with Re $\lambda \geq 0$, $|\lambda|$ large. Hence, (see [K.2, p. 135 and p. 140]), problem (6.0) has a unique classical solution u provided that $u_0 \in \mathcal{D}(\mathcal{G}) = \mathcal{D}(B) = \mathcal{D}(A^\alpha)$ and: either $f(t) \in \mathcal{D}(B)$, $0 \leq t \leq T$, $Bf(t)$ being continuous, or else f is continuously differentiable. However, less restrictive assumptions can be made when the number α is suitably restricted. To this end, we observe that estimate (6.1) remains true for all λ in the region

$$\Sigma = \{\lambda \in \mathbb{C}:\ \mathrm{Re}\,\lambda \geq -c\,(1 + |\mathrm{Im}\lambda|)^{1/\alpha}\},\ c > 0, \tag{6.2}$$

of the resolvent set of $\mathcal{G}$ (see [F-Y.1]). Hence we are allowed to apply [F-Y.1] under the assumption $1 < \alpha < 3/2$ and deduce the following

Theorem 6.1. Under assumption (H.1), (H.2), (H.3G), let $1 < \alpha < 3/2$. Then, given $2(\alpha - 1) < \gamma \leq 1$, for any $u_0 \in \mathcal{D}(B)$ and all $f \in C^\gamma([0,T]; X)$, (Hölder continuous of exponent γ) problem (6.0) has a unique classical solution □

We also quote [K.2, T.2] for similar results. A further existence result for (6.0) can be obtained by noticing that in our case the resolvent $R(\lambda, \mathcal{G})$ satisfies in fact the stronger condition

$$\|R(\lambda, \mathcal{G})\|_{\mathcal{L}(X)} \leq M\,(1 + \mathrm{Re}\lambda + |\mathrm{Im}\,\lambda|^{1/\alpha})^{-1},\quad \mathrm{Re}\,\lambda \geq 0, \tag{6.3}$$

Hence, in view of [K.2, Theorem 6.10, p. 142], we obtain

Theorem 6.2. Assume (H.1), (H.2), (H.3G), $1 < \alpha < 3/2$. If f is strongly continuous from $[0,T]$ into $\mathcal{D}(A^\omega)$, with $\omega > 2(\alpha - 1)$ and $u_0 \in \mathcal{D}(A^\gamma)$, $\gamma > 1$, then (6.0) has a unique classical solution □

Notice that f satisfies a time regularity assumption in Theorem 6.1 and a space regularity assumption in Theorem 6.2.

6.2. Second order equations which can be factored in first order terms with $\alpha > 1$.

In this subsection the object of our interest is the second order Cauchy problem

$$(\mathcal{S}) \qquad \begin{cases} u''(t) + 2\rho B u'(t) + Au(t) = f(t) & 0 < t \leq T \\ u(0) = u_0,\ u'(0) = u_1 \end{cases} \tag{6.4}$$

subject to the assumptions below. The goal is to obtain *classical* solutions to $(\mathcal{S})$ as an application of Theorems 6.1 and 6.2 on classical solutions to Cauchy problems for first order equations of type (6.0), to which $(\mathcal{S})$ will be reduced by suitable factorization in two first order terms. Standing assumptions on (6.4) are as follows:

a_1) the operators A and B are positive self-adjoint on the complex Hilbert space H;

a_2) A and B commute (in the sense that $A^{-1}B^{-1} = B^{-1}A^{-1}$);

a_3)

$$\mathcal{D}(B^\alpha) \subset \mathcal{D}(A^{1/2}) \subset \mathcal{D}(B) \qquad \alpha > 1 \tag{6.5a}$$

equivalently

$$A^{1/2}B^{-\alpha} \in \mathcal{L}(H), \qquad BA^{-1/2} \in \mathcal{L}(H) \tag{6.5b}$$

in turn equivalent, because of a_2), to

$$\mathcal{D}(B^{2\alpha}) \subset \mathcal{D}(A) \subset \mathcal{D}(B^2) \tag{6.5c}$$

a_4) the constant ρ satisfies

$$0 < \rho < 1/\sqrt{k}, \quad \text{where } k \equiv \|BA^{-1/2}\| \tag{6.6}$$

We first notice that since for any $u \in \mathcal{D}(A^{1/2}) \subset \mathcal{D}(B)$

$$((A - \rho^2 B^2)u, u) = \|A^{1/2}u\|^2 - \rho^2 \|BA^{-1/2} A^{1/2}u\| \tag{6.7}$$

(by (6.6)) $$\geq (1 - \rho^2 k^2)\|A^{1/2}u\| \geq \frac{(1-\rho^2k^2)}{\|A^{-1/2}\|^2} \|u\|^2$$

then $A - \rho^2 B^2$ is a positive self-adjoint operator with domain equal to $\mathcal{D}(A)$. Thus, in particular, the positive square root $(A - \rho^2 B^2)^{1/2}$, with domain equal to $\mathcal{D}(A^{1/2})$, is well-defined as a self-adjoint operator on H. The key observation of this subsection is that, if α is suitably chosen — more precisely, $1 < \alpha < 3/2$ — then the Cauchy problem for the second order problem $(\mathcal{S})$ = (6.4) admits a factored version

$$(\mathcal{F}) \begin{cases} \left(\frac{d}{dt} + \rho B + i(A - \rho^2B^2)^{1/2}\right)\left(\frac{d}{dt} + \rho B - i(A-\rho^2B^2)^{1/2}\right) u(t) = f(t), \ 0 < t \leq T & (6.8) \\ u(0) = u_0, \ u'(0) + (\rho B - i(A - \rho^2B^2)^{1/2})\, u_0 = u_1 + (\rho B - i(A-\rho^2B^2)^{1/2})u_0 = v_0 & (6.9) \end{cases}$$

to which Theorems 6.1 and 6.2 can be applied. To this end, we set

$$A_0 = \rho B - i(A - \rho^2B^2)^{1/2} : \ \mathcal{D}(A_0) = \mathcal{D}(A^{1/4}) \to \mathcal{D}(A^{1/2}) \tag{6.10}$$

$$A_1 = \rho B + i(A - \rho^2B^2)^{1/2} : \ \mathcal{D}(A_1) = \mathcal{D}(A^{1/2}) \to H \tag{6.11}$$

and we consider the following problem $(\mathcal{P})$ in $E \equiv \mathcal{D}(A^{1/2}) \times X$:

$$(\mathcal{P}) \quad \begin{cases} u'(t) + A_0 u(t) = v(t) \\ v'(t) + A_1 v(t) = f(t), \quad 0 < t \leq T \\ u(0) = u_0, \ v(0) = v_0 \end{cases} \tag{6.12}$$

rewritten as a system as

$$\frac{d}{dt}\begin{vmatrix} u(t) \\ v(t) \end{vmatrix} = -\mathcal{A}\begin{vmatrix} u(t) \\ v(t) \end{vmatrix} + \begin{vmatrix} 0 \\ f(t) \end{vmatrix}; \tag{6.13}$$

$$-\mathcal{A} = \begin{vmatrix} -A_0 & I \\ 0 & -A_1 \end{vmatrix}; \quad (\lambda + \mathcal{A})^{-1} = \begin{bmatrix} (\lambda + A_0)^{-1} & (\lambda + A_0)^{-1}(\lambda + A_1)^{-1} \\ 0 & (\lambda + A_1)^{-1} \end{bmatrix}$$

$$\mathcal{D}(\mathcal{A}) = \mathcal{D}(A_0) \times \mathcal{D}(A_1) \to E \tag{6.14}$$

We note that the operators $-A_i$ in (6.10), (6.11) are of the canonical type (1.1) of this paper, with $\{\mathcal{G}, A, B\}$ and the space X in (1.1) corresponding to $\{-A_i, -\rho B, \pm(A - \rho^2 B^2)^{1/2}\}$ and the space E in (6.10), (6.11). Thus, the condition (6.5a): $\mathcal{D}(B^\alpha) \subset \mathcal{D}(A^{1/2})$, $\alpha > 1$, for the operators

A_i corresponds to the condition (4.1) (or, with equality, the condition of Remark 1.1.3) for the operator $\mathcal{G}$ in (1.1). Accordingly, because of $\mathcal{D}(B^{\alpha}) \subset \mathcal{D}(A^{1/2}), \alpha > 1$ as well as the commutativity property in a_2), we are in a position to apply Remark 1.4, or Remark 4.1, and obtain the estimates

$$|\lambda|^{1/\alpha} \|\lambda + A_i\|_{\mathcal{L}(H)} \leq C, \ i = 0, 1; \quad \text{Re}\,\lambda > 0, \ |\lambda| > 1 \tag{6.15}$$

(see e.g. estimate (4.3) with $\gamma = 1$ as guaranteed by Remark 4.1) and the operators A_i are generators of s.c. contractions semigroups of Gevrey class $> \alpha$ for all $t > 0$ on H. The same holds true for the operator A_0, this time with domain $\mathcal{D}(A^{1/4})$, on the space $\mathcal{D}(A^{1/2})$. Moreover, (6.14) (left) shows, by perturbation, that $-\mathcal{A}$ is the generator of a s.c. semigroup on E. Finally we estimate the resolvent of $-\mathcal{A}$ on E. For $\{f, g\} \in E \equiv \mathcal{D}(A_0) \times H$, $\mathcal{D}(A_0) = \mathcal{D}(A^{1/2})$, (6.14) (right) shows that

$$(\lambda + \mathcal{A})^{-1} \begin{vmatrix} f \\ g \end{vmatrix} = \begin{vmatrix} u \\ v \end{vmatrix}; \quad A_0 u = (\lambda + A_0)^{-1} A_0 f + [1 - \lambda(\lambda + A_0)^{-1}](\lambda + A_1)^{-1} g \tag{6.16a}$$

$$v = (\lambda + A_1)^{-1} g \tag{6.16b}$$

Hence, by (6.15), we estimate from (6.16a)

$$\begin{aligned} \|u\|_{\mathcal{D}(A_0)} = \|A_0 u\|_H &\leq \frac{C}{|\lambda|^{1/\alpha}} \|f\|_{\mathcal{D}(A_0)} + (1 + C|\lambda|^{1-1/\alpha}) \frac{c}{|\lambda|^{1/\alpha}} \|g\| \\ &\leq C \left[\frac{1}{|\lambda|^{1/\alpha}} + \frac{1}{|\lambda|^{2/\alpha - 1}} \right] \|\{f, g\}\|_{\mathcal{D}(A_0) \times H = E} \\ &\leq C \frac{1}{|\lambda|^{2/\alpha - 1}} \|\{f, g\}\|_E \end{aligned} \tag{6.17}$$

where as the last step we have used $1 < \alpha < 2$ so that $1/\alpha > 2/\alpha - 1 > 0$. From (6.17), and (6.16b) via (6.15), we obtain, that for $1 < \alpha < 2$:

$$|\lambda|^{2/\alpha - 1} \|(\lambda + \mathcal{A})^{-1}\|_{\mathcal{L}(E)} \leq C, \quad \text{Re}\,\lambda > 0, \quad |\lambda| > 1. \tag{6.18}$$

In view of Krein's results quoted in section 6.1, we can solve $(\mathcal{P}) = (6.12)$ for any continuously differentiable function f from [0, T] into H, and for any $u_0 \in \mathcal{D}(A)$, $v_0 \in \mathcal{D}(A^{1/2})$. On the other hand, if

$$\begin{cases} 1 < \alpha < 2 \ \text{and} \\ 3(2/\alpha - 1) > 2 \end{cases} \quad \text{i.e. for } 1 < \alpha < 6/5$$

then for all of $f \in C^{\gamma}([0, T]; H)$, $4\dfrac{\alpha - 1}{2 - \alpha} < \gamma \leq 1$, and any $u_0 \in \mathcal{D}(A)$, $v_0 \in \mathcal{D}(A^{1/2})$, there is a unique *classical* solution to problem $(\mathcal{P}) = (6.12)$. The key point of this discussion is that, in this case, we *have found an effective classical solution to the original second order problem* $(\mathcal{S}) =$ (6.4). In fact, $u' \in C((0, T]; \mathcal{D}(A_0)) = C((0, T]; \mathcal{D}(A^{1/2}))$, $A_0 u \in C((0, T]; \mathcal{D}(A^{1/2})) = C((0, T]; \mathcal{D}(A_0))$, that is $u \in C((0, T]; \mathcal{D}(A))$. But $u' \in C((0, T]; \mathcal{D}(A_0))$ implies that $A_0 u \in C^1((0, T]; H)$ and $\frac{d}{dt}(A_0 u) = A_0 u'$. Since v has a derivative $v' \in C((0, T]; H)$, it follows that $u'(t) = v(t) - A_0 u(t)$ has a derivative continuous on (0, T]. Hence

$$u''(t) + A_0 u'(t) = - A_1 v(t) + f(t) = - A_1 [u'(t) + A_0 u(t)] + f(t)$$
$$= - A_1 u'(t) - A_1 A_0 u(t) + f(t), \quad 0 < t \le T.$$

Also, if $u_0 \in \mathcal{D}(A)$ and $u_1 \in \mathcal{D}(A^{1/2})$, then $v_0 = u_1 + (\rho B - i(A - \rho^2 B^2)^{1/2}) u_0 \in \mathcal{D}(A^{1/2})$. We have thus shown

Theorem 6.3. Assume $a_1) - a_3)$ as well as $1 < \alpha < 6/5$. Then for all $f \in C^\gamma(([0, T]; H)$, $4 \dfrac{\alpha - 1}{2 - \alpha} < \gamma \le 1$, and any $u_0 \in \mathcal{D}(A)$, $v_0 \in \mathcal{D}(A^{1/2})$, there exists a unique classical solution to the original problem $(\mathcal{S}) = (6.4)$.

Remark 6.3 In the case $B = A^\sigma$, $0 < \sigma < 1$, then assumption $a_3)$, i.e. (6.5) holds true with $\alpha = \dfrac{1}{2\sigma}$ if and only if $0 < \sigma < 1/2$. However, the further restriction $1 < \alpha < 6/5$ means equivalently $5/12 < \sigma < 1/2$ for Theorem 6.3 to hold □

If one, instead, aims at results for problem $(\mathcal{F})$; i.e.(6.8) and (6.9), [equivalently $(\mathcal{P}) = (6.12)$], arguments as those in section 6.1 yield the following

Theorem 6.4. Assume $a_1) - a_3)$ as well as $1 < \alpha < 3/2$. Then for all $f \in C^\gamma([0, T]; H)$, $2(\alpha - 1) < \gamma \le 1$, all $u_0 \in \mathcal{D}(A)$ and all $u_1 \in \mathcal{D}(A^{1/2})$, there is a unique $u \in C([0, T]; H)$, $u' \in C((0, T[; H)$

$$u \in C((0, T]; \mathcal{D}(A^{1/2})), \quad u' + A_0 u \in C((0,T]; \mathcal{D}(A^{1/2})) \cap C^1((0, T]; H)$$

satisfying problem $(\mathcal{F})$ □

Remark 6.4. If $B = A^\sigma$, $0 < \sigma < 1$, and with reference to Remark 6.3, then condition $1 < \alpha < 3/2$ holds true with $\alpha = \dfrac{1}{2\sigma}$ if and only if $1/3 < \sigma < 1/2$ □

APPENDIX

1) Let E_λ by the resolution of the identity associated with the self-adjoint operator B on the complex Hilbert space X, where λ is real:

$$Bx \equiv \int_{-\infty}^{\infty} \lambda \, dE_\lambda x; \quad \mathcal{D}(B) = \{x \in X: \int_{-\infty}^{\infty} |\lambda|^2 (dE_\lambda x, x) < \infty\} \tag{A.1}$$

$$B^+ = \int_{-\infty}^{\infty} |\lambda| \, dE_\lambda x = \text{non-negative square root of} \tag{A.2}$$

$$B^2 x = \int_{-\infty}^{\infty} \lambda^2 \, dE_\lambda x; \quad \mathcal{D}(B^2) = \{x \in X: \int_{-\infty}^{\infty} |\lambda|^4 (dE_\lambda x, x) < \infty\} \tag{A.3}$$

$$\mathcal{D}(B^+) = \mathcal{D}(B) \tag{A.4}$$

2) Define Π^- as the orthogonal projection

$$\Pi^- x = \int_{-\infty}^{0} dE_\lambda x, \quad x \in X, \tag{A.5}$$

so that

$$x = \int_{-\infty}^{\infty} dE_\lambda\, x = \Pi^- x + \int_0^{\infty} dE_\lambda x\,, \quad x \in X\,. \tag{A.6}$$

3) Writing from (A.2)

$$B^+ x = \int_{-\infty}^{0} (-\lambda)\, dE_\lambda x + \int_0^{\infty} \lambda dE_\lambda x\,, \quad x \in \mathcal{D}(B^+)\,, \tag{A.7}$$

we introduce the non-negative operator

$$\Pi^- B^+ x = \int_{-\infty}^{0} (-\lambda)\, dE_\lambda\, x\,, \tag{A.8}$$

$$\mathcal{D}(\Pi^- B^+) = \{x \in X\colon \int_{-\infty}^{0} |\lambda|^2\, (dE_\lambda\, x, x) < \infty\} \supset \mathcal{D}(B^+) = \mathcal{D}(B)\,. \tag{A.9}$$

4) Also, adding and subtracting, we write recalling (A.2)

$$\begin{aligned} Bx &= \int_{-\infty}^{\infty} \lambda\, dE_\lambda\, x + \int_0^{\infty} \lambda\, dE_\lambda\, x + \int_{-\infty}^{0} (-\lambda)\, dE_\lambda\, x + \int_{-\infty}^{0} \lambda\, dE_\lambda x \\ &= B^+ x + 2 \int_{-\infty}^{0} \lambda\, dE_\lambda x = B^+ x - 2\Pi^- B^+ x, \quad x \in \mathcal{D}(B^+) \end{aligned} \tag{A.10}$$

5) We notice

$$(B^+)^{1/2}\, x = \int_{-\infty}^{\infty} |\lambda|^{1/2}\, dE_\lambda\, x; \quad \mathcal{D}((B^+)^{1/2}) = \{x \in X\colon \int_{-\infty}^{\infty} |\lambda|\, (dE_\lambda\, x, x) < \infty\} \tag{A.11}$$

$$(\Pi^- B^+)^{1/2}\, x = \int_{-\infty}^{0} (-\lambda)^{1/2}\, dE_\lambda x\,; \quad \mathcal{D}((\Pi^- B^+)^{1/2}) = \{x \in X\colon \int_{-\infty}^{0} (-\lambda)\, (dE_\lambda\, x, x) < \infty\} \tag{A.12}$$

$$\mathcal{D}((\Pi^- B^+)^{1/2}) \supset \mathcal{D}((B^+)^{1/2}) \tag{A.13}$$

6) Finally, we show — as needed in section 3.3 — that for $0 < \theta \le 1$ (and B^+ positive):

$$B\colon \mathcal{D}((B^+)^{1-\theta/2}) \text{ onto } [\mathcal{D}((B^+)^{\theta/2})]' \tag{A.14}$$

In fact, with $x \in \mathcal{D}((B^+)^{1-\theta/2})$, we rewrite by (A.10)

$$Bx = (B^+)^{\theta/2}\, (B^+)^{1-\theta/2} x - 2\, (B^+)^{\theta/2}\, (\Pi^- B^+)\, (B^+)^{-1}\, (B^+)^{1-\theta/2}\, x \in [\mathcal{D}((B^+)^{\theta/2})]'$$

since $y = (B^+)^{-1} (B^+)^{1-\theta/2}\, x \in \mathcal{D}(B^+) \subset \mathcal{D}(\Pi^- B^+)$ by (A.9) and (A.14) is proved.

REFERENCES

[D-S.1] N. Dunford and J. T. Schwartz, Linear Operators, I, Wiley-Interscience, New York 1958.

[C-T.1] S. Chen and R. Triggiani, Proof of two conjectures by G. Chen and D. L. Russell on structural damping for elastic systems, *Springer-Verlag Lectures Notes in Mathematics,* n. 1354, 234-256, Proceedings of Conference on Approximation and Optimization held at the University of Havana, Cuba, January 1987.

[F-L-T] F. Flandoli, I. Lasiecka and R. Triggiani, Algebraic Riccati equations with non-smoothing observation arising in hyperbolic and Euler-Bernoulli equations, *Ann. Matem. Pura and Appl.,* Vol. CLIII (1988), 307-382.

[F-Y.1] A. Favini and A. Yagi, Multivalued linear operator and degenerate evolution equations, *Annali Mat. Pura Appl.,* to appear.

[K.1] T. Kato, Perturbation theory for linear operators, Springer-Verlag 1966.

[K.2] S. G. Krein, Linear differential equations in Banach space, *American Math. Soc.,* Translations of Math. Monographs Vol 29, 1971.

[L-M.1] J. L. Lions and E. Magenes, Non-homogeneous boundary value problems, Springer-Verlag 1970.

[L.T.1] I. Lasiecka and R. Triggiani, Riccati equations arising from systems with unbounded input-solution operator: applications to boundary control problems for wave and plate problems, *J. of Non Linear Analysis,* to appear.

[L-T.2] I. Lasiecka and R. Triggiani, Riccati equations for hyperbolic partial differential equations with $L_2(0,T;L_2(P))$ - Dirichlet boundary terms, *SIAM J. Control & Options* Vol. 24 (1986), 884-924.

[L-T.3] I. Lasiecka and R. Triggiani, Feedback semigroups and cosine operators for boundary feedback and hyperbolic parabolic equations, *J. Diff. Eqts.,* 47 (1983), 246-272.

[P.1] A. Pazy, Semigroups of linear operators and applications to partial differential equations, Springer-Verlag, Berlin - Heidelberg - Tokyo, 1986.

[Ta.1] S. Taylor, Ph.D. thesis, Chapter 1 Gevrey class semigroups, School of Mathematics University of Minnesota, 1989.

[T.2] K. Taira, The theory of semigroups with weak singularity and its applications to partial differential equations, *Tsakuba J. Math.* 13 (1989), 513-562.

[Tr.1] H. Triebel, Interpolation theory, function spaces, differential operators, North-Holland 1978.

[W.1] J. Weidmann, Linear Operators in Hilbert Spaces, Springer-Verlag, 1980.

The Kompaneets Equation

JEROME A. GOLDSTEIN Department of Mathematics, Tulane University, New Orleans, LA 70188, U.S.A. and Department of Mathematics, Louisiana State University, Baton Rouge, LA 70803, U.S.A.

1. INTRODUCTION

The Kompaneets equation is a spatially degenerate nonlinear parabolic partial differential equation arising in plasma physics. It was introduced by A.S. Kompaneets [11] in 1957. For an exposition of known results about this equation and further references see Caflisch and Levermore [2].

Of concern is the radiation field in a fully ionized plasma. The effect of Compton scattering is described by a nonlinear Fokker-Planck diffusion equation which in this case is the Kompaneets equation. Its nondimensionalized form is

$$\frac{\partial u}{\partial t} = \frac{1}{\beta(x)} \frac{\partial}{\partial x}\left[\alpha(x)\left\{\frac{\partial u}{\partial x} + u + \gamma(u)\right\}\right] \tag{1}$$

for $0 < x, t < \infty$. The initial condition and boundary conditions are

$$u(0, x) = f(x), \tag{2}$$

$$\alpha(x)\left\{\frac{\partial u}{\partial x} + u + \gamma(u)\right\} \to \quad \text{as } x \to 0, \infty \tag{3}$$

The most important case is $\alpha(x) = x^4, \beta(x) = x^2$, and $\gamma(u) = u^2$. The radiation density $u(t,x)$ is nonnegative and gives the total photon number, namely

$$N = \int_0^\infty u(t,x)x^2 dx.$$

Here $x = h\nu/\theta$ is the normalized photon energy; h is Planck's constant, ν is the frequency and θ is the temperature. Also, the x^2 factor is a geometric factor expressing spherical symmetry.

Well-posedness for the problem (1)-(3) is open. We shall approach this problem by the method of nonlinear semigroups. The problem is very degenerate at the origin because of the fact that the diffusion coefficient α/β is such that its reciprocal is not integrable in a neighborhood of $x = 0$. We shall attempt to apply the Crandall-Ligett theorem, which involves checking both dissipativity and a range condition. Our effort will only be partially successful; we shall find the right space for the problem and establish dissipativity of the relevant operator. We shall then indicate the difficulty in the range condition computation.

2. FORMULATION

Let α, β, γ be continuous functions on $[0,\infty)$ to $[0,\infty)$ which are positive on $(0,\infty)$. Suppose also that γ is nondecreasing. Let $X_p = L^p((0,\infty); \beta(x)dx)$ consist of the real measurable functions u on $(0,\infty)$
with finite norm

$$||u||_p = \left\{\int_0^\infty |u(x)|^p \beta(x)dx\right\}^{1/p},$$

$1 \le p < \infty$. One can modify this definition in the usual way to obtain X_∞. DefineA_p, an operator on X_p, as follows. $u \in$ Dom (A_p) iff $u \in X_p \cap W^{2,p}_{\text{loc}}(0,\infty); u \ge 0; w = \alpha(u' + u + \gamma(u)) \in W^{1,p}_{\text{loc}}(0,\infty)$; $w(x) \to 0$ as $x \to 0, \infty$; and $A_p u := \frac{1}{\beta} w' \in X_p$.

The semigroup strategy is to write the problem (1) -(3) as

$$du/dt = Au, \ u(0) = f \tag{4}$$

which is an ordinary differential equation in a Banach space X. We must check two condition:

(i) A is dissipative, that is, for

$$u_i - \lambda A u_i = h_i,\ i = 1,2$$

with $\lambda > 0$, then $||u_1 - u_2|| \leq ||h_1 - h_2||$. In other words, $||(I - \lambda A)^{-1}||_{\text{Lip}} \leq 1$.

(ii) A satisfies the range condition, that is, there is a dense set D in $\overline{\text{Dom}(A)}$ such that for (small) $\lambda > 0$ and $h \in D$, the problem $u - \lambda A u = h$ has a solution.

If (i) and (ii) hold, then by the Crandall-Liggett theorem [6], (4) is governed by a semigroup T given by

$$u(t) = T(t)f = \lim_{n\to\infty} (I - \frac{t}{n}\bar{A})^{-n} f$$

where $\bar{A}$ is the closure of A (viewed as a graph in $X \times X$). One has $||T(t)||_{\text{Lip}} \leq 1$ and u is the unique mild solution of (4) when $f \in \overline{\text{Dom}(A)}$. For more on semigroups see for example [1], [8], [5].

3. DISSIPATIVITY

Our main result here is the following result.

THEOREM 1. A_1 is dissipative on X_1.

We remark that A_p is dissipative on X_p iff $p = 1$. Moreover, A defined by

$$Au = \beta^{-1}[\alpha(u' + u + \gamma(u))]'$$

(with the boundary condition (3)) can be studied on other spaces; but for example it is not dissipative on $L^p((0,\infty); dx)$ for any p. We believe that X_1 is the right space for the study of A.

Here is the proof of the theorem. Let $\lambda > 0, h_i \in X, 0 \leq u_i \in \text{Dom}(A_1)$,

$$u_i - \lambda A_1 u_i = h_i \tag{5}$$

for $i = 1, 2$. Since $u := u_1 - u_2$ is continuous on $(0, \infty)$, the sets

$$[u_1 > u_2] = \{x \in (0, \infty) : u_1(x) > u_2(x)\}\,,\ [u_2 > u_1]$$

are open. Thus they can be decomposed as unions of maximal open intervals:

$$[u > 0] = \bigcup_{n \in J} (a_n, b_n)\,,\ [u < 0] = \bigcup_{n \in K} (c_n, d_n)\,.$$

Let

$$\delta := \int_0^\infty (A_1 u_1 - A_1 u_2)\ \text{sign}_0(u)\beta(x)dx,$$

where $\text{sign}_0(r)$ is 1,0, or -1, according as r is positive, zero or negative. Then by (5),

$$\begin{aligned} ||u_1 - u_2||_1 &= \int_0^\infty u\ \text{sign}_0(u)\beta(x)dx \\ &= \lambda\delta + \int_0^\infty (h_1 - h_2)\ \text{sign}_0(u)\beta(x)dx \\ &\leq \lambda\delta + ||h_1 - h_2||_1 \leq ||h_1 - h_2||_1 \end{aligned}$$

provided $\delta \leq 0$. Thus it suffices to show $\delta \leq 0$. Now, using obvious notation,

$$\begin{aligned} \delta = \int_0^\infty &= \int_{[u>0]} + \int_{[u<0]} + \int_{[u=0]} \\ &= \sum_{n \in J} \int_{a_n}^{b_n} + \sum_{n \in K} \int_{c_n}^{d_n} + 0. \end{aligned}$$

It suffices to show that each term of the form $\int_{a_n}^{b_n}$ or $\int_{c_n}^{d_n}$ is nonpositive. We shall consider $\int_{a_n}^{b_n}$, the other term being similar. If $0 < a_n < b_n < \infty$, then the graph of $u = u_1 - u_2$ looks like this:

Thus $u(a_n) = 0 = u(b_n)\,, u'(a_n) \geq 0 \geq u'(b_n)$. Consequently

$$\begin{aligned}
&\int_{a_n}^{b_n} (A_1 u_1 - A_1 u_2)\ \mathrm{sign}_0\,(u_1 - u_2)\,\beta dx \\
&\qquad = \int_{a_n}^{b_n} \frac{1}{\beta(x)} [\alpha(x)(u_1'(x) + u_1(x) \\
&\qquad + \gamma(u_1(x)) - u_2'(x) - u_2(x) - \gamma(u_2(x))]\beta(x)dx \\
&\qquad = \alpha\{u' + u + \gamma(u_1) - \gamma(u_2)\}|_{a_n}^{b_n} \\
&\qquad = \alpha u'|_{a_n}^{b_n} \ \text{since}\ u(a_n) = u(b_n) = 0 \\
&\qquad \leq 0 \ \text{since}\ u'(b_n) \leq 0 \leq u'(a_n).
\end{aligned}$$

We must also consider other cases, such as $u > 0$ in all of $(0, \infty)$, or $(a_n, b_n) = (0, b_n)$ with $b_n < \infty$, and so on. All of the arguments are similar. We treat the case of $0 = a_n, b_n < \infty$. Then, as above,

$$\begin{aligned}
&\int_{a_n}^{b_n} (A_1 u_1 - A_1 u_2)\ \mathrm{sign}_0\,(u_1 - u_2)\,\beta dx \\
&\qquad = \alpha\{u' + u + \gamma(u_1) - \gamma(u_2)\}|_0^{b_n} \\
&\qquad = I_1 + I_2.
\end{aligned}$$

Now $I_1 = \alpha\{u' + u + \lambda(u_1) - \lambda(u_2)\}|_{b_2} \leq 0$ by the argument given above. Also, $u_1, u_2 \in$ Dom (A) implies

$$\lim_{x \to 0} \alpha\{u_i' + u_i + \gamma(u_i)\} = 0$$

for $i = 1, 2$ whence $I_2 = 0$ (see (3)). Theorem 1 follows. ■

4. RANGE CONDITION

To simplify matters, throughout this section we take $\alpha(x) = x^4, \beta(x) = x^2$, and $\gamma(u) = u^2$. Let $\lambda > 0, 0 \leq h \in C_c^1(0, \infty)$, the subscript c denoting compact support. We want to solve

$$u - \lambda A u = h,$$

that is

$$(6)\qquad \begin{cases} w(x) := x^4[u'(x) + u(x) + u(x)^2], \\ w(x) \;\to 0 \text{ as } x \to 0, \infty, \\ u(x) \;-\; \lambda x^{-2} w'(x) = h(x) \end{cases}$$

in the sense of distributions on $(0, \infty)$.

Let

$$v(x) := x^2 u(x), k(x) := x^2 h(x)(\geq 0).$$

Then $v' = 2xu + x^2 u'$, so that

$$w = x^2 v + x^2 v' - 2xv + v^2.$$

The homogeneous Dirichlet conditions for w imply

$$v(0) = v(\infty) = 0.$$

Thus (6) reduces to solving

$$w = x^2 v + x^2 v' - 2xv + v^2,$$

$$v - \lambda w' = k,$$

$$v(0) = v(\infty) = 0$$

for a $v \geq 0$. This is a problem of the form

$$v(x) - \lambda x^2 v''(x) + F(x, v(x), v'(x)) = 0,$$

$$v(0) = v(\infty) = 0.$$

This can be converted from a problem on $[0, \infty]$ to a problem on [0,1] by introducing the new independent variable $y = (2/\pi) \tan^{-1} x$. Then we get, for $v(x) = z(y)$,

$$(7)\qquad z - \lambda \eta(y) z'' + G(y, z, z') = 0$$

where ' stands for d/dy, and $\eta(y)/y^2 \to \pi^2/4$ as $y \to 0$ and $\eta(y) \to \infty$ as $y \to 1$.

This problem is closely related to recent joint work of the writer and C.-Y.Lin [9],[10]. We considered certain problems of the form

$$u_t = \varphi(u, u_x) u_{xx} + \psi(x, u, u_x)$$

for $t \geq 0$ and $0 \leq x \leq 1$ with (nonlinear) Wentzell boundary conditions. We applied the Crandall-Liggett theorem in $X = C[0,1]$. The range condition reduced to solving

$$u - \lambda\varphi(x,u')u'' = \lambda\psi(x,u,u') = k,$$
$$u(0) = u(1) = 0 \tag{8}$$

for $k \in C_c(0,1)$ and $\lambda > 0$. The motivation was an early result of Feller, put into final form by Clément and Timmermans [3]. Consider the linear problem

$$u - \lambda[\varphi_0(x)u'' + \varphi_1(x)u' + \varphi_2(x)u] = k \tag{9}$$

where $\varphi_i \in C(0,1)$, $\varphi_2 \in C[0,1]$, $\varphi_0 > 0$ on (0,1), and $\lambda > 0$ is sufficiently small (and $k \in C_c(0,1)$.) Let

$$Q(x) := \exp\left\{-\int_{1/2}^{x} \frac{\varphi_1(s)}{\varphi_0(s)}ds\right\} \tag{10}$$

If Q is integrable in a neighborhood of $x = 0$ and $x = 1$, then (9) is solvable. Moreover (10) is close to being a necessary and sufficient condition for the solvability of (9); see [3] or [9] for precise details. In our case when (8) reduces to (7), $\varphi(x,u') = \varphi_0(x)$ becomes a positive constant times $\sin^2 y \sec^2 y$ (with $y = (2/\pi)\tan^{-1}x$). Think of $G(y,z,z')$ as $G_0(y) + G_1(y)z + G_2(y)z' + \cdots$, and try to apply the Feller-Clément-Timmermans criteria with $\varphi_1 = G_2$, $\varphi_2 = G_1$, and $k = -G_2$. This is the approach Lin and I applied in [10]. Lin and I have obtained some extensions of our earlier results [9],[10], but not of a general enough nature to apply to the very singular Kompaneets ordinary differential equation (7). This is an open problem we hope to solve in the future.

5. *CONCLUDING REMARKS*

Suppose that (7) can be solved, so that the Kompaneets equations is governed by a contraction semigroup T on X_1. This T preserves the lattice structure, i.e., if $0 \leq f \leq g$ in X_1, then $0 \leq T(t)f \leq T(t)g$ for all $t \geq 0$; this follows from an abstract result of Crandall and Tartar [7]. Next, if $f \in D \subset X_1 \cap X_p$ for $1 \leq p \leq \infty$, then for $t > 0, T(t)f \in X_1 \cap X_p$ and $||T(t)f||_p$ is nonincreasing in t. Here is the proof. Let $p > 1$. By standard approximation procedures (for instance, using Yosida approximations) it suffices to consider f such that $T(t)f$ is a strong solution of $u' = Au, u(0) = f$. Then

$$\begin{aligned}\frac{d}{dt}||T(t)f||_p^p &= \int_0^\infty \frac{\partial}{\partial t}|T(t)f(x)|^p x^2\,dx \\ &= p\int_0^\infty |T(t)f|^{p-1}(AT(t)f)x^2\,dx,\end{aligned}$$

and it suffices to show

$$\int_0^\infty u(x)^{p-1}(Au(x))x^2\,dx \leq 0$$

for $0 \leq u$ in Dom $(A)(=$ Dom $(A_1)) \cap$ Dom $(A_p))$. But

$$\begin{aligned}\int_0^\infty u^{p-1}(Au)x^2\,dx &= \int_0^\infty u^{p-1}[x^4(u' + u + u^2)]'\,dx \\ &= -(p-1)\int_0^\infty [x^4(u' + u + u^2)]u^{p-2}u'\,dx \\ &= -(p-1)\int_0^\infty x^4u^{p-2}(u') - (p-1)\sum_{j=0}^{1}\int_0^\infty x^4u^{p-j}u'\,dx \\ &= (\leq 0) + (p-1)\sum_{j=0}^{1} -(p+1-j)^{-1}\int_0^\infty 4x^3u^{p+1-j}\,dx \leq 0.\end{aligned}$$

and the result follows.

6. ACKNOWLEDGEMENTS

I am very grateful to Dave Levermore who taught me about the Kompaneets equation in the fall of 1990 at MSRI in Berkeley. It was fun being Dave's student. I am grateful to Enrico Obrecht and the other organizers of the wonderful conference at Bologna at which this report was made. Finally I gratefully acknowledge the partial support of an NSF grant.

REFERENCES

1. Ph. Bénilan, M.G. Crandall and A. Pazy, *Nonlinear Evolution Governed by Accretive Operators,* in preparation.

2. R. E. Caflisch and C.D. Levermore, *Equilibrium for radiation in a homogeneous plasma,* Phys. Fluids 29 (1986), 748-752.

3. Ph. Clément and C. Timmermans, *On* (C_0) *semigroups generated by operators satisfying Ventcel's boundary conditions,* Indag. Math . 89 (1986), 379-387.

4. G. Cooper, *Compton Fokker-Planck equation for hot plasmas,* Phys. Rev. D3 (1971), 2312-2316.

5. M. G. Crandall, *An introduction to evolution governed by accretive operators, in Dynamical Systems,* Vol. 1 (ed. by L. Cesari, J.L. Hale and J.P. LaSalle), Academic (1976), 131-165.

6. M. G. Crandall and T. M. Liggett, *Generation of semigroups of nonlinear transormations on general Banach spaces,* Amer. J. Math. 93 (1971), 265-298.

7. M. G. Crandall and L. Tartar, *Some relations between nonexpansive and order preserving mappings,* Proc. Amer. Math. Soc. 78 (1980), 385-390.

8. J. A. Goldstein,*Semigroups of Nonlinear Operators and Applications,* in preparation.

9. J. A. Goldstein and C.-Y. Lin, *Highly degenerate parabolic boundary value problems,* Diff. Int. Eqns. 2 (1989), 216-227.

10. J. A. Goldstein and C.-Y. Lin, *Parabolic problems with strong degeneracy at the spatial boundary,* in Semigroup Theory and Evolution Equations (ed. by Ph. Clément, E. Mitidieri and B. de Pagter), Marcel Dekker (1991), 181-191.

11. A. S. Kompaneets, *The establishment of thermal equilibrium between quanta and electrons,* Soviet Phys. JETP 4 (1957), 730-737.

Multiplicative Perturbation of Resolvent Positive Operators

ALBRECHT HOLDERRIETH Mathematisches Institut, Universität Tübingen, Auf der Morgenstelle 10, D-W-7400 Tübingen, Germany

Multiplicative perturbations of generators of strongly continuous semigroups of bounded linear operators have been considered by many authors (see, e.g., [Do], [L1], [L2], [Do-H]) mostly assuming the contractivity of the original and of the perturbed semigroup. Here we present an approach taking into account an order structure of the underlying Banach space.
To this purpose we use the concept of resolvent positiv operators on a Banach lattice as introduced by W. Arendt in [Ar1] and we study the multiplicative perturbation of such operators by an element of the center of the Banach lattice.

In the following we always assume E to be a real Banach lattice with positive cone E_+. By $Z(E)$, the center of E, we denote all bounded linear operators on E which belong to some order interval of the form $[-nI, nI]$, $n \in \mathbb{N}$, where I denotes the identity. For more details on Banach lattices see [Me], [Sch]. If A is a linear operator we denote its domain by $D(A)$, the resolvent set by $\rho(A)$ and the resolvent at a point λ by $R(\lambda, A)$.

Now we are ready to recall the definition of a resolvent positive operator.

Definition ([Ar1]). An operator $(A, D(A))$ is called *resolvent positive* if there exists $\omega \in \mathbb{R}$ such that $(\omega, \infty) \subset \rho(A)$ and $R(\lambda, A) \geq 0$ for all $\lambda > \omega$.

The notion of resolvent positive operators is a useful tool in the investigation of abstract Cauchy problems (see [Na], [Ar1]). Especially every generator of a strongly continuous semigroup of positive operators is resolvent positive. In [Ar1] it was shown that every resolvent positive operator is the generator of a twice integrated semigroup and if the domain of the operator is dense, of a once integrated semigroup. (For the theory of integrated semigroups see [Ar2] or [Ne]). The following Proposition therefore yields a perturbation result for generators of integrated semigroups with positive resolvent.

Proposition. Let $(A, D(A))$ be a resolvent positive operator, with $(0, \infty) \subset \rho(A)$ such that the resolvent satisfies there an estimate

$$\|R(\lambda, A)\| \leq \frac{M}{\lambda}$$

for a suitable $M > 0$. If $0 \leq B \in Z(E)$ fulfills

$$I - B^{-1} \geq 0 \qquad \text{and} \qquad \|I - B^{-1}\| < \frac{1}{M},$$

then the operator BA with domain $D(BA) = D(A)$ is resolvent positive with $(0, \infty) \subset \rho(BA)$.

Proof. Let $\delta := \|I - B^{-1}\|M < 1$. For $\lambda > 0$ we have

$$\lambda - BA = B[B^{-1}\lambda - a] = B[(B^{-1} - I)\lambda + \lambda - A] = B[I - (I - B^{-1})\lambda R(\lambda, A)](\lambda - A).$$

From the assumption it follows the existence of $\delta < 1$ such that

$$\|(I - B^{-1})\lambda R(\lambda, A)\| \leq \delta < 1$$

for all $\lambda > 0$. From the Neumann series we obtain that $R(\lambda, BA)$ exist for all $\lambda > 0$. Furthermore it is given by

$$R(\lambda, BA) = R(\lambda, A) \sum_{n=0}^{\infty} [(I - B^{-1})\lambda R(\lambda, A)]^n B^{-1}.$$

Since B is a positive element of the center its inverse is positive, too. Therefore using the assumption on $I - B^{-1}$ we have that $R(\lambda, BA) \geq 0$ for all $\lambda \in (0, \infty)$. □

Let us now apply the proposition to generators of strongly continuous semigroups of positive operators.

Corollary 1. Let the operator $(A, D(A))$ be the generator of a bounded positive semigroup $(T(t))_{t \geq 0}$ on the Banach lattice E. If for $B \in Z(E)$ there exists a $\delta > 0$ such that

$\delta I \leq B$ then the operator BA is resolvent positive and therefore the generator of an once integrated semigroup on E.

Proof. First we define a new norm $\|.\|_n$ on E by

$$\|x\|_n := sup_{t\geq 0}\{\|T(t)|x|\|\}.$$

It is easy to see that $\|.\|_n$ is a lattice norm which is equivalent to the original one. Furthermore the semigroup $(T(t))_{t\geq 0}$ is contractive for the new norm.
Now we multiply the operator B by a constant μ such that $I \leq \mu B$. Using the fact that the center of E is isomorphic (in order and norm) to a space of continuous functions on a compact set, we obtain that $I-(\mu B)^{-1} \geq 0$ and $\|I-(\mu B)^{-1}\|_n < 1$. Since A is the generator of a positive contraction semigroup we can apply the Proposition and conclude that the operator μBA is resolvent positive. But then the operator BA is resolvent positive, too. □

Remark. 1. Corollary 1 is best possible in the following sense. In [deL] deLaubenfels presents in a different context an example of a generator of a strongly continuous group of positive operators and a perturbing center operator, for which the perturbed operator generates an integrated semigroup but not a strongly continuous semigroup.
2. The corollary is not true if the operator $(A, D(A))$ is just the generator of a bounded strongly continuous semigroup (see for example [Kr-L], chap. 2.7.3).

In [Na], Chap. B- II, W. Arendt proved that the assumptions of the Corollary imply on a Banach lattice $C(K)$, K compact, that the perturbed operator is even the generator of a positive strongly continuous semigroup. In arbitrary Banach lattices we just obtain generators of once integrated semigroups, but the stronger assumptions in the following corollary also imply that the operator BA is a generator of a strongly continuous semigroup.

Corollary 2. Let the operators A and B satisfy the assumptions of the Corollary 1. If in addition there exist $\lambda_0 > 0$ and $k > 0$ such that

$$\|R(\lambda_0, A)x\| \geq k\|x\|$$

for all $x \in E_+$, then the operator BA is the generator of a strongly continuous semigroup of positive operators.

Proof. For a suitable $\mu > 0$ we have that

$$R(\lambda_0, \mu BA) = R(\lambda_0, A)\sum_{n\geq 0}^{\infty}[(I-(\mu B)^{-1})\lambda_0 R(\lambda_0, A)]^n(\mu B)^{-1}.$$

and that $I - (\mu B)^{-1} \geq 0$. For $x \in E_+$ taking just the first part of the sum yields

$$R(\lambda_0, \mu BA)x \geq R(\lambda_0, A)(\mu B)^{-1}x.$$

Since $B^{-1} \in Z(E)$ is strictly positive there exists a constant $r > 0$ such that $(\mu B)^{-1}x \geq rx$ for all $x \in E_+$. Therefore

$$\|R(\lambda_0, \mu BA)x\| \geq r\|R(\lambda_0, A)\| \geq rk\|x\|$$

for all $x \in E_+$. Now the assertion follows from Theorem 2.5 in [Ar1]. □

References.

[Ar1] Arendt. W., *Resolvent positive operators*, Proc. London Math. Soc. (3) **54** (1987), 321–349.

[Ar2] Arendt, W., *Vector valued Laplace transforms and Cauchy problems*, Israel J. Math. **59** (1987), 327–352.

[deL] deLaubenfels, R., *Bounded, commuting multiplicative perturbations of strongly continuous group generators*, Houston J. Math. **17** (1991), 299–310.

[Do] Dorroh, J.R., *Contraction semigroups in a function space*, Pacific J. Math. **19** (1966), 35–38.

[Do-H] Dorroh, J.R. and A. Holderrieth, *Multiplicative perturbation of semigroup generators*, Bolletino Unione Mat. It., to appear.

[Kr-L] Kreiss, H.-O., and J. Lorenz, *Initial-Boundary Value Problems and the Navier-Stokes Equations*, Pure and Applied Mathematics Vol. 136, Academic Press Inc. 1989.

[L1] Lumer,G., *Homotopy-like perturbation: General results and applications*, Arch. Math. **52** (1989), 551-561.

[L2] Lumer, G., *New singular multiplicative perturbation results via homotopy-like perturbation*, Arch. Math. **53** (1989), 52-60.

[Me] Meyer-Nieberg, P., *Banach Lattices*, Springer-Verlag, Berlin-Heidelberg-New York, 1991.

[Na] Nagel, R. (Ed.), *One-parameter Semigroups of Positive Operators*, Lect. Notes Math. **1184**. Springer-Verlag, Berlin-Heidelberg-New York-Tokyo, 1986.

[Ne] Neubrander, F., *Integrated semigroups and their applications to the abstract Cauchy problem*, Pacific J. Math. **135** (1988), 111–155.

[Sch] Schaefer, H. H., *Banach Lattices and Positive Operators*, Springer-Verlag, Berlin-Heidelberg-New York, 1974.

Uniform Decay Rates for Semilinear Wave Equations with Nonlinear and Nonmonotone Boundary Feedback, Without Geometric Conditions

I. LASIECKA Department of Applied Mathematics, University of Virginia, Charlottesville, VA 22903, U.S.A.

D. TATARU Department of Applied Mathematics, University of Virginia, Charlottesville, VA 22903, U.S.A.

1. INTRODUCTION

Consider the following semilinear equation

$$(1.1)\quad \begin{cases} y_{tt} = \Delta y - f_0(y) & \text{in } \Omega \times (0, \infty) \\ \dfrac{\partial y}{\partial \gamma} = -\, g(y_t|_\Gamma) - f_1(y|_\Gamma) & \text{on } \Gamma \times (0, \infty) \\ y(0) = y_0 \in H^1(\Omega); \ \ y_t(0) = y_1 \in L_2(\Omega)\,. \end{cases}$$

Here, Ω is a bounded open region in R^n $n \geq 1$, with a smooth boundary Γ and γ stands for an outer normal direction to the boundary Γ.

The following assumptions are made on the nonlinear functions f_i, $i = 0, 1$, and g.

(H–1) g (s) is a continuous monotone increasing function on R such that

(i) $g(s)\, s > 0$ for $s \neq 0$;

(ii) $M_2 s^2 \leq g(s)\, s \leq M_1 s^2$ for $|s| \geq 1$, for some

(H–2) $f_0(s)$ is a $W^{1,\infty}_{loc}$ (R), piecewise $C^1(R)$ function differentiable at s = 0 such that

(i) $f_0(s)\, s \geq 0$ for $s \in R$.

(ii) $|f_0(s)| \leq N\,(1 + |s|^{k-1})$, $1 < k < \dfrac{n}{n-2}$ for $s > N$

(H–3) $f_1(s)$ is a continous function differentiable at 0 such that

(i) $f_1(s) \le 0$ for $s \in R$

(ii) $|f_1(s)| \le M|s|^{k_1} + A|s|$ for $s \in R$; $k_1 < \frac{n-1}{n-2}$.

The main goal of this paper is to prove that under the above hypothesis, the solutions to (1.1) exist in $C(0, \infty; H^1(\Omega) \times L_2(\Omega))$ and, moreover, they decay with the uniform rates to zero when $t \to \infty$.

The problem of proving uniform decay rates for the solutions to the wave equation with a boundary dissipation has attracted a lot of attention in recent years. Indeed, the linear problem (i.e. when $g(y) = y$ and $f_i(y) \equiv 0$) has been treated by several authors, see for instance ([Ch.1], [L.1], [T.1]). The case when the dissipation on the boundary is nonlinear (i.e. g(s) is nonlinear) and other nonlinearities are not present (i.e. $f_i(y) = 0$), has been studied by [Z.1] (see also [L.2] where the plate equation is considered, and [C-L-M] and [A-B-C] where the one dimensional beam and the wave equations are treated respectively). Assuming (H-1) and the additional hypothesis

(1.2) $|g(s)| \le C|s|^p$, $\forall s \in R$ for some $p \ge 1$,

[Z.1] proved the uniform decay rates for the corresponding solutions. The case when conservative nonlinear terms $f_0(y)$ is added to the equation and $g(s) = s$, $f_1 = 0$, has been treated in [K-Z]. Assuming (H-2) and in addition

(1.3) $\exists \delta > 0$, $f_0(s)s \ge (2+\delta)F_0(s)$ $\forall s \in R$

where $F_0(s) = \int_0^s f_0(t)\,dt$, the authors established in [K-Z] exponential decay rates for the solutions.

The main novelty of our paper is twofold:

(i) first of all we consider a fully nonlinear problem i.e.: both the stabilizing feedback $g(y_t|_\Gamma)$ as well as the boundary conditions and the equation are allowed to be nonlinear;

(ii) secondly, the assumptions imposed on g(s) and $f_0(s)$ are much weaker than the ones considered before. Indeed, the function g(s) is not required to satisfy any growth condition at the origin (compare with (1.2)) and $f_0(s)$ is not subject to an "artificial" structural condition as in (1.3).

It should be noted that the presence of a nonlinear term $f_1(y|_\Gamma)$ on the boundary and the lack of assumptions (1.2) and (1.3) contribute to some major technical difficulties at the level of proving decay rates as well as the existence theory. Indeed, the usual Liapunov type of approach as used in [Z.1], [L.2], [K-Z] does not provide, in our case, adequate results. The reason for this is that the problem does not possess "enough" structure in order to prove the desired differential inequalities. To force this "structure", even in the special when $f_1(y) = 0$, the additional assumptions like (1.3) (resp. (1.2)) were used in [K-Z] (resp [Z.1]). In order to cope with these difficulties, we shall use a different (than Liapunov function) approach. In fact, this approach is based on certain integral estimates for the energy functional combined with nonlinear compactness argument; it is motivated by the multipliers techniques used for the linear problem in [T.1]. Another difficulty of the problem mentioned above is related to the existence and uniqueness of the solutions. In fact, in general we do not have uniqueness, (see Theorem 1). Thus, in order to prove the existence of the solutions to (1.1), we first consider an approximating problem (when $g(x) - \varepsilon x$ is still increasing for some $\varepsilon > 0$, and f_0, f_1 are Lipschite continuous, and the existence and uniqueness for this approximating problem follows from the nonlinear semigroup theory. Passing through the limit on the approximating problems yields the desired existence result for the problem (1.1). Since the solutions do not depend continuously on the initial data, regular solutions (needed for p.d.e estimates) cannot be asserted by restricting as usual to smooth initial data (and then extending the result by density). Therefore, a different approximating argument (see Lemma 2.2) is needed to assert the existence of "smooth" approximations of the solutions.

Below we state our main results.

Theorem 1. (i) Assume (H-1) - (H-3). Then, for each $(y_0, y_1) \in H^1(\Omega) \times L_2(\Omega)$, problem (1.1) has at least one solution $y \in C(0, \infty; H^1(\Omega)) \cap C^1(0, \infty; L_2(\Omega))$ and such that

$$(1.4) \quad y_t \in L_2(0, \infty, \Gamma), \quad \frac{\partial y}{\partial \gamma} \in L_2(0, \infty; \Gamma)).$$

Remark. Actually, Theorem 1 can be proved under somewhat more general hypotheses assumed on the functions $f_i(s)$. Indeed, modification of some arguments in the proof of Theorem 1 allows us to replace the conditions (H-2) - (H-3) (iii) by the requirement that f_0 (resp. f_1) are compact (as Nemycki's operators) from $H^1(\Omega)$ to $L_2(\Omega)$ (resp. $L_2(\Gamma)$).

Corollary 1. In addition to the hypothesis of Theorem 1, we assume that either $f_1 \equiv 0$ or else $[g(s_1) - g(s_2)][s_1 - s_2] \geq \alpha |s_1 - s_2|^2$, $\alpha > 0$, and f_1 is locally Lipschitz from $H^1(\Omega)$ into $L^2(\Gamma)$. Then the solution y, y_t is unique.

In order to state our stability result, we introduce some notation. Let $h(s)$ be a real valued function which is defined for $s \geq 0$, it is concave, strictly increasing, $h(0) = 0$ and it satisfies

$$(1.5) \quad h(sg(s)) \geq s^2 + g^2(s) \text{ for } s \leq 1.$$

Such a function can be always constructed by the virtue of hypothesis (H-1). Let

$$\tilde{h}(x) \equiv h\left[\frac{x}{\text{mes}\,\Sigma}\right], \quad x > 0$$

where $\Sigma = \Gamma \times (0, T)$.

Since $\tilde{h}$ is monotone increasing, for some $c \geq 0$, $c + \tilde{h}$ is invertible. Define

$$(1.6) \quad p(x) \equiv \gamma (cI + \tilde{h})^{-1}(x)$$

where a positive constant γ will be suitably selected.

Then p is a positive, continuous, strictly increasing function with $p(0) = 0$.

$$(1.7) \quad q(x) \equiv x - (I + p)^{-1}(x), \quad x > 0.$$

Since $p(x)$ is positive, increasing, so is $q(x)$. Let $E(t)$ denote the energy of the solution y, y_t, i.e.

(1.8) $E(t) = |\nabla y(t)|^2_{L_2(\Omega)} + |y_t(t)|^2_{L_2(\Omega)} + 2\int_\Gamma F_1(y|_\Gamma)\, d\Gamma_1 + 2\int_\Omega F_0(y)\, d\Omega$

where

$$F_i(s) \equiv \int_0^s f_i(t)\, dt .$$

It can be easily shown that E(t) remains bounded for the solutions in a bounded set of $H^1(\Omega) \times L_2(\Omega)$. We are ready to state our stabilization result. We shall assume the following hypothesis

(H-4) At least one of the following conditions holds:

(i) f_0 is linear;

(ii) $f_1(u)\, u \geq \varepsilon u^2$ for some $\varepsilon > 0$.

We note that condition (H-4) will be necessary only in the compactness - uniqueness argument.

Theorem 2. Assume hypotheses (H-1) - (H-4). Let (y, y_t) be a solution to (1.1), with the properties $y \in C[0, T; H^1(\Omega)] \cap C^1[0, T; L_2(\Omega)]$; $y_t \in L_2(\Sigma)$; $\frac{\partial y}{\partial \gamma} \in L_2(\Sigma)$. Then, there exists $T_0 > 0$ such that

(1.9) $E(t) \leq S(t - T_0)(E(0))$ for $t > T_0$

where S(t) is the solution (contraction semigroup) of the differential equation:

$$\begin{cases} \frac{d}{ds} S(t) + q(S(t)) = 0 \\ S(0) = E(0) \end{cases}$$

and q(s) is given by (1.7), (1.6) with the constant γ in (1.6) depending in general (unless k = 1) on E(0).

Corollary 2. Assume that for some $\alpha, \beta > 0$,

$$\beta s^2 \geq g(s)\, s \geq \alpha |s|^{p+1} \text{ for each real s for } |s| \leq 1 \text{ for some } p \geq 1 .$$

Then

$E(t) \leq C(E(0))\, e^{-\alpha t} E(0)$ if $p = 1$;

$E(t) \leq C(E(0))\, t^{2/1-p}$ if $p > 1$.

proof: It is enough to construct a function h with the property (1.5). Indeed, we can take $h(s) = \alpha^{-\frac{2}{p-1}} s^m$, $m = \frac{2}{p+1} \leq 1$. Then $p(s) = (cI + \tilde{h})^{-1}(s)$, i.e., $p(cs + s^m) = s$ where c is a suitable constant depending on E(0). Recall also that:

$$q(s) = s - (I + p)^{-1} (s)$$

Since asymptotically (for s small) we have

$$p(s) \sim \alpha\, s^{\frac{1}{m}} \quad \text{and therefore } q(s) \sim \alpha\, s^{\frac{1}{m}}$$

we obtain:

$$\text{and } s(t)x = \begin{cases} c_1 (t + c_2 x^{\frac{1-p}{2}})^{\frac{2}{1-p}} & \text{if } p > 1 \\ e^{-\alpha t} x & \text{if } p = 1 \end{cases}$$

where c_1, c_2 depend only on α, p. The conclusion now follows from Theorem 2. ■

Remarks

1. The result of [Z.1] is a very special case of Corollary 2. Indeed, [Z.1] treats only the *linear* equation with monotone feedback ($f_1 \equiv 0$). Moreover, the geometric conditions of "star shaped" type are assumed in [Z.1].
2. Theorem 2 may be easily extended to the case when the function f_1 is not Lipschitz at "0". However, in this case, our proof does not provide a computable rate of energy decay.

2. PROOF OF THEOREM 1

The proof of Theorem 1 follows through the following two step procedure: we shall first consider an auxiliary approximating problem for which the existence of the unique solution can be established by nonlinear semigroup theory. In the second step we obtain the solutions of the problem (1.1) as the limits of the approximating equation. The proof of Theorem 1 is technical and we refer the reader to [L-T]. In the process of proving Theorem 1 we obtain the following result which will be needed in the sequel:

Lemma 2.1. Let u be a solution to (1.1) given by Theorem 1. Then the following energy identity holds:

$$(2.1) \quad E(t) + \int_0^t \int_\Gamma g(y_t(s))\, y_t(s)\, d\Gamma\, d s = E(0)\,.$$

3. PROOF OF THEOREM 2

It follows through a sequence of auxiliary results.

Proposition 3.1

Assume the hypothesis (H-4) is fulfilled. Let $u \in C(0,T; H^1(\Omega)) \cap C^1[0,T; L_2(\Omega)]$ be such that $\frac{\partial u}{\partial \nu} \in L^2(\Sigma)$. Then

$$(3.1) \quad (1 - \varepsilon\,[E(0)]^{k-1}) \int_0^T E(t)\, dt \le C(E(0)) \left(\int_\Sigma [g^2(y_t) + y_t^2(t)]\, d\Sigma + C(\varepsilon) \left[\int_\Sigma y^2\, d\Sigma + \int_Q y^2\, dQ\right] + E(T)\right).$$

The proof of Proposition 3.1 employs multiplier methods combined with Sobolev Imbeddings and microlocal analysis estimates. Technical details are referred to [L-T].

The next step is to absorb the lower order terms on the RHS of (3.1). This will be accomplished by applying a suitable nonlinear version of a compactness argument.

Lemma 3.1. Assume the hypotheses (H-1)-(H-6). Let (y, y_t) be a solution to (1.1). Then for $T \geq T_0$ (where T_0 is sufficiently large) we have:

$$(3.2) \qquad \int_{\Sigma} y^2 \, d\Sigma + \int_{Q} y^2 \, dQ \leq C(E(0)) \left[\int_{\Sigma} (y_t^2 + g^2(y_t)) \, d\Sigma \right]$$

proof: We shall argue by contradiction. Let $y_n(t)$ be a sequence of solutions to (1.1) such that

$$(3.3) \qquad \lim_{n \to \infty} \frac{\int_{\Sigma} y_n^2 \, d\Sigma + \int_{Q} y_n^2 \, dQ}{\int_{\Sigma} y_{nt}^2 \, d\Sigma + \int_{\Sigma} g^2(y_{nt}) \, d\Sigma} = \infty$$

while the energy of the initial data $(y_n(0), y_{nt}(0))$, denoted by $E_n(0)$, remains uniformly (in n) bounded by, say, $E_n(0) \leq M$.

Since $E_n(0) \leq M$, by the energy identity (2.1) we have $E_n(t) \leq M$. Hence,

$$(3.4) \qquad \begin{cases} y_n \to y \ \text{weakly} & \text{in } H^1(Q) \\ y_n \to y \ \text{strongly} & \text{in } L_2(Q). \end{cases}$$

Case A. Assume that $y \neq 0$. Then, we notice first that Aubin's type compactness result (see [S-1]) gives

$$y_n \to y \ \text{strongly in } L_\infty(0,T; H^{1-\varepsilon}(\Omega)), \ \text{for } \varepsilon > 0.$$

Then, by hypotheses (H-2) and (H-3), it follows easily that

$$f_0(y_n) \to f_0(y) \ \text{strongly in } L_\infty(0,T; L_2(\Omega))$$

$$f_1(y_n) \to f_1(y) \ \text{strongly in } L_\infty(0,T; L_2(\Gamma)$$

Also by (3.3) y_{nt}, $g(y_{nt}) \to 0$ in $L_2(\Sigma)$. Then, passing to the limit in the equation, we get for y:

$$(3.5) \qquad \begin{cases} y_{tt} - \Delta y = - f_0(y) \\ \dfrac{\partial y}{\partial \gamma} = - f_1(y), \quad y_t = 0 \quad \text{on } \Gamma \end{cases}$$

and for $y_t = v$

$$\begin{cases} v_{tt} - \Delta v = - f_0' (y)\, v \\ \dfrac{\partial v}{\partial \gamma} = v = 0 \qquad \text{on } \Gamma \end{cases}$$

where the equalities hold in the sense of distributions.

Now, by using hypothesis (H-4) and the uniqueness result of [R.1] we obtain

$$v = y_t = 0 .$$

Hence, we have proved that $y_t = 0$. Then, by (3.5) we get for y the elliptic equation:

$$\begin{cases} \Delta y = f_0(y) , \\ \dfrac{\partial y}{\partial \nu} = - f_1(y) \quad \text{on } \Gamma . \end{cases}$$

Multiplying by y and integrating by parts, we get:

(3.6) $\displaystyle\int_\Omega (|\nabla y|^2 + y f_0(y))\, d\Omega + \int_\Gamma y\, f_1(y)\, d\Gamma = 0 .$

Then, we get $\nabla y = 0$. By (H-4) (ii) and (3.6) we also obtain that either $y = 0$ in Ω or $y = 0$ in Γ, therefore $y = 0$. This contradicts our assumption that $y \neq 0$.

Case B. Assume that $y = 0$.

Denote $C_n = (|y_n|^2_{L_2(\Sigma)} + |y_n|^2_{L_2(Q)})^{1/2}$, $\tilde{y}_n = \dfrac{1}{C_n} \cdot y_n$. Clearly,

(3.7) $|\tilde{y}_n|^2_{L_2(\Sigma)} + |\tilde{y}_n|_{L_2(Q)} = 1$

Also, because $y = 0$ we get:

$$C_n \to 0 \ \text{ as } n \to \infty .$$

By (3.3) we get:

(3.8) $\tilde{y}_{nt} \to 0 \quad \text{in } L_2(\Sigma) .$

On the other hand, from (3.1) and (2.1) we obtain, after taking ε suitable small:

$$[T - C(E(0))]\, E(0) \le C(E(0)) \left\{ \int_\Sigma [g^2(y_t) + y_t^2 + y^2]\, d\Sigma + \int_Q y^2\, dQ \right\},$$

recalling Theorem 2.1, and taking T suitable large yields

(3.9) $\displaystyle E(t) \le E(0) \le C_T(E(0)) \left\{ \int_\Sigma [g^2(y_t) + y_t^2 + y^2]\, d\Sigma + \int_Q y^2\, dQ \right\}.$

Dividing both sides of (3.9) (applied to the solution y_n) by $|y_n|^2_{L_2(\Sigma)} + |y_n|^2_{L_2(Q)}$ and invoking (3.3) yields

(3.10) $|\nabla \tilde{y}_n(t)|^2_{L_2(\Omega)} + |\tilde{y}_{n,t}(t)|^2_{L_2(\Omega)} \le C_T\,(E(0))\,;\quad 0 \le t \le T\,.$

By (H.4), we obtain in the L.H.S. of (3.8) the term $\varepsilon\, |\tilde{y}_n|^2_{L_2(\Omega)}$ or $\varepsilon\, |\tilde{y}_n|^2_{L_2(\Gamma)}$, therefore we still get boundedness of $\tilde{y}_n$ in $H^1(Q)$. Thus, on a subsequence we have:

$$
(3.11)\qquad \begin{cases} \tilde{y}_n \to \tilde{y} \text{ weakly in } H^1(Q) \\ \tilde{y}_n \to \tilde{y} \text{ strongly in } L_2(Q) \\ \tilde{y}|_\Gamma \to \tilde{y}|_\Gamma \text{ strongly in } L_2(\Sigma)\,. \end{cases}
$$

Moreover, $\tilde{y}_n$ satisfies the equation

$$
(3.12)\qquad \begin{cases} \tilde{y}_{ntt} = \Delta \tilde{y}_n - \dfrac{f_0(y_n)}{C_n} & \text{in } Q \\[2ex] \dfrac{\partial}{\partial \gamma}\,\tilde{y}_n = \dfrac{-g(\tilde{y}_{nt}|_\Gamma) - f_1(\tilde{y}_n)}{C_n} & \text{on } \Sigma\,. \end{cases}
$$

In order to pass to the limit in (3.12) we need to determine the limits of nonlinear terms. This will be done in:

Proposition 3.4

$$(3.13)\qquad \frac{g(\tilde{y}_n)}{C_n} \to 0 \text{ in } L^2(\Sigma) \text{ as } n \to \infty,$$

$$(3.14)\qquad \frac{f_0(y_n)}{C_n} \to f_0'(0)\,\tilde{y} \text{ in } L^2(Q) \text{ as } n \to \infty,$$

$$(3.15)\qquad \frac{f_1(y_n)}{C_n} \to f_1'(0)\,\tilde{y} \text{ in } L^2(\Sigma) \text{ as } n \to \infty.$$

Proof: Follows from Sobolev's Imbeddings (see [L-T]).

Applying the result of Proposition 3.4 to the equation (3.12) and passing through the limit as $n \to \infty$ gives

$$
(3.16)\qquad \begin{cases} \tilde{y}_{tt} = \Delta \tilde{y} - f_0'(0)\,\tilde{y} \text{ in } Q\,, \\[1ex] \dfrac{\partial}{\partial \gamma}\,\tilde{y} = -f_1'(0)\,\tilde{y} \quad \text{on } \Sigma\,, \end{cases}
$$

$\tilde{y}_t = 0 \quad \text{on } \Sigma\,.$

Thus $v = \tilde{y}_t \in C\,[0,T;\, L_2(\Omega)]$ satisfies

(3.17) $$\begin{cases} v_{tt} = \Delta v - f_0'(0)\, v, \\ \dfrac{\partial}{\partial \gamma} v = 0 \quad \text{on } \Sigma. \end{cases}$$

We are in a position to apply standard uniqueness results for the wave equation, which yields

(3.18) $v = \tilde{y}_t \equiv 0$.

Returning to (3.16) and exploiting (3.18) we obtain

(3.19) $$\begin{cases} \Delta \tilde{y} - f_0'(0)\, \tilde{y} = 0 & \text{in } \Omega; \\ \dfrac{\partial}{\partial \gamma} \tilde{y} = -f_1'(0)\, \tilde{y} & \text{on } \Sigma. \end{cases}$$

As in case A, multiplying the first equation in (3.19) by $\tilde{y}$ we get $\tilde{y} = 0$, which contradicts (3.7). ∎

Using the result of Lemma 3.1 in the inequality (3.1) with ε suitably small, and recalling once again (3.1), (see [L-T]), we obtain the following

Proposition 3.5. Let $T > 0$ be sufficiently large. Then

$$E(T) \leq C_T(E(0)) \int_\Sigma (y_t^2(t) + g^2(y_t))\, d\Sigma.$$

Our final estimate is

Lemma 3.2. With p(s) defined by (1.6) and $T > 0$ sufficiently large we have

$$p(E(T)) + E(T) \leq E(0).$$

proof: see [L-T].

To conclude the proof of our theorem we need

Lemma 3.3. Let p be a positive, increasing function such that $p(0) = 0$. Since p is increasing, we can define a function q such that $q(x) = x - (I + p)^{-1}(x)$. Notice also that q is an increasing function. Consider the following sequence of positive numbers.

(3.20) $s_{m+1} + p(s_{m+1}) \leq s_m$.

Then $s_m \leq S(m)$ where S(t) is a solution of a differential equation

(3.21) $$\begin{cases} \dfrac{d}{dt} S(t) + q(S(t)) = 0, \\ S(0) = s_0. \end{cases}$$

Moreover, if $p(x) > 0$ for $x > 0$ then $\lim_{t \to \infty} S(t) = 0$.

proof: see [L-T]

Final step in the proof of Theorem 2.

Applying the results of Lemma 3.2 we obtain

(3.22) $E(m(T+1)) + p_m(E(m(T+1))) \leq E(mT)$

for $m = 0, 1, \ldots$ where p_m is a function defined by (1.6) with a constant γ depending (increasing) on E (mT). Thus, by virtue of the energy inequality (3.6), we infer

(3.23) $p_m \geq p_0$,

and p_m in (3.22) can be replaced by p_0. Thus, we are in a position to apply the result of Lemma 3.3 with

$$s_m \equiv E(mT), \quad s_0 \equiv E(0).$$

This yields

$$E(mT) \leq s(mT), \quad m = 0, 1, 2 \ldots$$

Setting $t = mT + \tau$ and recalling the evolution property gives

$$E(t) \leq E(mT) \leq S(t - \tau) \leq S(t-T) \quad \text{for } t > T,$$

which completes the proof of Theorem 2. ■

REFERENCES

[A-B-C] B. d' Andrea-Novel, F. Boustany, F. Conrad, Control of an overhead crane: stabilization of flexibilities. *Proceedings IFIP conference on Boundary Control and Boundary Variations* Sophie-Antipolis Oct. 1990

[B.1] V. Barbu, Nonlinear Semigroups and Differential Equations in Banach Spaces, Noordhoff, 1976.

[B.2] H. Brezis, Problemes unilateraux, *J. Math. Pures et Appl.* 51, (1972) pp 1-168.

[Ch.1] G. Chen, Energy decay estimates and exact boundary value controllability for the wave equation in a bounded domain, *J. Math. Pures Appl.* 58 (1979) pp. 248-274.

[C-L-M] F. Conrad, J. Lebland, J. P. Marmoret, Stabilization of second order evolution equations by unbounded nonlinear feedback, *Proceedings of IFAC Conference* Perpignan 1989, pp 111-116.

[K.Z] V. Komornik - E. Zuazua, Stabilization frontiere de l'equation des ondes. Une methode directe. C. R. Acad. Sci. Paris Ser I Math. 305 (1987) pp 605-608.

[L.1] J. Lagnese, Decay of solutions of the wave equation in a bounded region with boundary dissipation, *J. Diff. Eq.* 50, (1983) pp. 163-182.

[L.2] J. Lagnese, Boundary Stabilization of Thin Plates, *SIAM Studies in Applied*

Mathematics, 1990.

[L-T] I. Lasiecka, D. Tataru, Uniform Boundary Stabilization of Semilinear wave equation with nonlinear boundary conditions, *Journal of Differential and Integral Equations,* to appear.

[R.1] A. Ruiz, Unique confirmation for weak solutions of the wave equation plus a potential. Preprint, Universided Autonoma Madrid, (1988).

[S.1] J. Simon, Compact sets in the space L^P (0T; B), *Annali di Mat. Pura et Applicate,* IV Vol. CXLVI (1987) pp. 65-96.

[T.1] R. Triggiani, Wave equation on a bounded domain with boundary dissipation: an operator approach, *J. Math. Anal. Appl.,* 137 (1989) pp. 438-461

[Z.1] E. Zuazua, Uniform stabilization of the wave equation by nonlinear boundary feedback, *SIAM J. on Control and Optimization,* Vol. 28, (1990) pp. 466-478.

Sharp Trace Estimates of Solutions to Kirchhoff and Euler–Bernoulli Equations

I. LASIECKA Department of Applied Mathematics, University of Virginia, Charlottesville, VA 22903, U.S.A.

R. TRIGGIANI Department of Applied Mathematics, University of Virginia, Charlottesville, VA 22903, U.S.A.

1. INTRODUCTION: A WAVE PROBLEM

Let Ω be an open bounded domain in R^n with smooth boundary Γ. Let $-A(x,\partial)$ be a second-order elliptic operator of order two with smooth, space variable coefficients. Consider the second-order hyperbolic equation in $w(t,x)$:

$$w_{tt} = A(x,\partial)w \quad \text{in } (0,T]\times\Omega = Q_T . \tag{1.1}$$

The following estimate on the tangential trace $\frac{\partial}{\partial \tau} w$ of (smooth) solutions of (1.1) was proved in [L-T.1, Lemma 7.2].

THEOREM 1.0. Let Γ_1 be a non-empty open portion of Γ, and let $0 < \alpha < T$ be arbitrary. Then, the following estimate holds true for solutions w of (1.1),

$$\int_\alpha^{T-\alpha}\int_{\Gamma_1}\left(\frac{\partial w}{\partial \tau}\right)^2 d\Gamma_1 dt \le C_{T\alpha\varepsilon}\left\{\int_0^T\int_{\Gamma_1}\left(\frac{\partial w}{\partial \nu}\right)^2 + w_t^2\right\} d\Gamma_1 dt + \|w\|^2_{H^{\frac{1}{2}+\varepsilon}(Q_T)} , \tag{1.2}$$

where ν is a unit outward vector normal to Γ_1. ∎

The proof of Theorem 1.0 is obtained by pseudo-differential operators techniques and microlocal analysis [L-T.1, Sect. 7]. Besides being of interest in itself, estimate (1.2) permits to eliminate altogether (unnatural) geometrical conditions of the type $h\cdot\nu \ge \text{const} > 0$ on Γ_1, $h(x)$ a suitable vector field, that had been made in prior literature on both uniform stabilization (e.g., [Lag.1], see also [Lag.2] for a case "const = 0") and exact controllability (e.g., [L-T.2]) of wave equations with $L_2(0,T;L_2(\Gamma_1))$-control in the Neumann boundary conditions, in the space $H^1(\Omega)\times L_2(\Omega)$. This was shown in [L-T.1]. These assumptions were caused precisely by a lack of information on the tangential derivative $\frac{\partial w}{\partial \tau}$ such as is given by Theorem 1.0.

Motivated by the wave equation case, the aim of this paper is to provide similar trace estimates for Kirchhoff-type, or Euler-Bernoulli-type, of equations, which are of fourth-order in space, see Section 2. The proofs are again by pseudo-differential and micro-local analysis and follow the conceptual strategy of [L-T.1] used to prove Theorem 1.0 for second-order hyperbolic equations; however, additional technical difficulties need to be overcome now; see Orientation below (4.1.3).

The implications of the results given here in eliminating geometrical conditions, such as those described above, also in the theories of exact controllability and uniform stabilization of plate-like equations are examined in Section 3.

2. KIRCHHOFF AND EULER-BERNOULLI PROBLEMS

2.1. The cases $\gamma > 0$: Kirchhoff problems

Canonical model #1

Let Ω be an open bounded domain in R^n with smooth boundary $\Gamma = \Gamma_0 \cup \Gamma_1$, Γ_i open, Γ_0 possibly empty while Γ_1 non-empty and $\bar{\Gamma}_0 \cap \bar{\Gamma}_1 = \emptyset$ [see Remark 2.4 below on this point]. We consider the following mixed problem in the unknown $w(t,x)$:

$$\begin{cases} w_{tt} - \gamma \Delta w_{tt} + \Delta^2 w = f & \text{in } (0,T]\times\Omega \equiv Q_T; & (2.1a) \\ w \equiv \frac{\partial w}{\partial \nu} \equiv 0 & \text{in } (0,T]\times\Gamma_0 \equiv \Sigma_{0T}; & (2.1b) \\ \Delta w + \beta_1 w = g_1 & \text{in } (0,T]\times\Gamma_1 \equiv \Sigma_{1T}; & (2.1c) \\ \frac{\partial \Delta w}{\partial \nu} + \beta_2 w - \gamma \frac{\partial w_{tt}}{\partial \nu} = g_2 & \text{in } \Sigma_{1T}; & (2.1d) \end{cases}$$

with constant $\gamma > 0$ and boundary operators β_1 and β_2 of second and third order, respectively,

$$\beta_1 = \sum_{i,j=1}^{n-1} k_{ij} \frac{\partial^2}{\partial \tau_i \partial \tau_j} + \frac{\partial}{\partial \nu} \sum_{i=1}^{n-1} k_i \frac{\partial}{\partial \tau_i} + \ell.o.t.; \qquad (2.2)$$

$$\beta_2 = \sum_{i,j,\ell=1} c_{ij\ell} \frac{\partial^3}{\partial \tau_i \partial \tau_j \partial \tau_\ell} + \frac{\partial}{\partial \nu} \sum_{i=1}^{n-1} c_{ij} \frac{\partial^2}{\partial \tau_i \partial \tau_j} + \frac{\partial^2}{\partial \nu^2} \sum_{i=1}^{n-1} c_i \frac{\partial}{\partial \tau_i} + \ell.o.t. \qquad (2.3)$$

Here, ν is the unit outward normal vector to Γ and $\tau = [\tau_1, \cdots, \tau_{n-1}]$ is a unit tangential vector to Γ, with smooth, space variable dependent functions $k_{ij} \cdots c_{ij\ell} \cdots$ on Γ_1. Also, ℓ.o.t. denotes lower-order terms.

REMARK 2.1. When n = 2, we obtain the classical model of Kirchhoff for a thin, isotropic, homogeneous plate in the deflection w(t,x), where β_1 and β_2 specialize then to

$$\beta_1 = -(1-\mu) \left[\frac{\partial^2}{\partial \tau^2} + k \frac{\partial}{\partial \nu}\right];$$

$$\beta_2 = (1-\mu) \left[\frac{\partial}{\partial \nu} \frac{\partial^2}{\partial \tau^2} - \frac{\partial}{\partial \tau}\left(k \frac{\partial}{\partial \tau}\right)\right]; \tag{2.4}$$

with $0 < \mu < 1$ and curvature k [Lag.1]. ∎

Our main result is an estimate on the second order traces of w.

THEOREM 2.1. Let $0 < \alpha < T$ and $\varepsilon > 0$ be arbitrary. Let $s_0 < ½$. Then the following inequality holds true for solutions w of the Kirchhoff problem (2.1) with $\gamma > 0$:

$$\int_\alpha^{T-\alpha} \int_{\Gamma_1} \left\{\left(\frac{\partial^2 w}{\partial \tau^2}\right)^2 + \left(\frac{\partial^2 w}{\partial \nu^2}\right)^2 + \left(\frac{\partial^2 w}{\partial \tau \partial \nu}\right)^2\right\} d\Gamma_1 dt$$

$$\le c_{T,\alpha,\varepsilon,\gamma}\left\{\|f\|^2_{H^{-s_0}(Q_T)} + \|g_1\|^2_{L_2(\Sigma_{1T})} + \|g_2\|^2_{H^{-1}(\Sigma_{1T})}\right.$$

$$\left. + \| \, |\nabla w_t| \, \|^2_{L_2(\Sigma_{1T})} + \|w\|^2_{L_2(0,T;H^{3/2+\varepsilon}(\Omega))} + \|w_t\|^2_{L_2(\Sigma_{1T})}\right\}. \tag{2.5}$$

where ∇ denotes the gradient in all variables, so that

$$\| \, |\nabla w_t| \, \|^2 = \| \, |\nabla_{tang.} w_t| \, \|^2 + \left\|\frac{\partial w_t}{\partial \nu}\right\|^2. \quad ∎$$

REMARK 2.3. The proof by pseudo-differential methods of Section 4 will also allow to have (i) γ smoothly depending on the space variable on Γ_1: $0 < \gamma_0 \le \gamma(\cdot) \le \gamma_M < \infty$, and (ii) $(-\Delta)$ in (2.1) replaced by a general uniformly elliptic operator of order 2. ∎

REMARK 2.4. The assumption $\overline{\Gamma}_0 \cap \overline{\Gamma}_1 = \emptyset$ is made for the sole purpose of having solutions w with the degree of smoothness required by the right hand side of (2.5). ∎

Canonical model #2

We next consider the problem

$$\begin{cases} w_{tt} - \gamma\Delta w_{tt} + \Delta^2 w = f & \text{in } Q_T; & (2.6a) \\ w|_\Sigma \equiv g_0 & \text{in } \Sigma_T; & (2.6b) \\ \Delta w + \beta_1 w = g_1 & \text{in } \Sigma_T. & (2.6c) \end{cases}$$

The proof of Theorem 2.1 essentially contains the proof of the following

THEOREM 2.2. Let $0 < \alpha < T$ and $\varepsilon > 0$ be arbitrary. Let $s_0 < \frac{1}{2}$. Then the following inequality holds true for the solutions w of the Kirchhoff problem (2.6) with $\gamma > 0$:

$$\int_\alpha^{T-\alpha} \int_\Gamma \left\{ \left[\frac{\partial^2 w}{\partial \tau^2}\right]^2 + \left[\frac{\partial^2 w}{\partial \nu^2}\right]^2 + \left[\frac{\partial^2 w}{\partial \tau \partial \nu}\right]^2 \right\} d\Gamma\, dt$$

$$\leq C_{T,\alpha,\varepsilon,\gamma} \left\{ \|f\|^2_{H^{-s_0}(Q_T)} + \|g_0\|^2_{L_2(0,T;H^2(\Gamma))} + \|g_1\|^2_{L_2(\Sigma_T)} \right.$$

$$\left. + \| \, |\nabla w_t| \, \|^2_{L_2(\Sigma_T)} + \|w\|^2_{L_2(0,T;H^{3/2+\varepsilon}(\Omega))} + \|w_t\|^2_{L_2(\Sigma_T)} \right\}. \blacksquare \qquad (2.7)$$

2.2. The cases $\gamma = 0$: Euler-Bernoulli problems

Canonical model #1

We return to problem (2.1), and we set $\gamma = 0$ in Eqns. (2.1a) and (2.1d).

THEOREM 2.3. Let $0 < \alpha < T$ and $\varepsilon > 0$ be arbitrary. Let $s_0 < \frac{1}{2}$. Then the following inequality holds true for the solutions w of the Euler-Bernoulli problem obtained from (2.1) by setting $\gamma = 0$ in Eqns. (2.1a) and (2.1d):

$$\int_\alpha^{T-\alpha} \int_{\Gamma_1} \left\{ \left[\frac{\partial^2 w}{\partial \tau^2}\right]^2 + \left[\frac{\partial^2 w}{\partial \nu^2}\right]^2 + \left[\frac{\partial^2 w}{\partial \tau \partial \nu}\right]^2 \right\} d\Gamma_1 dt$$

$$\leq C_{T,\alpha,\varepsilon} \left\{ \|f\|^2_{L_2(0,T;H^{-s_0}(\Omega))} + \|g_1\|^2_{L_2(\Sigma_{1T})} + \|g_2\|^2_{L_2(0,T;H^{-1}(\Gamma_1))} \right.$$

$$\left. + \|w\|^2_{L_2(0,T;H^{3/2+\varepsilon}(\Omega))} + \|w_t\|^2_{L_2(\Sigma_{1T})} + \left\|\frac{\partial}{\partial \nu} w_t\right\|^2_{L_2(0,T;H^{-1}(\Gamma))} \right\}. \blacksquare \qquad (2.8)$$

The proof will be given in Section 5.

Canonical model #2

We return to problem (2.6) and set $\gamma = 0$ in Eqn. (2.6a).

THEOREM 2.4. Let $0 < \alpha < T$ and $\varepsilon > 0$ be arbitrary. Let $s_0 < \frac{1}{2}$. Then the following inequality holds true for the solutions w of the Euler–Bernoulli problem obtained from (2.6a) by setting $\gamma = 0$ in Eqn. (2.6a):

$$\int_\alpha^{T-\alpha} \int_\Gamma \left\{ \left(\frac{\partial^2 w}{\partial \tau^2}\right)^2 + \left(\frac{\partial^2 w}{\partial \nu^2}\right)^2 + \left(\frac{\partial^2 w}{\partial \tau \partial \nu}\right)^2 \right\} d\Gamma\, dt$$

$$\leq C_{T,\alpha,\varepsilon} \left\{ \|f\|^2_{L_2(0,T;H^{-s_0}(\Omega))} + \|g_0\|^2_{L_2(0,T;H^2(\Gamma))} + \|g_0\|^2_{H^1(0,T;L_2(\Gamma))} \right.$$

$$+ \|g_1\|^2_{L_2(\Sigma_T)} + \|\frac{\partial}{\partial \nu} w_t\|^2_{L_2(\Sigma_T)} + \|w\|^2_{L_2(0,T;H^{3/2+\varepsilon}(\Omega))}$$

$$\left. + \|w_t\|^2_{L_2(\Sigma_T)} \right\}. \quad \blacksquare \qquad (2.9)$$

3. IMPLICATIONS OF TRACE THEOREMS: ELIMINATION OF GEOMETRICAL CONDITIONS IN THE THEORIES OF EXACT CONTROLLABILITY AND UNIFORM STABILIZATION FOR PLATE EQUATIONS

In this section--following the case of the wave equation described in the introduction--we shall show that the trace estimates of Section 2 permit likewise to eliminate (unnatural) geometrical conditions on the triplet $\{\Omega, \Gamma_0, \Gamma_1\}$ that had been made in exact controllability and uniform stabilization results of the literature for plate-like equations. We shall explicitly treat the more demanding uniform stabilization problem whose solution then implies a corresponding exact controllability result.

3.1. A two-dimensional Kirchhoff plate model: $\gamma > 0$

We return to the Kirchhoff problem (2.1) in the two-dimensional case $\dim \Omega = n = 2$, where now the operators β_1 and β_2 are given by Eqn. (2.4). In addition, we now take the control functions g_1 and g_2 in (2.1c) and (2.1d) to be given in feedback form by

$$g_1 = -\frac{\partial w_t}{\partial \nu} \qquad \text{in } (0,\infty)\times\Gamma_1; \qquad (3.1a)$$

$$g_2 = w_t - \frac{\partial}{\partial \tau}\left(\frac{\partial}{\partial \tau} w_t\right) + k_0 w \qquad \text{in } (0,\infty)\times\Gamma_1. \qquad (3.1b)$$

Here and hereafter, the constant k 0 is taken to be: (i) $k_0 \geq 0$ if Γ_0 is non-empty: $\Gamma_0 \neq \emptyset$; and (ii) $k_0 > 0$ if Γ_0 is empty: $\Gamma_0 = \emptyset$.

With the above feedback model (2.1a), (2.1b), (3.1a), (3.1b), we associate the energy function

$$E_\gamma(t) \equiv a(w,w) + \int_\Omega [w_t^2 + \gamma^2 |\nabla w_t|^2] d\Omega + k_0 \int_{\Gamma_1} w^2 \, d\Gamma_1; \qquad (3.2a)$$

$$a(w,w) = \int_\Omega [|\Delta w|^2 + (1-\mu)(2w_{xy}^2 - w_{xx}^2 - w_{yy}^2)] d\Omega. \qquad (3.2b)$$

We recall that the term: $a(w,w) + k_0 \int_{\Gamma_1} w^2 d\Gamma_1$ is equivalent to the $H^2(\Omega)$-norm for w. We can now state our uniform stabilization result.

THEOREM 3.1. Consider problem (2.1), along with (2.4) and (3.1).

Case (i): Let $\Gamma_0 \neq \emptyset$ and $k_0 \geq 0$. Assume that there exists a point $x_0 \in R^2$ such that

$$(x - x_0) \cdot \nu \leq 0 \quad \text{on } \Gamma_0, \qquad (3.3)$$

where the unit normal ν is outward.

Case (ii): Let $\Gamma_0 = \emptyset$, $k_0 > 0$. Then, in either case, there exist constants C, $\omega > 0$ such that the energy $E_\gamma(t)$ defined in (3.2) satisfies

$$E_\gamma(t) \leq C\, e^{-\omega t} E_\gamma(0). \quad \blacksquare \qquad (3.4)$$

REMARK 3.1. A similar two-dimensional Kirchhoff model was considered in [Lag.1, Chapt. 4] in the case $\Gamma_0 \neq \emptyset$ and $k_0 = 0$. Then [Lag.1, Theorem 5.1, p. 88] gives the desired energy decay (3.4), provided, however, that in addition to (3.3), the following geometric condition on Γ_1 is assumed for the same feedbacks (3.1a) and (3.2b): there exists a constant $\rho > 0$ such that

$$(x - x_0) \cdot \nu \geq \rho > 0 \quad \text{on } \Gamma_1 \qquad (3.5)$$

[or, as a variant, the constant ρ may be taken equal to zero, $\rho = 0$, provided that one takes the following modification of the feedbacks in (3.1a), (3.1b):

$$g_1 = -[(x-x_0)\cdot\nu]\,\frac{\partial w_t}{\partial\nu} \qquad \text{in } (0,\infty)\times\Gamma_1;$$

$$g_2 = [(x-x_0)\cdot\nu]w_t - \frac{\partial}{\partial\tau}\left[[(x-x_0)\cdot\nu]\,\frac{\partial w_t}{\partial\nu}\right] \qquad \text{in } (0,\infty)\times\Gamma_1.$$

Thus, the main contribution of our preceding Theorem 3.1 over Theorem 5.1 in [Lag.1, p. 88] is twofold: (i) our Theorem 3.1 dispenses altogether with geometric conditions such as (3.5) on the controlled part of the boundary Γ_1; (ii) our Theorem 3.1 allows Γ_0 to be empty.

Proof of Theorem 3.1. (Sketch)

Step 1. Using multiplier techniques as in [Lag. 1, pp. 81-84], one obtains the following estimate for T sufficiently large (as to absorb $E_\gamma(0)$) and any $0 < \alpha < T$:

$$\int_\alpha^{T-\alpha} E_\gamma(t)dt \le C_T\left\{\int_\alpha^{T-\alpha}\int_{\Gamma_1}(w_t^2+|\nabla w_t|^2)d\Sigma_{1\alpha}\right.$$
$$\left.+\int_\alpha^{T-\alpha}\int_{\Gamma_1}\left[\left|\frac{\partial^2 w}{\partial\tau}\right|^2+\left|\frac{\partial^2 w}{\partial\nu^2}\right|^2+\left|\frac{\partial^2 w}{\partial\nu\partial\tau}\right|^2\right]d\Sigma_{1\alpha}+\int_\alpha^{T-\alpha}\int_{\Gamma_1} w^2\,d\Sigma_{1\alpha}\right\}, \tag{3.6}$$

where $\Sigma_{1\alpha} = (\alpha, T-\alpha)\times\Gamma_1$.

Step 2. To eliminate the boundary terms in the second integral on the right hand side of (3.6), we now apply Theorem 2.1 with $f \equiv 0$ and with g_1 and g_2 given by (3.1a), (3.1b). As a result, we readily obtain via (2.5):

$$\int_\alpha^{T-\alpha}\int_{\Gamma_1}\left[\left|\frac{\partial^2 w}{\partial\tau}\right|^2+\left|\frac{\partial^2 w}{\partial\nu^2}\right|^2+\left|\frac{\partial^2 w}{\partial\tau\partial\nu}\right|^2\right]d\Sigma_{1\alpha}$$
$$\le \text{Const}_{T,\alpha,\varepsilon,\gamma}\left\{\int_{\Sigma_1}[|\nabla w_t|^2+w_t^2]\,d\Sigma_1+\|w\|^2_{L_2(0,T;H^{3/2+\varepsilon}(\Omega))}\right\}. \tag{3.7}$$

By using (3.7) in the second integral on the right side of (3.6) we obtain

$$\int_\alpha^{T-\alpha} E_\gamma(t)dt \le C_{T,\alpha,\varepsilon,\gamma}\left\{\int_0^T\int_{\Gamma_1}[w_t^2+|\nabla w_t|^2]d\Gamma_1 dt+\int_0^T\|w(t)\|^2_{H^{3/2+\varepsilon}(\Omega)}dt\right\}. \tag{3.8}$$

Step 3. We recall the energy identity for the present feedback problem (e.g., Lag. 1, (4.54)]),

$$E_\gamma(t) + \int_0^t \int_{\Gamma_1} [w_t^2+|\nabla w_t|^2] \, d\Gamma_1 dt \equiv E_\gamma(0), \quad \forall\, t \geq 0. \tag{3.9}$$

Using (3.9) twice, we obtain

$$\int_0^\alpha E_\gamma(t)dt + \int_{T-\alpha}^T E_\gamma(t)dt \leq 2\alpha\, E_\gamma(0)$$

$$= 2\alpha\, E_\gamma(T)+2\alpha \int_0^T \int_{\Gamma_1} [w_t^2+|\nabla w_t|^2]d\Gamma_1\, dt. \tag{3.10}$$

Combining (3.10) with (3.8) yields the right hand inequality of

$$TE_\gamma(T) \leq \int_0^T E_\gamma(t)dt \leq C_{T,\alpha,\varepsilon,\gamma}\left\{\int_0^T \int_{\Gamma_1} [w_t^2+|\nabla w_t|^2]d\Gamma_1 dt + \int_0^T \|w(t)\|^2_{H^{3/2+\varepsilon}(\Omega)} dt\right\} + 2\alpha\, E_\gamma(T), \tag{3.11}$$

while the left-hand inequality of (3.11) uses that, from (3.9), $\dot{E}_\gamma(t) \leq 0$ and hence $E_\gamma(t)$ is decreasing. Next, we choose T such that $T-2\alpha > 0$, obtain from (3.11) an estimate for $E_\gamma(T)$ in terms of the bracket term $\{\ \ \}$, and use this estimate for $E_\gamma(T)$ on the last term of (3.11) to arrive at

$$\int_0^T E_\gamma(t)dt \leq C_{T,\alpha,\varepsilon,\gamma}\left\{\int_0^T \int_{\Gamma_1} [w_t^2+|\nabla w_t|^2]d\Gamma_1 dt+\int_0^T \|w(t)\|^2_{H^{3/2+\varepsilon}(\Omega)} dt\right\}. \tag{3.12}$$

Step 4. Using compactness-uniqueness arguments, one finally eliminates the lower-order terms on the right-hand side of inequality (3.12) to obtain the final estimate

$$TE_\gamma(T) \leq \int_0^T E_\gamma(t)dt \leq C_{T,\alpha,\varepsilon,\gamma}\left\{\int_0^T \int_{\Gamma_1} [w_t^2+|\nabla w_t|^2]d\Gamma_1 dt\right\}. \tag{3.13}$$

Step 5. Recalling identity (3.9) on the right side of (3.13) yields

$$TE_\gamma(T) \le C_{T,\alpha,\varepsilon,\gamma}[E_\gamma(0)-E_\gamma(T)],$$

from which we finally obtain

$$E_\gamma(T) \le \frac{C_T E_\gamma(0)}{T+C_T} = \rho_T E_\gamma(0), \tag{3.14}$$

with $\rho_T < 1$. A well-known semigroup result then yields the exponential decay (3.4) as usual. The proof of Theorem 3.1 is complete. ∎

3.2. A two-dimensional Euler-Bernoulli plate model: $\gamma = 0$

We return to the two-dimensional Euler-Bernoulli plate model with $n = \dim \Omega = 2$ obtained from problem (2.1) by setting $\gamma = 0$ in (2.1a) and (2.1d). Again, the boundary operators β_1 and β_2 are given by (2.4). Now, however, the control functions g_1 and g_2 in (2.1c) and (2.1d) are given in feedback form by

$$g1 = -\frac{\partial w_t}{\partial \nu}; \quad g_2 = w_t + k_0 w \quad \text{on } (0,\infty)\times\Gamma_1, \tag{3.15}$$

where the constants k_0 will be subject to the same conditions as in §3.1: $k_0 \ge 0$ if $\Gamma_0 \ne \emptyset$; and $k_0 > 0$ if $\Gamma_0 = \emptyset$. With the above feedback model (2.1a) with $\gamma = 0$, along with (3.15), we associate the energy functions (same as (3.2a) with $\gamma = 0$):

$$E_0(t) \equiv a(w,w) + \int_\Omega w_t^2 d\Omega + k_0 \int_{\Gamma_1} w^2 d\Gamma_1, \tag{3.16}$$

with $a(w,w)$ defined by (3.2b). Then we have the following uniform stabilization result.

THEOREM 3.2. Consider problem (2.1) with $\gamma = 0$ along with (2.4) and (3.15). Assume the same hypothesis of Case (i) and Case (ii) for Γ_0 and k_0 as in Theorem 3.1. Then, in either case, there exist constants C, w > 0 such that the energy $E_0(t)$ defined by (3.16) satisfies

$$E_0(t) \le C\, e^{-\omega t} E_0(0), \quad t \ge 0. \quad ∎ \tag{3.17}$$

REMARK 3.2. A similar two-dimensional Euler-Bernoulli model was considered in [Lag.3], [Lag.1, Chapt 4] in both cases $\Gamma_0 \ne \emptyset$ and $\Gamma_0 = \emptyset$.

If $\Gamma_0 \ne 0$, then [Lag.1, §6.1, p. 90] gives the energy decay (3.17) with controls

$$g_1 = 0; \quad g_2 = (x-x_0)w_t \quad \text{in } (0,\infty)\times\Gamma_1 , \tag{3.18}$$

for x_0 as in R^2, provided that, in addition to condition (3.3), the geometrical assumption (3.5) with $\rho = 0$ is also fulfilled on the controlled part Γ_1.

By contrast, our Theorem 3.1 for $\Gamma_0 \neq 0$ does not require the geometrical condition (3.5) with $\rho = 0$ but, as stated, uses also an active control $g_1 = -\frac{\partial w_t}{\partial\nu}$, see (3.15). It should be possible, however, to eliminate such control action on g_1 and take simply $g_1 = 0$ on Γ_1 by a further argument, similar to that which is carried out for Euler-Bernoulli plates with second- and third-order boundary conditions in [H.2]. This analysis consists in comparing the $L_2(\Sigma)$-norm of $\frac{\partial w_t}{\partial\nu}$ with the $L_2(\Sigma)$-norm of $\frac{\partial\Delta w}{\partial\nu}$ and is beyond the scope of the present paper.

If $\Gamma_0 = \emptyset$, then [Lag.1, Theorem 6.1, p. 96] gives the energy decay (3.17) by using a feedback control g_2 more complicated then our g_2 given by (3.15), and which moreover requires some additional restrictions on constants α and β which enter into the definition of g_1 and g_2, see [Lag.1, Theorem 6.1, p. 96].

Proof of Theorem 3.2. (Sketch)

Step 1. Using multiplier techniques as in [Lag.3], one obtains the following estimate for T sufficiently large (as to absorb E(0)) and any $0 < \alpha < T$:

$$\int_\alpha^{T-\alpha} E_0(t)dt \le C_T \left\{\int_\alpha^{T-\alpha}\int_{\Gamma_1} \left[w_t^2 + \left|\frac{\partial w_t}{\partial\nu}\right|^2\right] d\Sigma_{1\alpha} \right.$$
$$\left. + \int_\alpha^{T-\alpha}\int_{\Gamma_1} [|\Delta w|^2 + 2(1-\eta)(w_{xy}^2 - w_{xx}w_{yy})]\, d\Sigma_{1\alpha} + \int_\alpha^{T-\alpha}\int_{\Gamma_1} w^2\, d\Sigma_{1\alpha}\right\}. \tag{3.19}$$

Step 2. To eliminate the boundary terms in the second integral on the right-hand side of (3.19), we now apply Theorem 2.3 with $f \equiv 0$ and with g_1 and g_2 given by (3.15). As a result, we readily obtain via (2.8):

$$\int_\alpha^{T-\alpha}\int_{\Gamma_1}[|\Delta w|^2+2(1-\eta)(w_{xy}^2-w_{xx}w_{yy})]\,d\Sigma_{1\alpha}$$
$$\le C_{T,\alpha,\varepsilon}\left\{\int_0^T\int_{\Gamma_1}\left[w_t^2+\left|\frac{\partial w_t}{\partial\nu}\right|^2\right]d\Sigma_{1\alpha}+\|w\|^2_{L_2(0,T;H^{3/2+\varepsilon}(\Omega))}\right\}. \tag{3.20}$$

By using (3.20) into the second integral on the right of (3.19) we obtain

$$\int_\alpha^{T-\alpha}E_0(t)dt\le C_{T,\alpha,\varepsilon}\left\{\int_0^T\int_{\Gamma_1}\left[w_t^2+\left|\frac{\partial w_t}{\partial\nu}\right|^2\right]d\Sigma_{1\alpha}+\int_0^T\|w\|^2_{H^{3/2+\varepsilon}(\Omega))}dt\right\}. \tag{3.21}$$

which is the counterpart of (3.8) in the present case.

Step 3. We recall the energy identity for the present feedback problem

$$E_0(t)+\int_0^t\int_{\Gamma_1}\left[w_t^2+\left|\frac{\partial w_t}{\partial\nu}\right|^2\right]d\Gamma_1dt=E_0(0),\qquad\forall\ t\ge 0 \tag{3.22}$$

counterpart of (3.9). From here on, the proof proceeds exactly as in Steps 3 through 5 of Theorem 3.1. ∎

4. KIRCHHOFF EQUATION. PROOF OF THEOREM 2.1

4.1. Preliminaries

We return to the Kirchhoff problem (2.1a-d), which we rewrite here for convenience

$$\begin{cases} w_{tt}-\gamma\Delta w_{tt}+\Delta^2 w=f & \text{in } (0,T]\times\Omega; & (4.1.1a)\\ w\equiv\frac{\partial w}{\partial\nu}\equiv 0 & \text{in } (0,T]\times\Gamma_0; & (4.1.1b)\\ \Delta w+\beta_1 w=g_1 & \text{in } (0,T]\times\Gamma_1; & (4.1.1c)\\ \frac{\partial\Delta w}{\partial\nu}+\beta_2 w-\gamma\frac{\partial w_{tt}}{\partial\nu}=g_2 & \text{in } (0,T]\times\Gamma_1; & (4.1.1d)\end{cases}$$

with constant $\gamma>0$ and boundary operators β_1 and β_2 of second and third order, respectively, as in (2.1d), a canonical specialization thereof is given by:

$$\beta_1 = k_{11}\frac{\partial^2}{\partial\tau^2} + k_{12}\frac{\partial}{\partial\tau}\frac{\partial}{\partial\nu}; \tag{4.1.2}$$

$$\beta_2 = c_{21}\frac{\partial^3}{\partial\tau^3} + c_{22}\frac{\partial^2}{\partial\tau^2}\frac{\partial}{\partial\nu} + c_{23}\frac{\partial}{\partial\tau}\frac{\partial^2}{\partial\nu^2}; \tag{4.1.3}$$

where the k_{1j} and c_{2j} are smooth boundary functions on Γ_1 and $\partial/\partial\tau$ denotes tangential derivative.

Orientation. The proof in this case follows the strategy of the proof in [L-T.1, Section 7] for the wave problem with additional technical difficulties. In fact, now, both time-localization and dual variables location yield higher order terms ($\tilde{D}_x^3$ and $D_y\tilde{D}_x^2$ in (4.7.1)) which cannot be estimated by direct *a-priori* regularity. Instead, use of the equation and/or boundary conditions is now required for the analysis of these higher order terms. This is carried out in Sections 4.8 and 4.9 to prove the critical Propositions 4.7 and 4.11 respectively. We note at the outsets that the most critical term to estimate is Δw_t, which arises from the time commutator and which yields in particular the term a $\tilde{D}_x^2 D_t(\psi' w)$ in (4.3.9a). See Remark 4.1.

4.2. Half-space problems

Henceforth we write $\Omega = R^1_{x^+} \times R^{n-1}_y$ and $\Gamma = R^{n-1}_y = \Omega|_{x=0}$. We consider the version of problem (4.1.1a,c,d) on the half-space Ω, via partition of unity and local coordinates. Problem (4.1.1a,c,d) becomes (renaming the functions g_1 and g_2 in (4.1.1c-d) modulo a constant factor)

$$\begin{cases} Pw = f & \text{in } Q_T = (0,T]\times\Omega; & (4.2.1a)\\ B_1 w = g_1 & \text{in } \Sigma_T = (0,T]\times\Gamma; & (4.2.1b)\\ B_2 w = g_2 & \text{in } \Sigma_T; & (4.2.1c)\end{cases}$$

where (see also Sections 6 and 7 in [L-T.1]) $D_t = -\sqrt{-1}\,\frac{\partial}{\partial t}$, etc., and

$$\begin{cases} P(x,y;D_t,D_x,D_y) = -aD_t^2 - \gamma a D_t^2[\tilde{D}_x^2+\tilde{D}_y^2] + [\tilde{D}_x^2+\tilde{D}_y^2]^2; & (4.2.2)\\ \tilde{D}_x = D_x + \sum_{j=1}^{n-1} a_{nj}D_{y_j};\quad \tilde{D}_y^2 = \sum_{i,j=1}^{n-1} a_{ij}D_{y_i}D_{y_j} - \left(\sum_{j=1}^{n-1} a_{nj}D_{y_j}\right)^2; & (4.2.3)\end{cases}$$

with coefficients a, a_{ij} depending on x and y and constant outside a compact set of $R^n_{x^+y}$ and with symbols, respectively,

$$\begin{cases} \text{symb}\{P\} = p(x,y;s,\xi,\eta) = -as^2-\gamma as^2\left[\tilde{\xi}^2+\frac{d}{a^2}\right]+\left[\tilde{\xi}^2+\frac{d}{a^2}\right]^2; & (4.2.4) \\ \text{symb}\{\tilde{D}_x\} = \tilde{\xi} = \xi+\sum_{j=1}^{n-1} a_{nj}\eta_j;\ s^2 = (\sigma-ir_0)^2; & (4.2.5) \\ \text{symb}\{\tilde{D}_y^2\} = \sum_{i,j=1}^{n-1} a_{ij}\eta_i\eta_j-\left(\sum_{j=1}^{n-1} a_{nj}\eta_j\right)^2 = \frac{d(x,y;\eta)}{a^2(x,y)}. & (4.2.6) \end{cases}$$

$\tilde{D}_x|_{x=0}$ being the co-normal operator and

$$B_1 = [\tilde{D}_x^2+\tilde{D}_y^2]+\beta_1 \qquad \text{at } x = 0; \qquad (4.2.7)$$

$$B_2 = \tilde{D}_x[\tilde{D}_x^2+\tilde{D}_y^2]+\beta_2+\gamma a\tilde{D}_x D_t^2 \qquad \text{at } x = 0, \qquad (4.2.8)$$

where letting D_y be the gradient on Γ with symbol η, we may take

$$\begin{cases} \beta_1 = b_{11}D_y^2+b_{12}D_y\tilde{D}_x + \ell.o.t., & \\ b_{ij} \text{ incorporating pseudo-differential operators of order zero in } y; & (4.2.9a) \\ \text{symb}\{\beta_1\} = \mathcal{O}(|\eta|^2+|\eta|\tilde{\xi}); & (4.2.9b) \end{cases}$$

$$\beta_2 = b_{21}D_y^3+b_{22}D_y^2\tilde{D}_x+b_{23}D_y\tilde{D}_x^2 + \ell.o.t. \qquad (4.2.10)$$

A more convenient expression for $\beta_2 w$ is obtained by invoking (4.2.1b) and (4.2.7) to express $\tilde{D}_x^2 w$ in terms of g_1 and substituting into (4.2.10). This will be done, when needed, in (4.3.6) below for a different problem.

Our goal is to show the following result which then contains Theorem 2.1 on a finite domain (the two terms with $\tilde{D}_x w_t$ and $D_y w_t$ in (4.2.11)) below combine into the gradient $\nabla w_t = \nabla_{tang} w_t + \frac{\partial w_t}{\partial \nu}$ in (2.5)).

MAIN THEOREM 4.0. Let $0 < \alpha < T$ and $\varepsilon > 0$ be arbitrary. Then

$$\|D_y^2 w\|_{L_2(\alpha,T-\alpha;L_2(\Gamma_1))} + \|D_y\tilde{D}_x w\|_{L_2(\alpha,T-\alpha;L_2(\Gamma_1))}$$

$$+ \|\tilde{D}_x^2 w\|_{L_2(\alpha,T-\alpha;L_2(\Gamma_1))}$$

$$\le C_{T,\alpha,\varepsilon,\gamma}\Big\{\|f\|_{H^{-s_0}(Q_T)} + \|g_1\|_{L_2(\Sigma_{1T})} + \|g_2\|_{H^{-1}(\Sigma_{1T})}$$

$$+ \|w\|_{L_2(0,T;H^{3/2+\varepsilon}(\Omega))} + \|\tilde{D}_x w_t\|_{L_2(\Sigma_{1T})}$$

$$+ \|D_y w_t\|_{L_2(\Sigma_{1T})} + \|w_t\|_{L_2(\Sigma_{1T})}\Big\}. \quad \blacksquare \tag{4.2.11}$$

4.3. Localization in time

As in Section 7 of [L-T.1], let $\psi(t)$ be a $C_0^\infty(R)$-function, which we specify to be identically 1 in $[\alpha,T-\alpha]$ and to vanish outside $(\alpha/2, T-\alpha/2)$. We then apply ψ to problem (4.2.1) after setting $w_c(t,\cdot) = \psi(t)w(t,\cdot)$ (c stands for 'cut') and obtain (where [,] denotes commutator):

$$\begin{cases} Pw_c = \psi f + [P,\psi]w & \text{in } (0,\infty)\times\Omega = Q_\infty; \quad (4.3.1a) \\ B_1 w_c = \psi g_1 & \text{in } (0,\infty)\times\Gamma = \Sigma_\infty; \quad (4.3.1b) \\ B_2 w_c = \psi g_2 + [B_2,\psi]w & \text{in } \Sigma_\infty. \quad (4.3.1c) \end{cases}$$

Applying $[\tilde{D}_x^2+\tilde{D}_y^2]$ and the co-normal $\tilde{D}_x$ (half-space versions of $-\Delta$ and $\frac{\partial}{\partial\nu}$) to the identity

$$\psi w_{tt} = (\psi w)_{tt} - 2(\psi' w)_t + \psi'' w$$

yields, respectively

$$[P,\psi]w = i2\gamma\, a[\tilde{D}_x^2+\tilde{D}_y^2]D_t(\psi' w) + \ell.o.t. \quad \text{in } Q_\infty; \tag{4.3.2}$$

$$[B_2,\psi]w = i2\gamma\, a\tilde{D}_x D_t(\psi' w) + \ell.o.t. \quad \text{in } \Sigma_\infty; \tag{4.3.3}$$

where we note that $\psi' = 0$ identically, except on $(\alpha/2,\alpha)$ and $(T-\alpha,T-\alpha/2)$. For further use, we note the following properties:

(i) that (4.3.3) implies

$$\|[B_2,\psi]w\|_{H^{-1}(R^1_t;L_2(\Gamma))} \le C\gamma\|\tilde{D}_x(\psi'w)\|_{L_2(\alpha/2,T-\alpha/2;L_2(\Gamma))}$$

$$\le C_\varepsilon\gamma\|w\|_{L_2(\alpha/2,T-\alpha/2;H^{3/2+\varepsilon}(\Omega))} \qquad (4.3.4)$$

by trace theory, $\forall\ \varepsilon > 0$.

(ii) Eqn. (4.3.1b) with B_1 defined by (4.2.7) becomes explicitly

$$\tilde{D}^2_x w_c = \psi g_1 - \tilde{D}^2_y w_c - \beta_1 w_c. \qquad (4.3.5)$$

We can then use (4.3.5) in the definition of β_2 in (4.2.10), thus obtaining the more convenient expression

$$\beta_2 w_c = b_{21}D^3_y w_c + b_{22}D^2_y\tilde{D}_x w_c - b_{23}D_y\tilde{D}^2_y w_c - b_{23}D_y\beta_1 w_c + b_{23}D_y\psi g_1. \qquad (4.3.6)$$

Recalling (4.2.9b) for β_1, we arrive at the estimate

$$[\text{symb}\{\beta_2\}]w_c = \mathcal{O}([|\eta|^3+|\eta|^2\tilde{\xi}]w_c) + \mathcal{O}(|\eta|g_1). \qquad (4.3.7)$$

(iii) We can rewrite (4.3.2) as

$$[P,\psi]w = i2\gamma a\tilde{D}^2_x D_t(\psi'w) + i2\gamma a\tilde{D}^2_y D_t(\psi'w) + \ell.o.t. \text{ in } Q_\infty; \qquad (4.3.9a)$$

$$\text{symb}\{[P,\psi]\}w = \gamma\mathcal{O}(s\tilde{\xi}^2+|\eta|^2 s)(\psi'w). \qquad (4.3.9b)$$

Since w_c has compact support in $(\alpha/2, T-\alpha/2)$, we may view $s = \sigma - ir_0$ as Laplace transform variable of D_t (i.e., $\frac{\partial}{\partial t} \to r_0 + i\sigma$). Moreover, η is the Fourier variable corresponding to D_y (or $\frac{\partial}{\partial y} \to i\eta$).

4.4. Localization in the dual (Laplace/Fourier) variables

As in Sections 6 and 7 of [L-T.1], it will suffice to consider only the quarter space $R^{2n}(+)$ where $\sigma > 0$ and $\eta_j > 0$ and define mutually disjoint regions (cones)

$$\mathcal{R}_1 = \{(x,y;\sigma,\eta) \in R^{2n}(+) : \sigma < c_0|\eta|\}; \qquad (4.4.1a)$$

$$\mathcal{R}_{tr} = \{(x,y;\sigma,\eta) \in R^{2n}(+) : c_0|\eta| \le \sigma < 2\,c_0|\eta|\}; \qquad (4.4.1b)$$

$$\mathcal{R}_2 = \{(x,y;\sigma,\eta) \in R^{2n}(+) : 2\,c_0|\eta| \le \sigma\}. \qquad (4.4.1c)$$

$$\mathcal{E}_1 = \mathcal{R}_1 \cup \mathcal{R}_{tr}; \qquad \mathcal{E}_2 = \mathcal{R}_2 \cup \mathcal{R}_{tr}. \tag{4.4.2}$$

It is readily seen from (4.2.4) that with $\gamma > 0$, by selecting the positive constant c_0 in (4.4.1) suitably small (depending on the smooth coefficients $a_{ij}(x,y)$), then the symbol $p(x,y;\sigma,\xi,\eta)$ is elliptic of order 4 in the cone $\mathcal{E}_1$:

$$p \geq \delta\{\tilde{\xi}^4+\sigma^4+|\eta|^4\} \quad \text{in } \mathcal{E}_1, \quad \text{for some } \delta > 0 \tag{4.4.3}$$

(where δ may be taken independent of $\gamma > 0$). With reference to the cones in (4.4.1), let, as in Sections 6, 7 of [L-T.1], $\chi = \chi(x,y;\sigma,\eta) \in S^0(R^{n+1}_{tyx})$ be a (homogeneous) symbol of order zero of localization, defined as C^∞-function in all variables such that

$$\chi(x,y;\sigma,\eta) = \begin{cases} 1 & \text{in } \mathcal{R}_1 \\ 0 & \text{in } \mathcal{R}_2 \end{cases}, \quad \text{supp}\,\chi \subset \mathcal{R}_1 \cup \mathcal{R}_{tr} = \mathcal{E}_1. \tag{4.4.4}$$

Let $\chi \in OPS^0(R^{n+1}_{tyx})$ be the corresponding pseudo-differential operator. Applying χ to problem (4.3.1) yields

$$\begin{cases} P(\chi w_c) = \chi\psi f+\chi[P,\psi]w+[P,\chi]w_c \equiv \chi\psi f+F & \text{in } Q_\infty; & (4.4.5a) \\ B_1(\chi w_c) = \chi\psi g_1+[B_1,\chi]w_c \equiv \hat{g}_1 & \text{in } \Sigma_\infty; & (4.4.5b) \\ B_2(\chi w_c) \equiv \chi\psi g_2+\chi[B_2,\psi]w+[B_2,\chi]w_c \equiv \hat{g}_2 & \text{in } \Sigma_\infty; & (4.4.5c) \end{cases}$$

$$F \equiv \chi[P,\psi]w+[P,\chi]w_c. \tag{4.4.6}$$

We next analyze the new commutators appearing in problem (4.4.5). To this end, for the purpose of avoiding non-essential notation, we *shall henceforth indicate generically by* Φ_{ij} *a pseudo-differential operator corresponding to the (generic) smooth symbol* $\phi_{ij}(\cdot;\sigma,\eta)$ *of order* i^{th} *in* σ *and* j^{th} *in* η. The symbols ϕ_{ij} will be all supported in $\mathcal{E}_1$.

LEMMA 4.1. We have
(i)

$$[P,\chi] = [\Phi_{00}\tilde{D}_x^3+\Phi_{00}D_y\tilde{D}_x^2+\Phi_{00}D_y^2\tilde{D}_x+\Phi_{00}D_y^3]$$

$$+ \gamma[\Phi_{0,-1}\tilde{D}_x^2D_t^2+\Phi_{00}\tilde{D}_xD_t^2+\Phi_{00}\tilde{D}_yD_t^2]w_c + \ell.o.t.; \tag{4.4.7a}$$

$$\text{symb}\{[P,\chi]\} = [\phi_{00}(x,y;\sigma,\eta)\tilde{\xi}^3+\phi_{01}(x,y;\sigma,\eta)\tilde{\xi}^2$$
$$+ \phi_{02}(x,y;\sigma,\eta)\tilde{\xi}+\phi_{03}(x,y;\sigma,\eta)]$$
$$+ \gamma s^2[\phi_{0,-1}(x,y;\sigma,\eta)\tilde{\xi}^2+\phi_{00}(x,y;\sigma,\eta)\tilde{\xi}$$
$$+ \phi_{01}(x,y;\sigma,\eta)] + \ell.o.t. \text{ in } \mathcal{E}_1; \tag{4.4.7b}$$

with all ϕ_{ij} supported in $\mathcal{E}_1$. Thus, using (4.4.1a) for s whereby σ is dominated by $|\eta|$ in $\mathcal{E}_1$, we obtain:

$$\begin{cases}\text{symb}\{[P,\chi]\} = (1+\gamma)\mathcal{O}(\tilde{\xi}^3+|\eta|\tilde{\xi}^2+|\eta|^2\tilde{\xi}+|\eta|^3) \text{ in } \mathcal{E}_1; & (4.4.7c)\\ \text{supp symb}\{[P,\chi]\} \subset \mathcal{E}_1 = \mathcal{R}_1 \cup \mathcal{R}_{tr}. & (4.4.7d)\end{cases}$$

(ii)

$$[B_1,\chi] = \phi_{00}\tilde{D}_x+\phi_{00}D_y + \ell.o.t.; \tag{4.4.8a}$$

$$\text{symb}\{[B_1,\chi]\} = \phi_{00}(y;\sigma;\eta)\tilde{\xi}+\phi_{01}(y;\sigma,\eta) + \ell.o.t.; \tag{4.4.8b}$$

with all ϕ_{ij} supported by $\mathcal{E}_1$. Thus

$$\begin{cases}\text{symb}\{[B_1,\chi]\} = \mathcal{O}(\tilde{\xi}+|\eta|) \text{ in } \mathcal{E}_1; & (4.4.8c)\\ \text{supp symb}\{[B_1,\chi]\} \subset \mathcal{E}_1 = \mathcal{R}_1 \cup \mathcal{R}_{tr}. & (4.4.8d)\end{cases}$$

(iii)

$$[B_2,\chi] = \phi_{00}\tilde{D}_x^2+\phi_{00}D_y\tilde{D}_x+\phi_{00}D_y^2+\gamma\phi_{00}D_t^2 + \ell.o.t.; \tag{4.4.9a}$$

$$\text{symb}\{[B_2,\chi]\} = \phi_{00}(y;\sigma;\eta)\tilde{\xi}^2+\phi_{01}(y;\sigma,\eta)\tilde{\xi}+\phi_{02}(y;\sigma,\eta)$$
$$+ \gamma s^2\phi_{00}(y;\sigma,\eta) + \ell.o.t.; \tag{4.4.9b}$$

with all ϕ_{ij} supported in $\mathcal{E}_1$. Thus, using again (4.4.1a) for s,

$$\begin{cases}\text{symb}\{[B_2,\chi]\} = (1+\gamma)\mathcal{O}(\tilde{\xi}+|\eta|\tilde{\xi}+|\eta|^2) \text{ in } \mathcal{E}_1; & (4.4.9c)\\ \text{supp symb}\{[B_2,\chi]\} \subset \mathcal{E}_1 = \mathcal{R}_1 \cup \mathcal{R}_{tr}. \quad \blacksquare & (4.4.9d)\end{cases}$$

Proof. One first proves the expressions (4.4.7b), (4.4.8b), (4.4.9b) for the symbols by using the asymptotic expansions of symbols as in [H.1, p. 70] (thus obtaining explicit expressions of them). That all symbols ϕ_{ij} are supported in $\mathcal{E}_1$ is a consequence of the fact that they are sums of products containing derivatives of χ in x, y, or η. From the symbol expressions in Equations "b", one deduces the expressions for the corresponding pseudo-differential operators in Equations "a". ∎

4.5. Elliptic estimates for χw_c in $\mathcal{E}_1$

We have already noted in (4.4.3) that the symbol p of the operator P is elliptic of order 4 in the cone $\mathcal{E}_1$. Thus, problem (4.4.5) is elliptic and satisfies elliptic estimates in all variables. From [L-M.1], [H.1], we obtain

LEMMA 4.2. With reference to problem (4.4.5) we have: There exists a constant C such that

$$\|\chi w_c\|_{H^{5/2}(Q_\infty)} + \|\chi w_c\|_{H^2(\Sigma_\infty)} + \|\tilde{D}_x(\chi w_c)\|_{H^1(\Sigma_\infty)}$$

$$\leq C\Big\{\|\chi\psi f\|_{H^{-s_0}(Q_\infty)} + \|F\|_{L_2(R^1_{x+};H^{-3/2}(R^n_{ty}))} + \|\hat{g}_1\|_{L_2(\Sigma_\infty)}$$

$$+ \|\hat{g}_2\|_{H^{-1}(\Sigma_\infty)} + \|w\|_{L_2(Q_T)}\Big\}, \quad (4.5.1)$$

where we recall that $s_0 < \frac{1}{2}$. ∎

In the next sections we estimate the terms on the right hand side of (4.5.1).

4.6. Estimates of boundary terms of problem (4.4.5) for χw_c

LEMMA 4.3. With reference to (4.4.8) we have $\forall\ \varepsilon > 0$:

$$\|[B_1,\chi]w_c\|_{L_2(\Sigma_\infty)} \leq C_\varepsilon \|w_c\|_{L_2(0,T;H^{3/2+\varepsilon}(\Omega))}. \quad ∎ \quad (4.6.1)$$

Proof. We use (4.4.8a), where each term is estimated by trace theory:

$$\|D_y w_c\|_{L_2(\Sigma_\infty)} \leq c\|w_c\|_{L_2(R^1_t;H^1(\Gamma))} \leq c\|w_c\|_{L_2(R^1_t;H^{3/2}(\Omega))}; \quad (4.6.2)$$

$$\|\tilde{D}_x w_c\|_{L_2(\Sigma_\infty)} \leq c_\varepsilon \|w_c\|_{L_2(0,T;H^{3/2+\varepsilon}(\Omega))}, \quad \forall\ \varepsilon > 0. \quad ∎ \quad (4.6.3)$$

LEMMA 4.4. With reference to (4.4.9) we have $\forall\ \varepsilon > 0$:

$$\|[B_2,\chi]w_c\|_{H^{-1}(\Sigma_\infty)} \le c_\varepsilon\left\{\|w_c\|_{L_2(0,T;H^{3/2+\varepsilon}(\Omega))} + \|g_1\|_{H^{-1}(\Sigma_T)}\right\}. \quad \blacksquare \tag{4.6.4}$$

Proof. We use (4.4.9a). (i) We first show that $\forall\ \varepsilon > 0$:

$$\|\tilde{D}_x^2 w_c\|_{H^{-1}(\Sigma_\infty)} \le c_\varepsilon\left\{\|g_1\|_{H^{-1}(\Sigma_\infty)} + \|w_c\|_{L_2(0,T;H^{3/2+\varepsilon}(\Omega))}\right\}. \tag{4.6.5}$$

To see this, we return to Eqn. (4.3.5), where β_1 is then defined by (4.2.9). We then estimate as in (4.6.3) by trace theory

$$\begin{aligned}\|(\tilde{D}_y^2+\beta_1)w_c\|_{H^{-1}(\Sigma_\infty)} &\le c\left\{\|D_y^2 w_c\|_{H^{-1}(\Sigma_\infty)} + \|D_y\tilde{D}_x w_c\|_{H^{-1}(\Sigma_\infty)}\right\} \\ &\le c\left\{\|w_c\|_{L_2(0,T;H^1(\Gamma))} + \|\tilde{D}_x w_c\|_{L_2(\Sigma_\infty)}\right\} \\ &\le c_\varepsilon\left\{\|w_c\|_{L_2(0,T;H^{3/2+\varepsilon}(\Omega))}\right\}. \end{aligned} \tag{4.6.6}$$

Then using (4.6.6) in (4.3.5) yields (4.6.5).

(ii) The terms $D_y\tilde{D}_x$ and D_y^2 in (4.4.9a) are estimated in (4.6.6), while $\Phi_{00}D_t^2 w_c$ can be estimated by $\Phi_{00}D_y^2 w_c$, since the symbol ϕ_{00} is supported in $\mathcal{E}_1$ where σ is dominated by $|\eta|$. Thus (4.6.4) is proved. ∎

COROLLARY 4.5. With reference to $\hat{g}_1$ and $\hat{g}_2$ defined by (4.4.5b) and (4.4.5c), we have

$$\|\hat{g}_1\|_{L_2(\Sigma_\infty)} \le c_\varepsilon\left\{\|g_1\|_{L_2(\Sigma_T)} + \|w_c\|_{L_2(0,T;H^{3/2+\varepsilon}(\Omega))}\right\}; \tag{4.6.7}$$

$$\begin{aligned}\|\hat{g}_2\|_{H^{-1}(\Sigma_\infty)} \le c_\varepsilon\Big\{&\|g_2\|_{H^{-1}(\Sigma_T)} + \|g_1\|_{H^{-1}(\Sigma_T)} \\ &+ \|w_c\|_{L_2(0,T;H^{3/2+\varepsilon}(\Omega))} + \gamma\|w\|_{L_2(\alpha/2,T-\alpha/2;H^{3/2+\varepsilon}(\Omega))}\Big\}. \quad \blacksquare\end{aligned} \tag{4.6.8}$$

Proof. To prove (4.6.7), we use the assumption on g_1 and estimate (4.6.1) in (4.4.5b). To prove (4.6.8), we use the assumption on g_2 as well as estimate (4.6.4) and estimate (4.3.4) in (4.4.5c). ∎

4.7. Estimates of interior terms of problem (4.4.5) for χw_c

We begin with $[P,\chi]w_c$.

LEMMA 4.6. We can write $[P,\chi]$ in (4.4.7a) as

$$[P,\chi] = \Phi_{00}\tilde{D}_x^3+\Phi_{00}D_y\tilde{D}_x^2+G, \tag{4.7.1}$$

where the (good) operator G has symbol supported in $\mathcal{E}_1$ and satisfies

$$\|Gw_c\|_{L_2(R^1_{x^+};H^{-3/2}(R^n_{ty}))} \le C_\varepsilon \|w_c\|_{L_2(0,T;H^{3/2+\varepsilon}(\Omega))}. \quad ∎ \tag{4.7.2}$$

Proof. From (4.4.7a) and (4.4.7c) we deduce the explicit expression for G (not needed) and moreover that

$$\|Gw_c\|_{L_2(R^1_{x^+};H^{-3/2}(R^n_{ty}))}$$

$$\le C\left\{\|D_y^2\tilde{D}_x w_c\|_{L_2(R^1_{x^+};H^{-3/2}(R^n_{ty}))}+\|D_y^3 w_c\|_{L_2(R^1_{x^+};H^{-3/2}(R^n_{ty}))}\right\}$$

$$\le C\left\{\|w_c\|_{L_2(0,T;H^{3/2}(\Omega))}\right\}, \tag{4.7.3}$$

(4.7.2) is proved. Indeed,

$$\|D_y^3 w_c\|_{L_2(R^1_{x^+};H^{-3/2}(R^n_{ty}))} \le C\|D_y^3 w_c\|_{L_2(R^1_{x^+}\times R^1_t;H^{-3/2}(R^{n-1}_y))} \tag{4.7.4}$$

$$\le C\|w_c\|_{L_2(R^1_t;H^{3/2}(\Omega))}; \tag{4.7.5}$$

$$\|D_y^2\tilde{D}_x w_c\|_{L_2(R^1_{x^+};H^{-3/2}(R^n_{ty}))} \le C\|D_y^2\tilde{D}_x w_c\|_{L_2(R^1_{x^+}\times R^1_t;H^{-3/2}(R^{n-1}_y))}$$

$$\le C\|\tilde{D}_x w_c\|_{L_2(R^1_t;H^{1/2}(\Omega))}$$

$$\le C\|w_c\|_{L_2(0,T;H^{3/2}(\Omega))}. \quad ∎ \tag{4.7.6}$$

To complete the estimate of $[P,\chi]$ we need to analyze the first two terms in (4.7.1). These are the most demanding terms which cannot be estimated by the *a-priori* regularity analysis used for G in Lemma 4.7. Rather (as in the case of Lemma 4.4 for $[B_2,\chi]$), we shall need to use problem (4.3.1). The following proposition is critical.

PROPOSITION 4.7. With reference to the first two terms of (4.7.1) we have (recalling that $s_0 < \frac{1}{2}$):

$$\|\Phi_{00}\tilde{D}_x^3 w_c\|_{L_2(R^1_{x+};H^{-\frac{3}{2}}(R^n_{ty}))} + \|\Phi_{00}D_y\tilde{D}_x^2 w_c\|_{L_2(R^1_{x+};H^{-\frac{3}{2}}(R^n_{ty}))}$$

$$\leq C_\varepsilon\Big\{\|g_1\|_{L_2(\Sigma_T)} + \|g_2\|_{L_2(0,T;H^{-1}(\Gamma))} + \|f\|_{H^{-s_0}(Q_T)}$$

$$+ \|w_c\|_{L_2(0,T;H^{\frac{3}{2}+\varepsilon}(\Omega))} + \|w\|^2_{L_2(0,T;H^{\frac{3}{2}+\varepsilon}(\Omega))}$$

$$+ \|\Phi_{00}\tilde{D}_x^2(\psi' w)\|_{L_2(R^1_{x+};H^{-\frac{3}{2}}(R^n_{ty}))}\Big\}, \qquad (4.7.7)$$

where Φ_{00} is, as usual, a pseudo-differential operator with symbol ϕ_{00} of order zero in σ and η supported in $\mathcal{E}_1$. ∎

Proof. The proof is given in Section 4.8. ∎

COROLLARY 4.8. With reference to (4.7.1) we have with $s_0 < \frac{1}{2}$:

$$\|[P,\chi]w_c\|_{L_2(R^1_{x+};H^{-\frac{3}{2}}(R^n_{ty}))}$$

$$\leq C_\varepsilon\Big\{\|g_1\|_{L_2(\Sigma_T)} + \|g_2\|_{L_2(0,T;H^{-1}(\Gamma))} + \|f\|_{H^{-s_0}(Q_T)}$$

$$+ \|w_c\|_{L_2(0,T;H^{\frac{3}{2}+\varepsilon}(\Omega))} + \|w\|^2_{L_2(0,T;H^{\frac{3}{2}+\varepsilon}(\Omega))}$$

$$+ \|\Phi_{00}\tilde{D}_x^2(\psi' w)\|_{L_2(R^1_{x+};H^{-\frac{3}{2}}(R^n_{ty}))}\Big\}. \qquad (4.7.8)$$

Proof. Combine (4.7.7) of Proposition 4.7 with (4.7.1), (4.7.2) of Lemma 4.6. ∎

We now turn to $\chi[P,\psi]w$. From (4.3.2) we write

$$\chi[P,\psi]w = i2\gamma\chi(a\tilde{D}_x^2 D_t(\psi' w)) + i2\gamma\chi(a\tilde{D}_y^2 D_t(\psi' w)) + \ell.o.t. \tag{4.7.9}$$

LEMMA 4.9. With reference to (4.7.8) we have
(i)

$$\|\chi(a\tilde{D}_y^2 D_t(\psi' w)\|_{L_2(R^1_{x^+};H^{-3/2}(R^n_{ty}))} \le C\|w\|_{L_2(0,T;H^{3/2}(\Omega))}; \tag{4.7.10}$$

(ii) hence, by (4.7.9) and (4.7.10),

$$\|\chi[P,\psi]w\|_{L_2(R^1_{x^+};H^{-3/2}(R^n_{ty}))} \le C\left\{\|w\|_{L_2(0,T;H^{3/2}(\Omega))} + \|\chi(a\tilde{D}_x^2 D_t(\psi' w))\|_{L_2(R^1_{x^+};H^{-3/2}(R^n_{ty}))}\right\}. \quad \blacksquare \tag{4.7.11}$$

Proof. (i) Proceeding as in going from (4.7.4) to (4.7.5) we have

$$\|\chi(a\tilde{D}_y^2 D_t(\psi' w))\|^2_{L_2(R^1_{x^+};H^{-3/2}(R^n_{ty}))}$$

$$\le C \int_{R^1_{x^+}} \int_{\mathcal{E}_1} \frac{(\sigma|\eta|^2|\psi' w|)^2}{\sigma^3 + |\eta|^3}\, d\sigma d\eta dx \le C \int_{R^1_{x^+}} \int_{\mathcal{E}_1} \frac{\sigma^2|\eta|\ \ |\psi' w|^2}{(\frac{\sigma}{|\eta|})^3 + 1}\, d\sigma d\eta dx$$

$$\le C \int_{R^1_{x^+}} \int_{\mathcal{E}_1} |\eta|^3 |\psi' w|^2 d\sigma d\eta dx = C\|\Phi_{00}(\psi' w)\|^2_{L_2(R^1_{x^+}\times R^1_t;H^{3/2}(R^{n-1}_y))}$$

$$\le C\|w\|^2_{L_2(0,T;H^{3/2}(\Omega))},$$

where Φ_{00} is, as usual, a pseudo-differential operator with smooth symbol ϕ_{00} of order zero in σ and η supported in $\mathcal{E}_1$. $\blacksquare$

COROLLARY 4.10. With reference to F defined by (4.4.6) we have

$$\|F\|_{L_2(R^1_{x^+};H^{-3/2}(R^n_{ty}))} \le C\Big\{\|g_1\|_{L_2(\Sigma_T)} + \|g_2\|_{L_2(0,T;H^{-1}(\Gamma))}$$

$$+ \|f\|_{H^{-s_0}(Q_T)} + \|w_c\|_{L_2(0,T;H^{3/2+\varepsilon}(\Omega))} + \|w\|^2_{L_2(0,T;H^{3/2+\varepsilon}(\Omega))}$$

$$+ \|\chi(a\tilde{D}^2_x D_t(\psi' w))\|_{L_2(R^1_{x^+};H^{-3/2}(R^n_{ty}))}\Big\}. \quad ■ \tag{4.7.12}$$

Proof. We return to (4.4.6) and combine (4.7.8) for $[P,\chi]w_c$ with (4.7.11) for $\chi[P,\psi]w$, where the term $\Phi_{00}\tilde{D}^2_x(\psi' w)$ of the former is dominated in norm by the term $\chi(a\tilde{D}^2_x D_t(\psi' w))$ of the latter. ■

REMARK 4.1. Thus, it is the last term $\chi(a\tilde{D}^2_x D_t(\psi' w))$ that causes additional difficulties. Notice that this term originates from the time-commutator $[P,\psi]$ in the original equation. Also, $\psi' \equiv 0$ except on $(\alpha/2,\alpha) \cup (T-\alpha, T-\alpha/2)$. ■

In the next critical proposition, we estimate the last term in (4.7.12).

PROPOSITION 4.11. We have

$$\|\chi(a\tilde{D}^2_x D_t(\psi' w))\|_{L_2(R^1_{x^+};H^{-3/2}(R^n_{ty}))}$$

$$\le C_\varepsilon\Big\{\|g_1\|_{L_2(\Sigma_T)} + \|g_2\|_{L_2(0,T;H^{-1}(\Gamma))} + \|w\|_{L_2(0,T;H^{3/2+\varepsilon}(\Omega))}\Big\}. \tag{4.7.13}$$

Proof. The proof is given in Section 4.9. ■

COROLLARY 4.12. With reference to (4.7.9) we have with $s_0 < 1/2$:

$$\|F\|_{L_2(R^1_{x^+};H^{-3/2}(R^n_{ty}))} \le C_\varepsilon\Big\{\|g_1\|_{L_2(\Sigma_T)} + \|g_2\|_{L_2(0,T;H^{-1}(\Gamma))}$$

$$+ \|f\|_{H^{-s_0}(Q_T)} + \|w\|^2_{L_2(0,T;H^{3/2+\varepsilon}(\Omega))}\Big\}. \tag{4.7.14}$$

Proof. We use (4.7.13) in (4.7.12) and recall that $\|w_c\| \le C\|w\|$. ■

4.8. Proof of Proposition 4.7

Step 1. We return to the w_c-problem (4.3.1). Recalling (4.2.2) for P, we rewrite Eqn. (4.3.1a) explicitly as

$$[\tilde{D}_x^2+\tilde{D}_y^2]^2 w_c - \gamma a D_t^2[\tilde{D}_x^2+\tilde{D}_y^2]w_c = \psi f + aD_t^2 w_c + [P,\psi]w.$$

Then, recalling also (4.2.7) and (4.2.8) for B_1 and B_2, rewrite problem (4.3.1) as

$$\begin{cases} \tilde{D}_x^4 w_c + [2\tilde{D}_y^2 - \gamma a D_t^2]\tilde{D}_x^2 w_c = \psi f + U w_c + [P,\psi]w & \text{in } Q_\infty; \quad (4.8.1a) \\ \tilde{D}_x^2 w_c\big|_{x=0} = \psi g_1 - \tilde{D}_y^2 w_c - \beta_1 w_c & \text{in } \Sigma_\infty; \quad (4.8.1b) \\ \tilde{D}_x^3 w_c\big|_{x=0} = \psi g_2 - \tilde{D}_x\tilde{D}_y^2 w_c - \beta_2 w_c - \gamma a \tilde{D}_x D_t^2 w_c + [B_2,\psi]w & \text{in } \Sigma_\infty. \quad (4.8.1c) \end{cases}$$

$$\begin{cases} U = aD_t^2 - \tilde{D}_y^4 - [\tilde{D}_x^2, \tilde{D}_y^2] + \gamma a D_t^2 \tilde{D}_y^2; & (4.8.2a) \\ \text{symb}\{U\} = (1+\gamma)\mathcal{O}(|\eta|^4) + \mathcal{O}(|\eta|^2\tilde{\xi}) \quad \text{in } \mathcal{E}_1, & (4.8.2b) \end{cases}$$

since $\text{symb}\{[\tilde{D}_x^2, \tilde{D}_y^2]\} = \mathcal{O}(|\eta|^2\tilde{\xi})$, and recalling the definition (4.4.1a-b), (4.4.2) of $\mathcal{E}_1$. We note now some relevant properties of (4.8.1).

(i) By (4.2.6),

$$\text{symb}\{2\tilde{D}_y^2 - \gamma a D_t^2\} = \frac{2d}{a^2} - \gamma a(\sigma^2 - ir_0)^2 = q + \ell.o.t.; \quad (4.8.3)$$

$$q(x,y;\sigma,\eta) \equiv \frac{2d(x,y;\eta)}{a^2(x,y)} - \gamma a(x,y)\sigma^2 \geq \delta_0|\eta|^2$$
$$\text{in } \mathcal{E}_1, \quad \delta_0 > 0; \quad (4.8.4)$$

by selecting the constant c_0 in the definition (4.4.1) of $\mathcal{E}_1$ suitably small.

(ii) By (4.2.6) and (4.2.9b)

$$\text{symb}\{\tilde{D}_y^2 - \beta_1\} = \mathcal{O}(|\eta|^2 + |\eta|\tilde{\xi}). \quad (4.8.5)$$

(iii) By (4.2.6), (4.3.7), and (4.4.1) on σ dominated by $|\eta|$:

$$\text{symb}\{\tilde{D}_x\tilde{D}_y^2 + \beta_2 + \gamma a \tilde{D}_x \tilde{D}_t^2\}w_c$$
$$= \gamma\mathcal{O}([|\eta|^3 + |\eta|^2\tilde{\xi}]w_c) + \mathcal{O}(|\eta|g_1) \text{ in } \mathcal{E}_1. \quad (4.8.6)$$

It is then natural to introduce a new variable

$$z(\sigma,x,\eta) \equiv \tilde{D}_x^2 w_c(\sigma,x,\eta), \quad [\sigma,\eta] \in \mathcal{E}_1 , \tag{4.8.7}$$

and consider the version of the problem (4.8.1) with all operators, except $\tilde{D}_x$, being replaced by their symbols. We thus arrive by (4.8.3)-(4.8.7) at the following problem for x > 0:

$$\begin{cases} \tilde{D}_x^2 z+qz = r; & (4.8.8a) \\ z|_{x=0} = \psi g_1+\mathcal{O}([|\eta|^2+|\eta|\tilde{\xi}]w_c); & (4.8.8b) \\ \tilde{D}_x z|_{x=0} = \psi g_2+\mathcal{O}([|\eta|^3+\gamma|\eta|^2\tilde{\xi}]w_c)+\mathcal{O}(|\eta|g_1) & (4.8.8c) \\ \qquad +\text{symb}\{[B_2,\psi]\}w \text{ in } \mathcal{E}_1; \end{cases}$$

with σ, η parameters in $\mathcal{E}_1$, where by (4.8.1a) and (4.8.2),

$$r = \psi f+(1+\gamma)\mathcal{O}(|\eta|^4 w_c)+\mathcal{O}(|\eta|^2\tilde{\xi} w_c)+\text{symb}\{[P,\psi]\}w$$
$$+ \ell.\text{o.t. in } \mathcal{E}_1 , \tag{4.8.9}$$

the last term being identified in (4.3.9b). Notice that (4.8.8b) follows from (4.8.7), (4.8.1b), and (4.8.5); while (4.8.8c) follows by (4.8.7), (4.8.1c), (4.8.6), with $[B_2,\psi]w$ given by (4.3.3).

Step 2. We next multiply Eqn. (4.8.8a) by z and integrate by parts in x over $R^1_{x^+}$: x > 0. We obtain for z vanishing at ∞,

$$\int_0^\infty |\tilde{D}_x z|^2 dx + \int_0^\infty q|z|^2 dx = \int_0^\infty rz\, dx - (\tilde{D}_x z|_{x=0})(z|_{x=0}), \tag{4.8.10}$$

where by (4.8.4) we have

$$\delta_0|\eta|^2\|z\|^2_{L_2(R^1_{x^+})} \le \int_0^\infty q|z|^2 dx \quad \text{in } \mathcal{E}_1, \tag{4.8.11}$$

and hence by (4.8.10), (4.8.11),

$$\|\tilde{D}_x z\|^2_{L_2(R^1_{x^+})} +\delta_0\| |\eta| z\|^2_{L_2(R^1_{x^+})} \le \int_0^\infty rz\, dx - (\tilde{D}_x z|_{x=0})(z|_{x=0}). \tag{4.8.12}$$

Step 3. Returning to (4.7.7), since by (4.8.7),

$$\|\Phi_{00}\tilde{D}_x^3 w_c\|^2_{L_2(R^1_{x^+};H^{-3/2}(R^n_{ty}))} + \|\Phi_{00}D_y\tilde{D}_x^2 w_c\|^2_{L_2(R^1_{x^+};H^{-3/2}(R^n_{ty}))}$$

$$\leq C \int_{\mathcal{E}_1} \frac{\|\tilde{D}_x z\|^2_{L_2(R^1_{x^+})} + |\eta|^2 \|z\|^2_{L_2(R^1_{x^+})}}{\sigma^3 + |\eta|^3} \, d\sigma d\eta \, , \tag{4.8.13}$$

as the symbols of Φ_{00} are supported in $\mathcal{E}_1$, we see that our goal is to show that the integral term on $\mathcal{E}_1$ on the right of (4.8.13) is dominated by the terms on the right side of the sought after (4.7.7). In view of (4.8.12), this goal is *a-fortiori* accomplished by the following two lemmas.

LEMMA 4.13. (i) With reference to (4.8.9), we have $\forall\, \varepsilon_1 > 0$ and $\alpha < \frac{1}{2}$:

$$\int_0^\infty rz\, dx \leq \varepsilon_1 \|z\|^2_{H^{s_0}(R^1_{x^+})} + \frac{C}{\varepsilon_1}\|f\|^2_{H^{-s_0}(R^1_{x^+})} + \varepsilon_1 \|\,|\eta|z\|^2_{L_2(R^1_{x^+})}$$

$$+ \mathcal{O}_{\varepsilon_1}\Big\{\gamma\|\,|\eta|^3 w_c\|^2_{L_2(R^1_{x^+})} + \|\,|\eta|\tilde{D}_x w_c\|^2_{L_2(R^1_{x^+})}$$

$$+ \|\,|\eta|^2(\psi' w)\|^2_{L_2(R^1_{x^+})} + \|\tilde{D}_x^2(\psi' w)\|^2_{L_2(R^1_{x^+})}\Big\} \quad \text{in } \mathcal{E}_1 \, . \tag{4.8.14}$$

(ii)

$$\int_{\mathcal{E}_1} \frac{(1-\varepsilon_1)\|\tilde{D}_x z\|^2_{L_2(R^1_{x^+})} + (\delta_0 - \varepsilon_1)\|\,|\eta|z\|^2_{L_2(R^1_{x^+})}}{\sigma^3 + |\eta|^3}\, d\sigma d\eta$$

$$\leq C_{\varepsilon_1}\Big\{\|f\|^2_{H^{-s_0}(Q_T)} + \gamma\|w_c\|^2_{L_2(0,T;H^{3/2}(\Omega))} + \|w\|^2_{L_2(0,T;H^1(\Omega))}$$

$$+ \|\tilde{D}_x^2(\psi' w)\|^2_{L_2(R^1_{x^+});H^{-3/2}(\mathcal{E}_1))}\Big\}$$

$$+ \int_{\mathcal{E}_1} \frac{|(\tilde{D}_x z|_{x=0})(z|_{x=0})|}{\sigma^3 + |\eta|^3}\, d\sigma d\eta . \quad \blacksquare \tag{4.8.15}$$

REMARK 4.2. Notice that the third term on the right hand side of (4.8.15) contains w, not w_c, and that $\psi' \equiv 0$ except on $(\alpha/2,\alpha)$ and $(T-\alpha, T-\alpha/2)$. ∎

LEMMA 4.14. With reference to (4.8.8b) and (4.8.8c), we have

$$\int_{\mathcal{E}_1} \frac{|(\tilde{D}_x z|_{x=0})(z|_{x=0})|}{|\eta|^3}\, d\sigma d\eta \le C_\varepsilon \Big\{ \|g_1\|^2_{L_2(\Sigma_T)} + \|g_2\|^2_{L_2(0,T;H^{-1}(\Gamma))} + \|w_c\|^2_{L_2(0,T;H^{3/2+\varepsilon}(\Omega))} + \|w\|^2_{L_2(0,T;H^{3/2+\varepsilon}(\Omega))} \Big\}. \tag{4.8.16}$$

Proof of Lemma 4.13. (i) Multiplying r in (4.8.9) by z and integrating in x over $R^1_{x^+}$ yields

$$\int_0^\infty rz\, dx = (\psi f, z)_{L_2(R^1_{x^+})} + (1+\gamma)\mathcal{O}((|\eta|^3 w_c, |\eta| z)_{L_2(R^1_{x^+})} + \mathcal{O}\Big[(|\eta|\tilde{D}_x w_c, |\eta| z)_{L_2(R^1_{x^+})}\Big] + \mathcal{O}\Big[(\tilde{D}^2_x(\psi' w), |\eta| z)_{L_2(R^1_{x^+})}\Big] + \mathcal{O}(|\eta|^2 \psi' w, |\eta| z)_{L_2(R^1_{x^+})}, \tag{4.8.17}$$

since, by (4.3.9b), $\text{symb}[P,\psi]w = \mathcal{O}(|\eta|\tilde{\xi}^2 + |\eta|^3](\psi' w)$ in $\mathcal{E}_1$ (see (4.4.1)). From (4.8.17) one then obtains (4.8.14) at once playing $2ab \le \varepsilon_1 a^2 + \frac{1}{\varepsilon_1} b^2$.

(ii) Insert (4.8.14) into (4.8.12); move to the left $\varepsilon_1 \||\eta| z\|^2_{L_2(R^1_{x^+})}$ and $\varepsilon_1 \|z\|^2_{H^{s_0}(R^1_{x^+})} \le \varepsilon_1 \|\tilde{D}_x z\|^2_{L_2(R^1_{x^+})}$; divide by $\sigma^3 + |\eta|^3$ and integrate over $\mathcal{E}_1$. We then readily obtain (4.8.15) (we may drop σ^3 in all terms except $\tilde{\xi}^2 \psi' w$). ∎

Proof of Lemma 4.14. We use (4.8.8b) and (4.8.8c) where, by (4.3.3), and (4.4.1), (4.4.2) where σ is dominated by $|\eta|$:

$$\text{symb}[B_2,\psi] = \mathcal{O}(\tilde{\xi} s) = \mathcal{O}(\tilde{\xi}|\eta|) \quad \text{in } \mathcal{E}_1, \quad x = 0, \tag{4.8.18}$$

and obtain

$$(\tilde{D}_x z|_{x=0})(z|_{x=0}) = \mathcal{O}\Big\{[\psi g_1+|\eta|^2 w_c+|\eta|\tilde{D}_x w_c][\psi g_2+|\eta| g_1 + |\eta|^3 w_c+\gamma|\eta|^2\tilde{\xi} w_c+|\eta|\tilde{D}_x(\psi' w)]\Big\} \quad \text{in } \mathcal{E}_1, \tag{4.8.19}$$

from which we obtain

$$\frac{(\tilde{D}_x z|_{x=0})(z|_{x=0})}{|\eta|^3} = \mathcal{O}\Bigg\{\frac{g_1}{|\eta|^2}\frac{g_2}{|\eta|} + \frac{g_1^2}{|\eta|^2} + g_1 w_c + \frac{g_1}{|\eta|}\tilde{D}_x w_c + g_1\frac{\tilde{D}_x(\psi' w)}{|\eta|^2} + \frac{g_2}{|\eta|} w_c + |\eta|^2 w_c^2$$
$$+ (|\eta| w_c)(\tilde{D}_x w_c) + w_c(\tilde{D}_x(\psi' w)) + \frac{g_2}{|\eta|}\frac{\tilde{D}_x w_c}{|\eta|} + w_c(\tilde{D}_x w_c) + (\tilde{D}_x w_c)^2 + \frac{(\tilde{D}_x w_c)}{|\eta|}(\tilde{D}_x(\psi' w))\Bigg\}$$
$$\text{in } \mathcal{E}_1, \quad \text{at } x = 0. \tag{4.8.20}$$

Integrating (4.8.20) over $\mathcal{E}_1$ readily gives (4.8.16). The most critical term for w_c is:

$$\int_{\mathcal{E}_1} [(\tilde{D}_x w_c)^2+(|\eta| w_c)(\tilde{D}_x w_c)]_{x=0} d\sigma d\eta$$
$$\leq C_\varepsilon\left\{\|w_c\|^2_{L_2(0,T;H^{3/2+\varepsilon}(\Omega))} + \|w_c\|^2_{L_2(R^1_{x^+}\times R^1_t;H^1(\Gamma))}\right\}$$
$$\leq C_\varepsilon\left\{\|w_c\|^2_{L_2(0,T;H^{3/2+\varepsilon}(\Omega))}\right\} \tag{4.8.21}$$

by trace theory. Also, the most critical terms for w are:

$$\int_{\mathcal{E}_1}\left[w_c(\tilde{D}_x(\psi' w)) + \frac{\tilde{D}_x w_c}{|\eta|}(\tilde{D}_x(\psi' w))\right]_{x=0} d\sigma d\eta$$
$$\leq C_\varepsilon\left\{\|w_c\|^2_{L_2(R^1_t;H^{1/2+\varepsilon}(\Omega))} + \|\psi' w\|^2_{L_2(R^1_t;H^{3/2+\varepsilon}(\Omega))} + \|w_c\|^2_{L_2(R^1_t;H^{3/2+\varepsilon}(\Omega))} + \|\psi' w\|^2_{L_2(R^1_t;H^{3/2+\varepsilon}(\Omega))}\right\}$$

$$\leq C_\varepsilon \left\{ \|w_c\|^2_{L_2(0,T;H^{3/2+\varepsilon}(\Omega))} + \|w\|^2_{L_2(0,T;H^{3/2+\varepsilon}(\Omega))} \right\}. \quad \blacksquare \tag{4.8.22}$$

To complete the proof of Proposition 4.7, we use estimate (4.8.16) into estimate (4.8.15), and we notice that

$$\|\tilde{D}_x^2(\psi' w)\|^2_{L_2(R^1_{x+};H^{-3/2}(\mathcal{E}_1))} = \int_{\mathcal{E}_1} \frac{\|\tilde{D}_x^2(\psi' w)\|^2_{L_2(R^1_{x+})}}{\sigma^3 + |\eta|^3} \, d\sigma d\eta \tag{4.8.23}$$

is equivalent to $\|\Phi_{00}\tilde{D}_x^2(\psi' w)\|^2_{L_2(R^1_{x+};H^{-3/2}(R^n_{ty}))}$ with Φ_{00}, as usual, denoting a pseudo-differential operator with smooth symbol ϕ_{00} of order zero in σ and η, supported in $\mathcal{E}_1$. This way, (4.8.15) gives (4.7.7). ■

4.9. Proof of Proposition 4.11

This proof is somewhat similar to that of the preceding section, to which we shall often refer. Since ψ' (in the estimate (4.7.13)) satisfies $\psi' \equiv 0$ except on $(\alpha/2,\alpha) \cup (T-\alpha, T-\alpha/2)$, we introduce a second time localization which picks these two desired time intervals. Let $\pi(t) \in C_0^\infty(R)$ with $\pi(t) \equiv 1$ on $(\alpha/2,\alpha) \cup (T-\alpha,T-\alpha/2)$, while $\pi(t) \equiv 0$ for $t < \alpha/4$ and $t > T-\alpha/4$, so that supp $\pi \subset (\alpha/4, T-\alpha/4)$. Define (the second time localization) $w_{cc}(t,\cdot) = \pi(t)w(t,\cdot)$. Then, w_{cc} solves the analog of problem (4.3.1), and hence of problem (4.8.1), with w_c and ψ there replaced by w_{cc} and π now. The present counterpart of Eqn. (4.8.8) is now for $x > 0$:

$$\begin{cases} \tilde{D}_x^4 w_{cc} + q\tilde{D}_x^2 w_{cc} = \rho; & (4.9.1a) \\ \tilde{D}_x^2 w_{cc}|_{x=0} = \pi g_1 + \mathcal{O}([|\eta|^2 + |\eta|\tilde{\xi}]w_{cc}); & (4.9.1b) \\ \tilde{D}_x^3 w_{cc}|_{x=0} = \pi g_2 + \mathcal{O}([|\eta|^3 + \gamma|\eta|^2\tilde{\xi}]w_{cc}) + \mathcal{O}(|\eta|g_1) \\ \qquad + \mathcal{O}(\tilde{\xi}|\eta|\pi' w) \text{ in } \mathcal{E}_1; & (4.9.1c) \end{cases}$$

see (4.3.3) for symb$[B_2,\psi]w$, with parameters σ and η in $\mathcal{E}_1$, where ρ is the present counterpart of r in (4.8.9), i.e.,

$$\rho = \pi f+(1+\gamma)\mathcal{O}(|\eta|^4 w_{cc})+\mathcal{O}(|\eta|^2\tilde{\xi} w_{cc})+\mathcal{O}(\tilde{\xi}^2|\eta|(\psi' w))$$

$$+ \mathcal{O}(|\eta|^3(\psi' w))+\ell.o.t. \quad \text{in } \mathcal{E}_1, \qquad (4.9.2)$$

recalling symb[P,ψ] from (4.3.9b). Our goal now is to show the sought-after estimate (4.7.13) of the term $[|\eta|\tilde{\xi}^2(\psi' w)]$ in the norm of $L_2(R^1_{x^+};H^{-3/2}(\mathcal{E}_1))$ (as σ is dominated by $|\eta|$ in $\mathcal{E}_1$) [while the goal in Section 4.8 for Proposition 4.9 involved an estimate for the term $\phi_{00}\tilde{\xi}^3 w_c$ in the norm of $L_2(R^1_{x^+};H^{-3/2}(\mathcal{E}_1))$. This is the basic difference between the two proofs.

Equivalently, to prove (4.7.13) of Proposition 4.11 we must establish the following

LEMMA 4.15. We have

$$\|\Phi_{00}\tilde{D}_x^2(\psi' w_{cc})\|_{L_2(R^1_{x^+};H^{-1/2}(R^n_{ty}))}$$

$$\text{equivalently } \|\tilde{\xi}^2\psi' w_{cc}\|_{L_2(R^1_{x^+};H^{-1/2}(\mathcal{E}_1))}$$

$$\leq C_\varepsilon\left\{\|g_1\|_{L_2(\Sigma_T)}+\|g_2\|_{L_2(0,T;H^{-1}(\Gamma))}+\|w\|_{L_2(0,T;H^{3/2+\varepsilon}(\Omega))}\right\}. \qquad (4.9.3)$$

Proof. We now multiply Eqn. (4.9.1a) by w_{cc} (while in Section 4.8 we multiplied Eqn. (4.8.8a) by $z = \tilde{D}_x^2 w_c$) and as before we integrate by parts in $x > 0$ this time twice. We obtain the counterpart of (4.8.10), i.e.,

$$\int_0^\infty |\tilde{D}_x^2 w_{cc}|^2 dx + \int_0^\infty q|\tilde{D}_x w_{cc}|^2 dx = \int_0^\infty \rho w_{cc} dx - i(\tilde{D}_x^3 w_{cc}|_{x=0})(w_{cc}|_{x=0})$$

$$+ i(\tilde{D}_x^2 w_{cc}|_{x=0})(\tilde{D}_x w_{cc}|_{x=0}) + \ell.o.t. \qquad (4.9.4)$$

(where the ℓ.o.t. involve: $\int_0^\infty (\tilde{D}_x w_{cc})(\tilde{D}_x q) w_{cc} dx$, $q(\tilde{D}_x w_{cc}|_{x=0})(w_{cc}|_{x=0})$). In view of (4.8.4) we drop the positive second term on the left of (4.9.4), and obtain

$$\|\tilde{D}_x^2 w_{cc}\|^2_{L_2(R^1_{x+})} \leq \int_0^\infty \rho w_{cc}\,dx + |(\tilde{D}_x^3 w_{cc}|_{x=0})(w_{cc}|_{x=0})|$$

$$+ |(\tilde{D}_x^2 w_{cc}|_{x=0})(\tilde{D}_x w_{cc}|_{x=0}) + \ell.o.t. \tag{4.9.5}$$

Interior term. The interior term (which in Section 4.8 was responsible for problems) can now be readily estimated. From (4.9.2) we find the counterpart of (4.8.17), i.e.,

$$\int_0^\infty \rho w_{cc}\,dx = \mathcal{O}\Big[\|f\|_{H^{-s_0}(R^1_{x+})} \|w_{cc}\|_{H^{s_0}(R^1_{x+})}$$

$$+ (1+\gamma)\mathcal{O}((|\eta|^2 w_{cc}, |\eta|^2 w_{cc})_{L_2(R^1_{x+})}\Big]$$

$$+ \mathcal{O}\Big[(\tilde{D}_x w_{cc}, |\eta|^2 w_{cc})_{L_2(R^1_{x+})}\Big] + \mathcal{O}\Big[(\tilde{D}_x(\psi' w), |\eta|\tilde{D}_x w_{cc})_{L_2(R^1_{x+})}\Big]$$

$$+ \mathcal{O}\Big[(|\eta|^{3/2}\psi' w, |\eta|^{3/2} w_{cc})_{L_2(R^1_{x+})}\Big]. \tag{4.9.6}$$

To show (4.9.3), we see from (4.9.5) that we need to divide (4.9.6) by $\sigma+|\eta|$ and integrate over $\mathcal{E}_1$, thus obtaining

$$\int_{\mathcal{E}_1} \frac{\int_0^\infty \rho w_{cc}\,dx}{\sigma+|\eta|}\, d\sigma\, d\eta = \mathcal{O}\Big[\|f\|_{H^{-s_0}(Q_T)} + \|w_{cc}\|^2_{L_2(0,T;H^{s_0}(\Omega))}\Big]$$

$$+ \mathcal{O}\Big[\|w_{cc}\|^2_{L_2(0,T;H^{3/2}(\Omega))} + \|w\|^2_{L_2(0,T;H^1(\Omega))}\Big]. \tag{4.9.7}$$

Boundary terms. Recalling (4.9.1c) we have

$$\frac{(\tilde{D}_x^3 w_{cc}|_{x=0})(w_{cc}|_{x=0})}{\sigma+|\eta|} = \mathcal{O}\Big\{\frac{g_2}{|\eta|} w_{cc} + (|\eta| w_{cc})^2 + \gamma(\tilde{D}_x w_{cc})(|\eta| w_{cc})$$

$$+ g_1 w_{cc} + (\tilde{D}_x(\pi' w) w_{cc})\Big\} \text{ in } \mathcal{E}_1,\ x = 0, \tag{4.9.8}$$

from which using trace theory from the boundary to the interior we get

$$\int_{\mathcal{E}_1} \frac{|(\tilde{D}_x^3 w_{cc}|_{x=0})(w_{cc}|_{x=0})|}{\sigma+|\eta|}\, d\sigma d\eta$$

$$= \mathcal{O}\Big[\|g_2\|^2_{L_2(0,T;H^{-1}(\Gamma))} + \|w_{cc}\|_{L_2(0,T;H^{3/2}(\Omega))} + \gamma\|w_{cc}\|^2_{L_2(0,T;H^{3/2+\epsilon}(\Omega))}$$

$$+ \|g_1\|^2_{L_2(\Sigma_T)} + \|w\|_{L_2(0,T;H^{3/2+\epsilon}(\Omega))}\Big]. \qquad (4.9.9)$$

Finally, recalling (4.9.1b) we have

$$\frac{(\tilde{D}_x^2 w_{cc}|_{x=0})(\tilde{D}_x w_{cc}|_{x=0})}{\sigma+|\eta|} = \mathcal{O}\Big(\frac{g_1}{|\eta|}\tilde{D}_x w_{cc} + (|\eta| w_{cc})(\tilde{D}_x w_{cc}) + (\tilde{D}_x w_{cc})^2\Big)$$

$$\text{in } \mathcal{E}_1, \quad x = 0, \qquad (4.9.10)$$

from which again by trace theory we obtain as in (4.9.9),

$$\int_{\mathcal{E}_1} \frac{|(\tilde{D}_x^2 w_{cc}|_{x=0})(\tilde{D}_x w_{cc}|_{x=0})|}{\sigma+|\eta|}\, d\sigma d\eta$$

$$= \mathcal{O}\Big[\|g_1\|^2_{L_2(\Sigma_T)} + \|w_{cc}\|^2_{L_2(0,T;H^{3/2+\epsilon}(\Omega))}\Big]. \qquad (4.9.11)$$

Using (4.9.7), (4.9.9), (4.9.11) in (4.9.5) yields (4.9.3), as desired. ■

To complete the proof of Proposition 4.11, we notice that (4.9.3) yields (4.7.13). In fact, we have that $w_{cc} \equiv w$ on the interval $(\alpha/2,\alpha) \cup (T-\alpha, T-\alpha/2)$, where $\psi' \not\equiv 0$, so that $\psi' w_{cc} \equiv \psi' w$ on R^1_t. Then, returning to (4.7.13), we can replace $\psi' w$ by $\psi' w_{cc}$ in its left hand side:

$$\text{L.H.S. of (4.7.13)} = \|\chi(a\tilde{D}_x^2 D_t(\psi' w_{cc}))\|_{L_2(R^1_{x^+};H^{-3/2}(R^n_{ty}))}$$

$$\text{equivalent to} \int_{\mathcal{E}_1} \frac{\|\sigma\tilde{D}_x^2(\psi' w_{cc})\|^2_{L_2(R^1_{x^+})}}{\sigma^3+|\eta|^3}\, d\sigma d\eta \qquad (4.9.12)$$

$$\leq \int_{\mathcal{E}_1} \frac{\|\tilde{D}_x^2(\psi' w_{cc})\|^2_{L_2(R^1_{x^+})}}{\sigma + \frac{|\eta|^3}{\sigma^2}} d\sigma d\eta \leq \int_{\mathcal{E}_1} \frac{\|\tilde{D}_x^2(\psi' w_{cc})\|^2_{L_2(R^1_{x^+})}}{\sigma+\eta} d\sigma d\eta$$

$$\text{equivalent to } \|\Phi_{00}\tilde{D}_x^2(\psi' w_{cc})\|^2_{L_2(R^1_{x^+};H^{-\frac{1}{2}}(R^n_{ty}))}, \tag{4.9.13}$$

with symbol ϕ_{00} of Φ_{00} being of order zero in σ and η and supported in $\mathcal{E}_1$. The right hand side of (4.9.13) is then estimated by (4.9.3). Thus Proposition 4.11 is proved. ■

4.10. Final estimate for χw_c

THEOREM 4.16. With reference to χw_c we have with $s_0 < \frac{1}{2}$,

$$\|\chi w_c\|_{H^{\frac{5}{2}}(Q_\infty)} + \|\chi w_c\|_{H^2(\Sigma_\infty)} + \|\tilde{D}_x(\chi w_c)\|_{H^1(\Sigma_\infty)} + \|\tilde{D}_x^2(\chi w_c)\|_{L_2(\Sigma_\infty)}$$

$$\leq C_{\epsilon\gamma}\left\{\|f\|_{H^{-s_0}(Q_T)} + \|g_1\|_{L_2(\Sigma_T)} + \|g_2\|_{H^{-1}(\Sigma_T)} + \|w\|_{L_2(0,T;H^{\frac{3}{2}+\epsilon}(\Omega))}\right\}. \tag{4.10.1}$$

Proof. We use estimates (4.6.7) and (4.6.8) of Corollary 4.5 for $\hat{g}_1$ and $\hat{g}_2$ along with estimate (4.7.14) of Corollary 4.12 for F into the elliptic estimate (4.5.1) of Lemma 4.2, thereby obtaining (4.10.1). ■

4.11. Estimate for $(1-\chi)w_c$

Here the analysis is far simpler

PROPOSITION 4.17. With reference to $(1-\chi)w_c$ we have:

(i)

$$\|D_y^2(1-\chi)w_c\|_{L_2(\Sigma_\infty)} \leq C\left\{\|D_y(w_c)_t\|_{L_2(\Sigma_T)} + \|D_y w_c\|_{L_2(\Sigma_T)}\right\}$$

$$\leq C\left\{\|D_y w_t\|_{L_2(\Sigma_T)} + \|w\|_{L_2(0,T;H^{\frac{3}{2}}(\Omega))}\right\}; \tag{4.11.1}$$

(ii)

$$\|D_y\tilde{D}_x(1-\chi)w_c\|_{L_2(\Sigma_\infty)}+\|\tilde{D}_xD_y(1-\chi)w_c\|_{L_2(\Sigma_\infty)}$$

$$\leq C_\varepsilon\left\{\|\tilde{D}_x(w_c)_t\|_{L_2(\Sigma_T)}+\|w_c\|_{L_2(0,T;H^{3/2+\varepsilon}(\Omega))}\right\}$$

$$\leq C_\varepsilon\left\{\|\tilde{D}_xw_t\|_{L_2(\Sigma_T)}+\|w\|_{L_2(0,T;H^{3/2+\varepsilon}(\Omega))}\right\}; \quad (4.11.2)$$

(iii)

$$\|\tilde{D}_x^2(1-\chi)w_c\|_{L_2(\Sigma_\infty)} \leq C_\varepsilon\left\{\|g_1\|_{L_2(\Sigma_T)}+\|D_yw_t\|_{L_2(\Sigma_T)}+\|\tilde{D}_xw_t\|_{L_2(\Sigma_T)}\right.$$

$$\left.+\ \|w\|_{L_2(0,T;H^{3/2+\varepsilon}(\Omega))}\right\}. \quad (4.11.3)$$

Proof. In the cone $\mathcal{E}_2$ defined by (4.4.1), (4.4.2) with $2c_0|\eta| \leq \sigma$, where $\text{supp}(1-\chi) \subset \mathcal{E}_2$, we have that $|\eta|$ is dominated by σ. Thus, with support in here, the tangential derivative D_y is dominated by the time derivative D_t. Thus:

(i)

$$\|(1-\chi)D_y^2w_c\|_{L_2(\Sigma_\infty)} \leq C\|(1-\chi)D_y(w_c)_t\|_{L_2(\Sigma_T)}, \quad (4.11.4)$$

and (4.11.1) follows via a lower order commutator $[D_y^2, 1-\chi]$ of order 1 responsible for D_yw_c in $L_2(\Sigma_T)$, and trace theory.

(ii)

$$\|(1-\chi)D_y\tilde{D}_xw_c\|_{L_2(\Sigma_\infty)} \leq C\|(1-\chi)\tilde{D}_x(w_c)_t\|_{L_2(\Sigma_T)}, \quad (4.11.5)$$

and (4.11.2) follows via a lower order commutator $[D_y\tilde{D}_x, 1-\chi]$ of order 1 responsible for $\tilde{D}_xw_c$ and D_yw_c in $L_2(\Sigma_T)$ and trace theory.

(iii) From (4.3.5) we obtain

$$(1-\chi)\tilde{D}_x^2w_c = (1-\chi)\psi g_1-(1-\chi)\tilde{D}_y^2w_c-(1-\chi)\beta_1w_c, \quad (4.11.6)$$

from which we obtain via (4.11.4) and (4.11.5), upon recalling the definition (4.2.9a) for β_1:

$$\|(1-\chi)\tilde{D}_x^2 w_c\|_{L_2(\Sigma_\infty)} \leq C_\varepsilon\Big\{\|g_1\|_{L_2(\Sigma_T)}+\|D_y(w_c)_t\|_{L_2(\Sigma_T)}$$

$$+\|\tilde{D}_x(w_c)_t\|_{L_2(\Sigma_T)}+\|w_c\|^2_{L_2(0,T;H^{3/2+\varepsilon}(\Omega))}\Big\}$$

$$\leq C_\varepsilon\Big\{\|g_1\|_{L_2(\Sigma_T)}+\|D_y w_t\|_{L_2(\Sigma_T)}+\|\tilde{D}_x w_t\|_{L_2(\Sigma_T)}$$

$$+\|w\|_{L_2(0,T;H^{3/2+\varepsilon}(\Omega))}\Big\}, \qquad (4.11.7)$$

from which (4.11.3) follows via a lower order commutator $[\tilde{D}_x^2, 1-\chi]$ of order one, responsible for $\tilde{D}_x w_c$, and trace theory. ■

4.12. Conclusion of the proof of Theorem 2.1

Since $w_c = \chi w_c+(1-\chi)w_c$, we invoke both estimate (4.10.1) of Theorem 4.17 for χw_c and estimates (4.11.1)-(4.11.3) of Proposition 4.18 for $(1-\chi)w_c$, to get the *final estimate* of w_c:

$$\|D_y^2 w_c\|_{L_2(\Sigma_T)}+\|D_y\tilde{D}_x w_c\|_{L_2(\Sigma_T)}+\|\tilde{D}_x^2 w_c\|_{L_2(\Sigma_T)}$$

$$\leq C_{\varepsilon\gamma}\Big\{\|f\|_{H^{-s_0}(Q_T)}+\|g_1\|_{L_2(\Sigma_T)}+\|g_2\|_{H^{-1}(\Sigma_T)}+\|w\|_{L_2(0,T;H^{3/2+\varepsilon}(\Omega))}$$

$$+\|w_t\|_{L_2(\Sigma_T)}+\|D_y w_t\|_{L_2(\Sigma_T)}+\|\tilde{D}_x w_t\|_{L_2(\Sigma_T)}\Big\}. \qquad (4.12.1)$$

Recalling then that $w \equiv w_c$ on $(\alpha, T-\alpha)$, we finally obtain estimate (4.2.12) of the Main Theorem 4.0 from (4.12.1). The proof of Theorem 4.0 is complete. ■

5. EULER-BERNOULLI PROBLEM. PROOF OF THEOREM 2.3

In this section we shall briefly indicate the modifications that need to be made on the proof of Section 4 in order to obtain Theorem 2.3 for the Euler-Bernoulli problem which is obtained by setting $\gamma = 0$ in Eqns. (4.1a) and (4.1d). This results, first of all, in the time localized problem (4.3.1) where now the *time commutators* $[P,\psi]$ *and* $[B_2,\psi]$ *are both zero.*

The first noteworthy modification occurs in the localization problem in the dual variables. Qualitatively, in the present case of the Euler-Bernoulli problem, one time derivative corresponds to two space derivatives (rather than to one space derivative as for the Kirchhoff problem). We set

$$\varsigma^2 = \sigma > 0; \qquad \varsigma = \sigma^{1/2} > 0, \tag{5.1}$$

and divide now the quarter space $R^{2n}(+)$ in the (x,y,ς,η)-variables (rather than (x,y,σ,η)-variable as in Section 4) accordingly,

$$\mathcal{R}_1 = \{(x,y;\varsigma,\eta) \in R^{2n}(+): 0 < \varsigma < \sqrt{c_0}\,|\eta| \iff \sigma < c_0|\eta|^2\}; \tag{5.2a}$$

$$\mathcal{R}_{tr} = \{(x,y;\varsigma,\eta) \in R^{2n}(+): \sqrt{c_0}\,|\eta| \le \varsigma < \sqrt{2c_0}\,|\eta| \iff c_0|\eta|^2 \le \sigma < 2c_0|\eta|^2\}; \tag{5.2b}$$

$$\mathcal{R}_2 = \{(x,y;\varsigma,\eta) \in R^{2n}(+): \sqrt{2c_0}\,|\eta| \le \varsigma \iff 2c_0|\eta|^2 \le \sigma\}. \tag{5.2c}$$

$$\mathcal{E}_1 = \mathcal{R}_1 \cup \mathcal{R}_{tr}; \qquad \mathcal{E}_2 = \mathcal{R}_2 \cup \mathcal{R}_{tr}. \tag{5.3}$$

The role of the 'time-scaling' $\sigma \to \sigma^{1/2} = \varsigma$ is to obtain conical regions in the (ς,η)-coordinates rather than parabolic sectors in the (σ,η)-coordinates. The symbol

$$p(x,y;\varsigma,\xi,\eta) = -a(\varsigma^2 - ir_0)^2 + [\tilde{\xi}^2 + \frac{d}{a^2}]^2 \tag{5.4}$$

satisfies now, by selecting a suitably small constant c_0 in (5.2), the ellipticity property

$$\text{Re}\ p \ge \delta[\tilde{\xi}^4 + |\eta|^4 + \varsigma^4] \quad \text{in } \mathcal{E}_1, \tag{5.5}$$

and so p is elliptic of order 4 in all variables $\{\varsigma,\xi,\eta\}$ in $\mathcal{E}_1$. Thus, we now let $\chi(x,y;\varsigma,\eta)$ be a smooth homogeneous symbol of order zero of localization satisfying: $\chi \equiv 1$ in $\mathcal{R}_1$; $\chi \equiv 0$ in $\mathcal{R}_2$; supp $\chi \subset \mathcal{E}_1$, as in (4.4.4). Let $\mathcal{X}$ be the corresponding pseudo-differential operator.

Estimate of $\mathcal{X}w_c$. The proof of Sections 4.5 through 4.10 goes through in a simplified version since now $[B_2,\Psi] = 0$ and $[P,\Psi] = 0$. Moreover, the dual variable σ there (corresponding to the primal variable t) is replaced now by the dual variable ς

(corresponding to another primal variable, say u). In fact, now estimate (4.3.4) and Lemma 4.10 are not needed.

Thus, the term $a\tilde{D}_x^2 D_t(\psi' w)$ of Sections 4.7 through 4.10 does not come into play now and it is replaced in the counterpart of Corollary 4.11 for F by the better behaved term $\Phi_{00}\tilde{D}_x^2(\psi' w)$ in the norm of $L_2(R^1_{x^+};H^{-3/2}(R^n_{uy}))$, which comes from Lemma 4.9.

As a result, the present final estimate for χw_c will be the counterpart of estimate (4.10.1) of Theorem 4.17 with the time regularity there being halved now. Thus, setting as usual,

$$H^{\alpha,\beta}(Q_\infty) = L_2(R^1_t;H^\beta(\Omega)) \cap H^\beta(R^1_t;L_2(\Omega)),$$

and likewise for $H^{\alpha,\beta}(\Sigma_\infty)$, we obtain from (4.10.1) the following final estimate of χw_c in the Euler-Bernoulli problem.

PROPOSITION 5.1. We have

$$\|\chi w_c\|_{H^{3/2,3/4}(Q_\infty)} + \|\chi w_c\|_{H^{2,1}(\Sigma_\infty)} + \|\tilde{D}_x(\chi w_c)\|_{H^{1,1/2}(\Sigma_\infty)} + \|\tilde{D}_x^2(\chi w_c\|_{L_2(\Sigma_\infty)}$$

$$\leq C_\varepsilon \Big\{ \|f\|_{H^{-s_0,-s_0/2}(Q_T)} + \|g_1\|_{L_2(\Sigma_T)} + \|g_2\|_{H^{-1,-1/2}(\Sigma_T)}$$

$$+ \|w\|_{L_2(0,T;H^{3/2+\varepsilon}(\Omega))} \Big\}. \quad (5.6)$$

Estimate of $(1-\chi)w_c$. Now, in the region $\mathcal{E}_2$, we have by (5.2) that the tangential derivatives D_y, respectively, D_y^2 are dominated in norm by the time derivatives $D_t^{1/2}$, respectively, D_t. Thus, with reference to Proposition 4.18 and its proof, we have now the following counterpart versions.

(i) The counterpart of (4.11.4) is now

$$\|(1-\chi)D_y^2 w_c\|^2_{L_2(\Sigma_\infty)} \leq C\int_{\mathcal{E}_2} ||\eta|^2 w_c|^2 d\sigma\, d\eta \leq C\int_{\mathcal{E}_2} |\sigma w_c|^2 d\sigma\, d\eta$$

$$\leq C\|D_t(w_c)\|^2_{L_2(\Sigma_T)}. \quad (5.7)$$

(ii) The counterpart of (4.11.5) is now

$$\|(1-\chi)D_y\tilde{D}_x^2 w_c\|^2_{L_2(\Sigma_\infty)} \le C\int_{\mathcal{E}_2} \big||\eta|\tilde{D}_x w_c\big|^2 d\sigma\, d\eta \le C\int_{\mathcal{E}_2} |\sigma^{1/2}\tilde{D}_x w_c|^2 d\sigma\, d\eta$$

$$\left[\begin{array}{ll} \text{equivalent to } \|\tilde{D}_x w_c\|^2_{H^{1/2}(R^1_t;L_2(\Gamma))} & (5.8) \\ \le C\int_{\mathcal{E}_2} \left|\dfrac{\sigma\tilde{D}_x w_c}{\eta}\right|^2 d\sigma\, d\eta \text{ equiv. to } \|\tilde{D}_x D_t w_c\|^2_{L_2(R^1_t;H^{-1}(\Gamma))}. & (5.9) \end{array}\right.$$

(iii) Using now (5.7) and (5.8), resp. (5.9), in (4.11.5), we obtain the counterpart of (4.11.7):

$$\|(1-\chi)\tilde{D}_x^2 w_c\|_{L_2(\Sigma_\infty)} \le C\Bigg\{\|g_1\|_{L_2(\Sigma_T)} + \|D_t(w_c)\|_{L_2(\Sigma_T)}$$

$$+ \begin{cases} \|\tilde{D}_x w_c\|_{H^{1/2}(R^1_t;L_2(\Gamma))}; & (5.10) \\ \|\tilde{D}_x D_t w_c\|_{L_2(R^1_t;H^{-1}(\Gamma))}. & (5.11) \end{cases}$$

Taking now into account commutators as in the proof of Proposition 4.18, we then obtain its counterpart in the following

PROPOSITION 5.2. We have

(i)

$$\|D_y^2(1-\chi)w_c\|_{L_2(\Sigma_\infty)} \le C\left\{\|D_t(w_c)\|_{L_2(\Sigma_T)} + \|w\|_{L_2(0,T;H^{3/2+\varepsilon}(\Omega))}\right\}; \tag{5.12}$$

(ii)

$$\|D_y\tilde{D}_x(1-\chi)w_c\|_{L_2(\Sigma_\infty)} + \|\tilde{D}_x D_y(1-\chi)w_c\|_{L_2(\Sigma_\infty)}$$

$$\le C\left\{\|w_c\|_{L_2(0,T;H^{3/2+\varepsilon}(\Omega))} + \begin{cases} \|\tilde{D}_x w_c\|_{H^{1/2}(R^1_t;L_2(\Gamma))} & (5.13) \\ \|\tilde{D}_x D_t w_c\|_{L_2(R^1_t;H^{-1}(\Gamma))} & (5.14) \end{cases}\right\}.$$

(iii)

$$\|\tilde{D}_x^2(1-\chi)w_c\|_{L_2(\Sigma_\infty)} \le \Big\{\|g_1\|_{L_2(\Sigma_T)}+\|D_t(w_c)\|_{L_2(\Sigma_T)}$$

$$+ \|w_c\|_{L_2(0,T;H^{3/2+\varepsilon}(\Omega))}$$

$$+ \left\{\begin{array}{l}\|\tilde{D}_x w_c\|_{H^{1/2}(R_t^1;L_2(\Gamma))} \qquad (5.15)\\ \|\tilde{D}_x D_t w_c\|_{L_2(R_t^1;H^{-1}(\Gamma))} \qquad (5.16)\end{array}\right\}. \quad \blacksquare$$

Thus, Proposition 5.1 and Proposition 5.2 (version (5.16)), combined in $w_c = \chi w_c+(1-\chi)w_c$ yield the final estimate for w_c.

PROPOSITION 5.3. We have

$$\|D_y^2 w_c\|_{L_2(\Sigma_T)}+\|D_y\tilde{D}_x w_c\|_{L_2(\Sigma_T)}+\|\tilde{D}_x^2 w_c\|_{L_2(\Sigma_T)}$$

$$\le C_\varepsilon\Big\{\|f\|_{H^{-s_0,-s_0/2}(Q_T)}+\|g_1\|_{L_2(\Sigma_T)}+\|g_2\|_{H^{-1,-1/2}(\Sigma_T)}$$

$$+ \|w\|_{L_2(0,T;H^{3/2+\varepsilon}(\Omega))}+\|D_t w_c\|_{L_2(\Sigma_T)}+\|\tilde{D}_x D_t w_c\|_{L_2(0,T;H^{-1}(\Gamma))}\Big\}. \quad \blacksquare \tag{5.17}$$

Finally, since $(w_c)_t = \psi w_t+\psi' w$, we have

$$\|D_t w_c\|_{L_2(\Sigma_T)}+\|\tilde{D}_x D_t w_c\|_{L_2(0,T;H^{-1}(\Gamma))}$$

$$\le C\Big\{\|D_t w\|_{L_2(\Sigma_T)}+\|\tilde{D}_x D_t w\|_{L_2(0,T;H^{-1}(\Gamma)}+\|w\|_{L_2(0,T;H^{3/2+\varepsilon}(\Omega))}\Big\}, \tag{5.18}$$

using trace theory on $\tilde{D}_x w$. Then using (5.18) into (5.17), and recalling $w \equiv w_c$ in $(\alpha, T-\alpha)$, we obtain Theorem 2.3.

REFERENCES

[H.1] L. Hormander, *The Analysis of Linear Partial Differential Operators*, Springer-Verlag.

[H.2] M. A. Horn, Uniform decay rates for the solutions to the Euler-Bernoulli plate equation with boundary feedback acting via bending moments, University of Virginia, 1991, Differential and Integral Equations, to appear.

[Lag.1] J. Lagnese, *Boundary Stabilization of Thin Plates*, SIAM Studies in Appl. Math., Philadelphia, 1989.

[Lag.2] J. Lagnese, A note on the boundary stabilization of wave equations, SIAM J. Control & Optimiz. 26 (1988), 1250-1256.

[Lag.3] J. Lagnese, Uniform boundary stabilization of homogeneous isotropic plates, Lecture Notes in Inform. & Control Sciences, vol. #102, Springer Verlag 19, Proceedings of Conference held in Vorau, Austria, July 1986.

[L-T.1] I. Lasiecka and R. Triggiani, Uniform stabilization of the wave equation with Dirichlet-feedback control without geometrical conditions, Appl. Math. & Optimiz., vol. 25 (1992), 189-224.

[L-T.2] I. Lasiecka and R. Triggiani, Exact controllability of the wave equation with Neumann boundary control, Appl. Math. & Optimiz. 19 (1989), 243-290.

[Lit.1] W. Littman, Near optimal time boundary controllability for a class of hyperbolic equations, Springer-Verlag Lecture Notes, LNICS #97, 1987, pp. 307-312.

[L-M.1] J. L. Lions and E. Magenes, *Non-homogeneous boundary value problems*. Springer-Verlag, 1972.

[L-L.1] J. Lagnese and J. L. Lions, *Modelling, analysis, and control of thin plates*, Masson, 1988.

Boundary Values of Holomorphic Semigroups, H^∞ Functional Calculi, and the Inhomogeneous Abstract Cauchy Problem

RALPH DELAUBENFELS Mathematics Department, Ohio University, Athens, Ohio 45701, U.S.A.

0. INTRODUCTION. I would like to discuss some joint work with Khristo Boyadzhiev. Section I discusses how to "regularize" the (usually) unbounded group $\{e^{isB}\}_{s\in\mathbf{R}}$, when $\{e^{zB}\}_{Re(z)>0}$ is a bounded strongly continuous holomorphic semigroup of angle $\frac{\pi}{2}$. In Section II, we will focus on the particular semigroup $\{A^{-z}\}_{Re(z)>0}$, the fractional powers of a fixed operator. Section III will discuss regularized H^∞ functional calculi; this produces an H^∞ functional calculus, hence bounded imaginary powers, on a subspace. Section IV discusses the inhomogeneous abstract Cauchy problem.

Except in the last section, we will give merely outlines of the proofs. This will be more satisfying for the reader, since the proofs consist of a construction, followed by a verification that the construction does what one would expect; the reader will be spared the relatively tedious details of these verifications.

All operators are linear, on a Banach space, X. We will write $\mathcal{D}(A)$ for the domain of the operator A, $B(X)$ for the space of bounded linear operators from X into itself.

I. BOUNDARY VALUES AND REGULARIZED SEMIGROUPS. By $\{e^{zB}\}_{Re(z)>0}$, we will mean a bounded strongly continuous holomorphic semigroup of angle $\frac{\pi}{2}$, generated by B. This terminology is somewhat ambiguous, in that $z \mapsto e^{zB}$ is bounded in $S_\theta \equiv \{z \mid |arg(z)| < \theta\}$ whenever $\theta < \frac{\pi}{2}$, but it may not be bounded in the right half-plane $S_{\frac{\pi}{2}}$;

$$M(\theta) \equiv \sup_{z\in S_\theta} \|e^{zB}\|$$

will, in general, diverge to infinity, as $\theta \to \frac{\pi}{2}$.

Note that iB generates a bounded strongly continuous group if and only if $M(\theta)$ is bounded as θ approaches $\frac{\pi}{2}$.

Example 1.1. Let $B \equiv \Delta$, the Laplacian, on $L^p(\mathbf{R}^n), 1 \leq p < \infty$, with maximal domain. Then $M(\theta) = \left(\frac{1}{\cos(\theta)}\right)^{n|\frac{1}{p}-\frac{1}{2}|}$.

By the *boundary values* of $\{e^{zB}\}_{Re(z)>0}$, we mean that we wish to let z approach the imaginary axis, and make sense out of $\{e^{isB}\}_{s\in\mathbf{R}}$. Example 1.1 demonstrates that this "sense" must be more general than strongly continuous semigroups. We will describe our results in the language of regularized semigroups. Thus we will briefly digress, to present the definition, and some motivating properties of, regularized semigroups.

Definition 1.2. Suppose C is a bounded, injective operator. The strongly continuous family of bounded operators $\{W(t)\}_{t\geq 0}$ is a *C-regularized semigroup* (introduced in [10]) or *C-semigroup* (introduced in [13]) if $W(0) = C$ and $CW(t+s) = W(t)W(s), \forall s, t \geq 0$.

The operator A *generates* $\{W(t)\}_{t\geq 0}$ if

$$Ax = C^{-1}\left(\lim_{t\to 0}\frac{1}{t}(W(t)x - Cx)\right),$$

with maximal domain, that is, $\mathcal{D}(A)$ equals the set of all x for which the limit exists and is in the image of C.

The generator is automatically closed.

We think of $W(t)$ as being $e^{tA}C = Ce^{tA}$. Many operators that do not generate strongly continuous semigroups generate regularized semigroups(see [14], [15], [16] and [17]).

What desirable information do we gain by knowing that an operator generates a regularized semigroup?

We use the following terminology. When C is a bounded, injective operator, $[Im(C)]$ will be the Banach space with norm $\|x\|_{[Im(C)]} \equiv \|C^{-1}x\|$. When Y and W are Frechet spaces, we will write $Y \hookrightarrow W$ to mean that Y is continuously embedded in W, that is, $Y \subseteq W$ and the identity map from Y into W is continuous. By the *abstract Cauchy problem* we will mean

$$\frac{d}{dt}u(t,x) = A(u(t,x))\,(t \geq 0),\; u(0,x) = x. \tag{1.3}$$

Proposition 1.4(see [16]). *Suppose A generates a C-regularized semigroup. Then*

(1) *(1.3) has a unique mild solution, $\forall x \in Im(C)$. The solutions $u(t, Cx_n) \to 0$, uniformly on compact sets, whenever $x_n \to 0$.*

(2) *$\exists Z$, a Frechet space, such that*

$$[Im(C)] \hookrightarrow Z \hookrightarrow X,$$

and $A|_Z$ generates a strongly continuous semigroup.

If the C-regularized semigroup is exponentially bounded, then Z may be chosen to be a Banach space.

Thus we obtain a lot of solutions, and, perhaps more importantly, well-posedness on a *subspace*. The choice of C measures how far from well-posedness we are. It also gives an approximation of the subspace on which the abstract Cauchy problem is well-posed.

Clearly, the goal is to find C whose image is as large as possible.

The following theorem is saying that the amount of regularizing required to make e^{isB} a strongly continuous family of bounded operators is equated with the rate of growth of $\|e^{zB}\|$ as z approaches the imaginary axis. Note that $\frac{|z|}{Re(z)} = \frac{1}{\cos(\theta)}$, where θ is the argument of z.

Theorem 1.5. *Suppose $\gamma \geq 0$. Then the following are equivalent.*

(a) $\forall r > \gamma, \exists M_r < \infty$ *such that*

$$\|e^{zB}\| \leq M_r\left(\frac{1+|z|}{Re(z)}\right)^r.$$

(b) $\forall r > \gamma, \exists C_r$ *such that* iB *generates a* $(1-B)^{-r}$*- regularized group* $\{W_r(s)\}_{s\in\mathbf{R}}$ *such that*

$$\|W_r(s)\| \leq C_r(1+|s|^r).$$

Proof: (a) $\rightarrow$ (b). For $r > 0$, we have the well-known resolvent formula,

$$(1-B)^{-r}x = \frac{1}{\Gamma(r)}\int_0^\infty e^{-t}t^{r-1}e^{tB}x\,dt,$$

thus, at least formally,

$$e^{isB}(1-B)^{-r}x = \frac{1}{\Gamma(r)}\int_0^\infty e^{-t}t^{r-1}e^{(t+is)B}x\,dt;$$

elementary calculations show that choosing $r > \gamma$ will give us a strongly continuous family of bounded operators $\{W_r(s)\}_{s\in\mathbf{R}} = \{e^{isB}(1-B)^{-r}\}_{s\in\mathbf{R}}$, with the polynomial growth condition of (b).

(b) $\rightarrow$ (a). Fix $r > \gamma$. For $z = t + is$, with $t > 0$,

$$\begin{aligned}\|e^{zB}\| &= \|(1-B)^r e^{tB}W_r(s)\|\\ &\leq \|(1-B)^r e^{tB}\|\|W_r(s)\|\\ &\leq K_r t^{-r}C_r(1+|s|^r) \leq (K_r)(C_r)\left(\frac{1+|z|}{Re(z)}\right)^r,\end{aligned}$$

for some constant K_r, since e^{zB} is a bounded strongly continuous holomorphic semigroup. ∎

The proof of (a) implies (b) actually gives us a better growth condition on the regularized group.

Theorem 1.6. *Suppose $\gamma \geq 0$ and $\exists M < \infty$ such that*

$$\|e^{zB}\| \leq M\left(\frac{1+|z|}{Re(z)}\right)^\gamma.$$

Then, $\forall r > \gamma, \exists C_{r,\gamma}$ *such that* iB *generates a* $(1-B)^{-r}$*- regularized group* $\{W_r(s)\}_{s\in\mathbf{R}}$ *such that*

$$\|W_r(s)\| \leq C_{r,\gamma}(1+|s|^\gamma).$$

Example 1.7. Let B be as in Example 1.1. By writing the heat semigroup e^{zB} as a convolution operator, a direct computation shows that

$$\|e^{zB}\|_1 = \left(\frac{|z|}{Re(z)}\right)^{\frac{n}{2}}.$$

It is also well-known, by the spectral theorem, that $\|e^{z\Delta}\|_2 = 1$, whenever $Re(z) > 0$. An application of the Riesz convexity theorem then shows that, for $1 \leq p \leq \infty$,

$$\|e^{zB}\|_p \leq \left(\frac{|z|}{Re(z)}\right)^{n|\frac{1}{p}-\frac{1}{2}|},$$

whenever $Re(z) > 0$.

Thus Theorem 1.6 gives us

Theorem 1.8. *Suppose $1 \leq p < \infty$. Then $\forall r > n|\frac{1}{p} - \frac{1}{2}|, i\Delta$, on $L^p(\mathbf{R}^n)$, generates a $(1-\Delta)^{-r}$-regularized group, $\{W_r(s)\}_{s\in\mathbf{R}}$ that is $O(1+|s|^{n|\frac{1}{p}-\frac{1}{2}|})$.*

On $C_0(\mathbf{R}^n)$ or $BUC(\mathbf{R}^n), i\Delta$ generates a $(1-\Delta)^{-r}$-regularized group, $\{W_r(s)\}_{s\in\mathbf{R}}$ that is $O(1+|s|^{\frac{n}{2}}), \forall r > \frac{n}{2}$.

In different mathematical languages, Theorem 1.8 has appeared in many places: in [46], in the language of Fourier multipliers, in [1], in the language of smooth distribution groups and, more recently, in [26], in the language of integrated semigroups, to name a few. We offer our proof as a much simpler, and conceptually more clear, method.

Example 1.9. By using more sophisticated norm estimates, we may apply Theorem 1.6 to obtain new results about the Schrödinger operator *with* potential, $i(\Delta - V)$, for appropriate potentials V(see also [2] and [39]).

Denote by K^n the *Kato class* of measurable functions on $\mathbf{R}^n$, as defined in [45, p. 453]. This includes, but is not limited to, $L^\infty(\mathbf{R}^n)$.

For $V \in K^n$, it is shown in [45, Theorem A.2.7] that

$$H \equiv \Delta - V,$$

defined as a quadratic form, is a self-adjoint operator on $L^2(\mathbf{R}^n)$.

The following theorem states that iH, with appropriate domain, generates an exponentially bounded $(\omega - H)^{-r}$-regularized group, for r twice as big as in Theorem 1.8.

Theorem 1.10. *Suppose $1 \leq p < \infty$, $V_+ \in K^n, V_- \in L^\infty(\mathbf{R}^n)$. Let H be as in Example 1.9. Then $\exists \omega \in \mathbf{R}$ such that $\forall r > 2n|\frac{1}{p} - \frac{1}{2}|$, $\{e^{isH}(\omega - H)^{-r}\}_{s\in\mathbf{R}}$ is an $(\omega - H)^{-r}$-regularized group on $L^p(\mathbf{R}^n)$ that is $O(1+|s|^{2n|\frac{1}{p}-\frac{1}{2}|})$.*

On $C_0(\mathbf{R}^n)$ or $BUC(\mathbf{R}^n), \{e^{isH}(\omega - H)^{-r}\}_{s\in\mathbf{R}}$ is an $(\omega - H)^{-r}$-regularized group that is $O(1+|s|^{\frac{n}{2}}), \forall r > n$.

The proof is essentially the same as the proof of Theorem 1.8, using the following([45, Theorem B.7.1] and [12, Theorem 9]; see also [39, Propositions 2.1 and 2.4]), with Δ replaced by $H + 1 - \omega$, where $\omega \equiv \max\{\|V_-\|_\infty, \mu\} + 1$.

Lemma 1.11. *Let H be as in Example 1.9. Then $\exists \mu, c, a \in \mathbf{R}^+$ and a kernel $\tilde{K}(z,x,y)$ such that*

$$e^{z(H-\mu)} f(x) = \int_{\mathbf{R}^n} \tilde{K}(z,x,y) f(y)\, dy,$$

for $Re(z) > 0, f \in L^p(\mathbf{R}^n)(1 \leq p \leq \infty)$ and

$$|\tilde{K}(z,x,y)| \leq c(Re(z))^{-\frac{n}{2}} \exp\left(-Re\left(\frac{|x-y|^2}{az}\right)\right),$$

for $Re(z) > 0, x, y \in \mathbf{R}^n$.

II. BOUNDED IMAGINARY POWERS AND SEMIGROUPS OF BOUNDED VARIATION.

In this section, we restrict ourselves to the semigroup of fractional powers of a fixed operator, $\{A^{-z}\}_{Re(z)>0}$, so that the boundary values are the imaginary powers $\{A^{-is}\}_{s\in\mathbf{R}}$. Of course, A^{-z} will not be bounded, for $Re(z) > 0$, unless $0 \in \rho(A)$, but a satisfactory theory of fractional powers exists without this hypothesis. We will assume that A is injective and has dense range.

Definition 2.1. We will write S_θ for $\{re^{i\phi} | r > 0, |\phi| < \theta\}$, H_ϵ for $\{z \| Im(z)| < \epsilon\}$.

Definition 2.2. Suppose Ω is an open subset of the complex plane whose closure is not the entire plane. We will say that A has an *$H^\infty(\Omega)$-functional calculus* if $sp(A) \subseteq \overline{\Omega}$ and $\exists$ a continuous algebra homomorphism, $f \mapsto f(A)$, from $H^\infty(\Omega)$ into $L(X)$, such that $(\lambda - A)^{-1} = f_\lambda(A), \forall \lambda \notin \overline{\Omega}$, where $f_\lambda(z) \equiv (\lambda - z)^{-1}$.

The recent papers [18], [19], [41] are concerned primarily with the consequences of $\{A^{-is}\}_{s\in\mathbf{R}}$ being a strongly continuous group of bounded operators. In this section, we give some simple necessary conditions and some sufficient conditions for this to occur, along with a characterization of a stronger condition, having an $H^\infty(S_\theta)$ functional calculus. We will write $\{e^{tB}\}_{t\geq 0}$ to mean a strongly continuous semigroup generated by B.

In [6], we also characterize operators with an $H^\infty(S_\theta)$ functional calculus, for $0 < \theta < \pi$, in terms of the variation of $t \mapsto A(t + A)^{-1}$.

Consider the following.

(1) A has an $H^\infty(S_\theta)$ functional calculus, for some $0 < \theta < \frac{\pi}{2}$.

(2) $\{A^{-is}\}_{s\in\mathbf{R}}$ is a strongly continuous group of bounded operators of exponential type less than $\frac{\pi}{2}$.

(3_q) $\{e^{-tA}\}_{t\geq 0}$ is holomorphic and $\exists \psi > 0$ such that

$$\int_0^\infty |tx^*(Ae^{-te^{i\phi}A}x)|^q \,\frac{dt}{t} < M_q\|x^*\|\|x\|,$$

$\forall x^* \in X^*, x \in X, |\phi| < \psi$.

(4_q) $\{e^{-tA}\}_{t\geq 0}$ is holomorphic and $\exists \psi > 0$ such that

$$\int_0^\infty \|tAe^{-te^{i\phi}A}x\|^q \,\frac{dt}{t} < M_q\|x\|,$$

$\forall x \in X, |\phi| < \psi$. .

(5_q) $\{e^{-tA}\}_{t\geq 0}$ is holomorphic and

$$\int_0^\infty \|tAe^{-tA}\|^q \,\frac{dt}{t} < \infty,$$

Theorem 2.3.

(a) $(1) \leftrightarrow (3_1)$.

(b) $(1) \rightarrow (2)$.

(c) *If X is a Hilbert space,* $(1) \leftrightarrow (2)$.

(d) $(2) \rightarrow (3_q), \forall q \geq 2$.

(e) $(4_2) \rightarrow (2)$.

(f) $(5_2) \leftrightarrow$ *A is bounded and $sp(A) \subseteq S_\theta$, for some $\theta < \frac{\pi}{2}$.*

Note that (3_1) is saying that $t \mapsto e^{-tA}$ is (uniformly weakly) a semigroup of bounded variation. Thus we would like to restate part (a), after some preliminary definitions.

Definition 2.4. We will say that the strongly continuous semigroup $\{e^{sA}\}_{s\geq 0}$ is *uniformly weakly of bounded variation* if $\exists$ a constant M such that

$$\int_0^\infty |< e^{rA}x, A^*x^* >| \,dr \leq M\|x\|\|x^*\|,$$

$\forall x \in X, x^* \in \mathcal{D}(A^*)$.

If $\{e^{zA}\}_{z\in S_\theta}$ is a strongly continuous holomorphic semigroup, we will say that it is a *strongly continuous holomorphic semigroup of angle θ that is uniformly weakly of bounded variation* , if $\forall\psi < \theta, \{e^{se^{i\psi}A}\}_{s\geq 0}$ is uniformly weakly of bounded variation, with constant M_ψ, and $\{M_\phi \,|\, |\phi| < \psi\}$ is bounded, that is, $\forall\psi < \theta$, $\exists M_\psi < \infty$ such that

$$\int_0^\infty |x^*(Ae^{te^{i\phi}A})x)|\, dt \leq M_\psi \|x^*\| \|x\|,$$

$\forall x^* \in X^*, x \in X, |\phi| < \psi$.

The following theorem may be considered to be in the spirit of the Spectral Theorem on a Hilbert space, which asserts an equivalence between an operator, A, having a $C_0(\mathbf{R})$ functional calculus, iA generating a bounded, strongly continuous group, and A (or the group it generates) having a certain integral representation. Being of bounded variation is analogous to being continuous, in the sense that it describes "good" behaviour of a function, but, unlike continuity, very little is known about semigroups of operators of bounded variation.

We will write $B \equiv logA$ to mean that A has bounded imaginary powers $\{A^{-is}\}_{s\in\mathbf{R}}$ and $-iB$ is the generator of $\{A^{-is}\}_{s\in\mathbf{R}}$.

We remark that the equivalence of (c) and (d) is true for any operator B; if iB generates a strongly continuous group, $-A$ may be defined to be the *analytic generator* of e^{isB}(see [9]), and it may then be shown that $B = -logA$.

Theorem 2.5. *Suppose $0 < \theta \leq \frac{\pi}{2}$. Then the following are equivalent.*

(a) *A has an $H^\infty(S_\phi)$-functional calculus, $\forall \frac{\pi}{2} > \phi > \frac{\pi}{2} - \theta$.*

(b) *$-A$ generates a bounded strongly continuous holomorphic semigroup of angle θ, that is uniformly weakly of bounded variation.*

(c) *$logA$ has an $H^\infty(H_\phi)$-functional calculus, $\forall \frac{\pi}{2} > \phi > \frac{\pi}{2} - \theta$.*

(d) *$iB \equiv -ilogA$ generates a strongly continuous group of bounded operators $\{e^{isB}\}_{s\in\mathbf{R}}$ and, for $\psi < \theta, j = 1,2, x^* \in X^*, x \in X, \exists$ signed measures $E_{j,x^*,x}$, on $\mathbf{R}$, of bounded variation, such that the map $(x^*,x) \mapsto E_{j,x^*,x}$ is bilinear, for $j = 1,2$, and continuously differentiable f_j such that $f_1(s), f_1'(s)$ are $O(e^{-(\frac{\pi}{2}-\psi)s})$, as $s \to -\infty$, $f_2(s), f_2'(s)$ are $O(e^{(\frac{\pi}{2}-\psi)s})$, as $s \to \infty$, with*

$$x^*(e^{isB}x) = f_j(s) \int_{\mathbf{R}} e^{isr}\, dE_{j,x^*,x}(r),$$

for $j = 1,2, x^* \in X^*, x \in X, s \in \mathbf{R}$.

Proof: (a) $\leftrightarrow$ (c) is clear by a change of variables: if $B \equiv -logA$, then given (c), we define, for $f \in H^\infty(S_\phi)$, $f(A) \equiv (f\circ g)(B)$, where $g(z) \equiv e^{-z}$, and given (a), we define, for $f \in H^\infty(H_\phi)$, $f(B) \equiv (f \circ g^{-1})(A)$. The resolvent requirement, in the definition of an H^∞ functional calculus, is verified by appropriate integral transforms.

(d) $\to$ (c). Fix $\phi > (\frac{\pi}{2} - \theta)$. Choose $\psi < \theta$ such that $\phi > (\frac{\pi}{2} - \psi)$.

Choose C^∞ h such that $h \equiv 1$ on $[1,\infty), \equiv 0$ on $(-\infty, 0]$, let $g_1 \equiv hf_1$, $g_2 \equiv (1-h)f_2$. Then $g_j, j = 1,2$, is continuously differentiable, with $g_j(s)$ and $g_j'(s)O(e^{(\frac{\pi}{2}-\psi)|s|})$, as $|s| \to \infty$, the support of g_1 is contained in $[0,\infty)$, support of g_2 is contained in $(-\infty, 1]$, and

$$x^*(e^{isB}x) = g_1(s)\left(\int_{\mathbf{R}} e^{isr}\, dE_{1,x^*,x}(r)\right) + g_2(s)\left(\int_{\mathbf{R}} e^{isr}\, dE_{2,x^*,x}(r)\right).$$

We wish to use the following functional calculus construction (see Davies [11], for this construction when $\{e^{isB}\}_{s\in\mathbf{R}}$ is bounded)

$$f(B) \equiv \int_{\mathbf{R}} e^{isB}(\mathcal{F}f)(s)\, ds \qquad (*)$$

Since $\|e^{isB}\|$ is unbounded, it is delicate to determine for which f this integral will converge. Being holomorphic in a strip is helpful, for the following reasons.

Write f_z for the translate of f, $f_z(w) \equiv f(z+w)$. For $f \in H^\infty(H_\phi)$, $\mathcal{F}f(s) = e^{\pm as}\mathcal{F}f_{\pm ai}(s)$, for a real, $|a| < \phi$. Thus, for any $f \in \mathcal{A}_\phi \equiv \{f \in H^\infty(H_\phi) \,|\, \exists M$ such that $|zf(z)| \leq M, \forall z \in H_\phi\}$, the Cauchy inequality implies that

$$\int_{\mathbf{R}} \|e^{isB}\| |\mathcal{F}f(s)|\, ds < \infty.$$

As in [11], (*) defines an algebra homomorphism from $\mathcal{A}_\phi$ into $L(X)$ and, if $h_{r,n}(s) \equiv (r-s)^{-n}$, then, whenever $|Im(r)| > \phi, n \in \mathbf{N}$,

$$h_{r,n}(B) = (r-B)^{-n}$$

for all such r, n.

By writing

$$\begin{aligned} x^*(f(B)x) &= \int_{\mathbf{R}} \left[\int_{\mathbf{R}} e^{isr} g_1(s)\mathcal{F}f(s)\, ds\right] dE_{1,x^*,x}(r) \\ &\quad + \int_{\mathbf{R}} \left[\int_{\mathbf{R}} e^{isr} g_2(s)\mathcal{F}f(s)\, ds\right] dE_{2,x^*,x}(r) \\ &= \int_{\mathbf{R}} \left[f_{-ai}(s) * \mathcal{F}^{-1}\left(g_1(s)e^{-as}\right)\right] dE_{1,x^*,x}(r) \\ &\quad + \int_{\mathbf{R}} \left[f_{ai}(s) * \mathcal{F}^{-1}\left(g_2(s)e^{as}\right)\right] dE_{2,x^*,x}(r), \end{aligned} \qquad (**)$$

we may show that, for $x^* \in X^*, x \in X$, $\exists M_{x^*,x} < \infty$ such that

$$|x^*(f(B)x)| \leq M_{x^*,x}\|f\|_\infty,$$

$\forall f \in \mathcal{A}_\phi$, where the supremum of f is taken over H_ϕ.

The Uniform Boundedness Theorem now implies that $\exists$ a constant M such that

$$\|f(B)\| \leq M\|f\|_\infty, \forall f \in \mathcal{A}_\phi.$$

We may now use (**) to define a bounded linear map, $f \mapsto f(B)$, from $H^\infty(H_\phi)$ into $L(X, X^{**})$, extending the map from $\mathcal{A}_\phi$ into $L(X)$. Using the fact that, for any $f \in H^\infty(H_\phi), \lambda > \phi$, $z \mapsto f(z)(i\lambda + z)^{-1} \in \mathcal{A}_\phi$, it can be shown that this is an algebra homomorphism into $L(X)$. ∎

(b) → (d). We use the following representation for fractional powers

$$x^*(A^{-is}x) = \frac{1}{\Gamma(1+is)} \int_0^\infty t^{is} x^*(Ae^{-tA}x)\, dt, \qquad (*)$$

$\forall x^* \in X^*, x \in D(A) \cap Im(A)$.

By the bounded variation of $\{e^{-tA}\}_{t\geq 0}$, $\{\|A^{-is}x\| \,|\, x \in D(A) \cap Im(A), \|x\| \leq 1\}$ is bounded, $\forall s \in \mathbf{R}$. Since $D(A) \cap Im(A)$ is dense, this implies that A^{-is} extends to a bounded operator, such that (*) holds.

We have shown that iB generates a strongly continuous group, given by

$$x^*(e^{isB}x) = \frac{1}{\Gamma(1+is)} \int_0^\infty t^{is} x^*(Ae^{-tA}x)\, dt = \frac{1}{\Gamma(1+is)} \int_{-\infty}^\infty e^{isr} x^*(Ae^{-e^r A}x)\, e^r dr.$$

Letting $g(s) \equiv \frac{1}{\Gamma(1+is)}, dE_{x^*,x}(r) \equiv x^*(Ae^{-e^r A}x)\, e^r$ *almost* gives us (d). All that's missing is the appropriate rate of growth of g; $\frac{1}{\Gamma(1+is)}$ is $O(e^{\frac{\pi}{2}|s|})$. To improve this rate of growth, we "rotate", that is, use the residue theorem to replace t by $te^{\pm i\psi}$ in the integral above.

(a) → (b). It is straightforward to show that A is densely defined and $\|w(w+A)^{-1}\|$ is bounded in appropriate sectors, so that $-A$ generates a bounded strongly continuous holomorphic semigroup of angle θ.

For $0 < \psi < \theta$, choose δ between $\frac{\pi}{2} - \psi$ and $\frac{\pi}{2} - \theta$. Since A has an $H^\infty(S_\delta)$- functional calculus, $\exists$ a constant M_ψ such that

$$\|f(A)\| \leq M_\psi \|f\|_{H^\infty(S_\delta)},$$

for all $f \in H^\infty(S_\delta)$.

For $x \in X, x^* \in X^*$, the map $f \mapsto x^*(f(A)x)$ is a linear functional on $H^\infty(S_\delta) \bigcap C_0(\overline{S_\delta})$, of norm less than or equal to $M_\psi\|x\|\|x^*\|$, thus there exists a signed measure $\mu_{x^*,x}$, of variation less than or equal to $M_\psi\|x^*\|\|x\|$, such that,

$$x^*(f(A)x) = \int_{S_\delta} f(z)\, d\mu_{x^*,x}(z), \forall f \in H^\infty(S_\delta) \bigcap C_0(\overline{S_\delta}).$$

We use this representation to directly estimate

$$\int_0^\infty |x^*(Ae^{-te^{i\phi}A}x)|\, dt = \int_0^\infty |\int_{S_\delta} ze^{-te^{i\phi}z}\, d\mu_{x^*,x}(z)|\, dt,$$

to get the desired inequality of Definition 2.4. ∎

Proof of Theorem 2.3 : We've just shown (a). (b) is straightforward and well-known.

(e). Similarly to (b) → (d) of Theorem 2.5, we construct fractional powers as follows. For $x^* \in X^*, x \in D(A^2)$,

$$\begin{aligned}
|x^*(A^{-is}x)| &= |\frac{1}{\Gamma(2+is)} \int_0^\infty t^{is+1} x^*(A^2 e^{-tA}x)\, dt| \\
&\leq \frac{1}{|\Gamma(2+is)|} |\int_0^\infty \left(\sqrt{t}A^* e^{-\frac{t}{2}A^*} x^*\right)(\sqrt{t}Ae^{-\frac{t}{2}A}x)\, dt| \\
&\leq \frac{1}{|\Gamma(2+is)|} \left(\int_0^\infty t\|A^* e^{-\frac{t}{2}A^*}x^*\|^2\, dt\right)^{\frac{1}{2}} \left(\int_0^\infty t\|Ae^{-\frac{t}{2}A}x\|^2\, dt\right)^{\frac{1}{2}} \\
&\leq \frac{1}{|\Gamma(2+is)|} M\|x^*\|\|x\|.
\end{aligned}$$

Since $D(A^2)$ is dense, this implies that A^{-is} extends to a bounded operator, $\forall s \in \mathbf{R}$.

By applying the same argument to $e^{i\delta}A$, for appropriate δ, we get the desired growth of the fractional powers. ∎

(d). In (b) → (d) of Theorem 2.5, we essentially constructed fractional powers as the Fourier transform of $r \mapsto e^r Ae^{-e^r A}$. Here, we will turn it around and write $r \mapsto e^r Ae^{-e^r A}$ as a Fourier transform of fractional powers,

$$\begin{aligned}
x^*(Ae^{-e^r A}x)e^r &= \int_{\mathbf{R}} \Gamma(1+is)x^*((e^r A)^{-is}x)\, \frac{ds}{2\pi}, \\
&= \int_{\mathbf{R}} e^{-isr}\Gamma(1+is)x^*(A^{-is}x)\, \frac{ds}{2\pi},
\end{aligned}$$

$\forall x \in X, x^* \in X^*, r \in \mathbf{R}$.

If $\frac{1}{p} + \frac{1}{q} = 1$, we apply the Hausdorf-Young theorem to

$$\int_0^\infty |tx^*(Ae^{-tA}x)|^q \frac{dt}{t} = \int_{\mathbf{R}} |e^r x^*(Ae^{-e^r A}x)|^q\, dr.$$

Careful choice of δ gives the same estimates for $e^{i\delta}A$, uniformly for δ in appropriate sectors. ∎

(c). When X is a Hilbert space, the converse follows from the (vector-valued) Plancherel theorem, just as (d) followed from the Hausdorf-Young theorem. ((c) first appeared in [36]) ∎

(f) Suppose A is bounded and $sp(A) \subseteq S_\theta, \theta < \frac{\pi}{2}$. Then $\{e^{-tA}\}_{t\geq 0}$ is a bounded holomorphic strongly continuous semigroup, $\|Ae^{-tA}\| \leq \|A\|\|e^{-tA}\|$ is exponentially decaying.

Conversely, given (5_2), as in the proof of (e), $\{A^{-is}\}_{s\in\mathbf{R}}$ is a strongly continuous group, given by

$$A^{-is} = \frac{1}{\Gamma(2+is)}\int_0^\infty t^{is+1}A^2e^{-tA}\,dt.$$

Since $\|t^{is+1}A^2e^{-tA}\| \leq |t^{is+1}|\|Ae^{-\frac{1}{2}A}\|^2$, dominated convergence now implies that $A^{-is} \to \int_0^\infty tA^2e^{-tA}\,dt = I$, in the operator norm, as $s \to 0$. This implies that the generator, $i\log A$, is bounded, hence A is bounded and $0 \in \rho(A)$. Since $\{e^{-tA}\}_{t\geq 0}$ is holomorphic, $sp(A) \subseteq \overline{S_\theta}$, for some $\theta < \frac{\pi}{2}$. ∎

III. H^∞ FUNCTIONAL CALCULUS ON SUBSPACES. When some desired behaviour is not found, it is natural to search for a subspace where it is found. Best of all is to find the largest possible such space. In this section, the desired behaviour is having an $H^\infty(S_\theta)$ functional calculus. This implies that A has bounded imaginary powers, and makes it easier to work with the imaginary powers. For example, if A has bounded imaginary powers and the spectrum of A is contained in $\overline{S_\theta}$, for some $\theta < \pi$, it is an open question if $\{A^{is}\}_{s\in\mathbf{R}}$ is of exponential type θ. However, if $-A$ has an $H^\infty(S_\phi)$ functional calculus, $\forall\phi > \theta$, then this growth condition is easy to verify.

The following first appeared in [17].

Definition 3.1. Let $\mathcal{H}$ be a Banach algebra of functions from a subset of the complex plane into the complex plane and suppose $C \in B(X)$ is injective. Suppose $f_0(z) \equiv 1 \in \mathcal{H}$ and $\exists r \in \rho(A)$ such that $g_r(z) \equiv (r-z)^{-1} \in \mathcal{H}$. A *$C$-regularized $\mathcal{H}$ functional calculus* for A is a continuous map $\Lambda : \mathcal{H} \to L(X)$ such that

(1) $\Lambda(f_0) = C$,

(2) $\Lambda f \Lambda g = C\Lambda(fg), \forall f, g \in \mathcal{H}$ and

(3) $\Lambda g_r = (r-A)^{-1}C$, whenever $g_r \in \mathcal{H}$ and $r \in \rho(A)$.

We think of Λf as being $f(A)C$, analogous to the definition of a C-regularized semigroup.

Proposition 3.2. *Suppose $\exists\lambda \in \rho(A)$ such that $g_\lambda \in \mathcal{H}$. Then the following are equivalent.*

(a) *A has a C-regularized $\mathcal{H}$ functional calculus.*

(b) *$\exists$ a Banach space Z such that*

$$[Im(C)] \hookrightarrow Z \hookrightarrow X$$

and $A|_Z$ has a $\mathcal{H}$ functional calculus.

If Λ is the C-regularized $\mathcal{H}$ functional calculus for A and $f \mapsto f(A|_Z)$ is the $\mathcal{H}$ functional calculus for $A|_Z$, then

$$\Lambda f = f(A|_Z)C, \forall f \in \mathcal{H}.$$

Z is maximal-unique in the following sense. If W is a Banach space as in (b), then $W \hookrightarrow Z$.

Proof: (a) $\to$ (b). Define the possibly unbounded operator $f(A)$, with domain equal to $\{x \mid (\Lambda f)x \in Im(C)\}$, $f(A)x \equiv C^{-1}(\Lambda f)x$.

Let $\mathcal{D} \equiv \bigcap_{f\in\mathcal{H}}\mathcal{D}(f(A))$.

If $x \in \mathcal{D}, \|x\|_Z \equiv \sup\{\|f(A)x\| \mid f \in \mathcal{H}, \|f\|_{\mathcal{H}} \leq 1\}$.

$Z \equiv \{x \in \mathcal{D} \mid \|x\| < \infty\}$.

(b) $\to$ (a). Let $\Lambda f \equiv f(A|_Z)C$. ∎

Definition 3.3. Suppose $0 \leq \theta < \pi$. We will say that the operator A is *of type θ* if $\forall\phi > \theta$, $sp(A) \subseteq \overline{S_\phi}$ and $\{\|w(w-A)^{-1}\| | w \notin \overline{S_\phi}\}$ is bounded.

Theorem 3.4. *Suppose A is of type $\theta < \pi$. Then $\forall \frac{1}{2} > r > 0, \pi > \phi > \theta, \exists$ an $A^r(1+A)^{-2r}$-regularized $H^\infty(S_\phi)$ functional calculus for A.*

Proof: Fix $r > 0, \pi > \phi > \theta$, and let $C \equiv A^r(1+A)^{-2r}$. Let $B \equiv A^r$. Then $-B$ generates a bounded strongly continuous holomorphic semigroup of angle $\delta \equiv (\frac{\pi}{2} - r\theta)$. For $|\epsilon| < \psi < \delta$,

$$\int_0^\infty |< Be^{-te^{i\epsilon}B}Cx, x^* > | \, dt \leq \|B^2(1+B)^{-2}\| \|x\| \|x^*\| \int_0^1 \|e^{-te^{i\epsilon}B}\| \, dt$$
$$+ M_\psi \|(1+B)^{-2}\| \int_1^\infty \frac{dt}{t^2},$$

since e^{-zB} is a bounded strongly continuous holomorphic semigroup.

The same arguments as in Theorem 2.5(b) $\rightarrow$ (a) show that A^r has a C-regularized $H^\infty(S_{r\phi})$ functional calculus. It is now easy to define a C-regularized $H^\infty(S_\phi)$ functional calculus for A, by

$$f(A)C \equiv f_r(A^r)C, \; f_r(z) \equiv f(z^{\frac{1}{r}});$$

note that $f \in H^\infty(S_\phi)$ if and only if $f_r \in H^\infty(S_{r\phi})$. ∎

Corollary 3.5. *Suppose A is of type $\theta < \pi$ and $0 \in \rho(A)$. Then $\forall \frac{1}{2} > r > 0, \pi > \phi > \theta, \exists$ an A^{-r}-regularized $H^\infty(S_\phi)$ functional calculus for A.*

IV. SEMIGROUPS OF BOUNDED VARIATION AND THE INHOMOGENEOUS ABSTRACT CAUCHY PROBLEM. Motivation for the study of bounded imaginary powers comes from the inhomogeneous abstract Cauchy problem

$$\frac{d}{dt}u(t,x,f) = A(u(t,x,f)) + f(t) \, (t \geq 0), \; u(0,x) = x. \tag{4.1}$$

For any $f \in L^1([0,T], X)$, when A generates a strongly continuous semigroup $\{e^{tA}\}_{t \geq 0}$, a mild solution

$$u(t,x,f) \equiv e^{tA}x + \int_0^t e^{(t-s)A} f(s) \, ds \tag{4.2}$$

exists on $[0,T]$. If $u \in C([0,\infty), [\mathcal{D}(A)]) \cap C^1([0,\infty), X)$, we will say it is a *strong solution*.

When X is UMD, A is of type θ less than $\frac{\pi}{2}$ and has bounded imaginary powers, and $1 < p < \infty$, it has been shown that u is a strong solution and Au is in $L^p([0,T], X)$ whenever f is (see [18], [19] and [41]), and the map $f \mapsto Au(t,0,f)$, from $L^p([0,T], X)$ into itself, is continuous.

An example of how much more trouble is possible for $p = 1$ or ∞ is the following result.

Definition 4.3. A function $\alpha : [a,b] \rightarrow B(X,Y)$ is *of bounded semivariation on* $[a,b]$ if $\exists$ a constant M such that

$$\| \sum_{i=1}^n [\alpha(t_i) - \alpha(t_{i-1})] \, x_i \| \leq M,$$

for any partition $\{t_i\}_{i=0}^n$ of $[a,b], \{x_i\}_{i=1}^n \in X$ such that $\|x_i\| \leq 1, 1 \leq i \leq n$.

Proposition 4.4(Proposition 3.1 in [47]). *The following are equivalent, if A is a closed operator.*

(a) *(4.1) has a unique strong solution on $[0,r]$, for any $f \in C([0,r], X)$.*

(b) *A generates a strongly continuous semigroup that is of bounded semivariation on $[0,r]$.*

In many ways, this is a negative result. For example, in a reflexive space, Proposition 4.4 implies that (a) holds if and only if A is bounded (see [47, Corollary 3.1]).

Thus, having the map $f \mapsto Au(t,0,f)$ be bounded from L^∞ into itself, or at least from $C([0,r],X)$ into itself, seems unlikely.

In this section, we would like to consider the case $p = 1$, on the infinite interval $[0,\infty)$, that is, we would like to consider in what sense the map $f \mapsto u(t,x,f)$, from $L^1([0,\infty),X)$ into itself, is bounded. Since we are using the entire positive real axis, these may be considered results about asymptotic behaviour.

We will state our results in the language of weak solutions (introduced in [3]), that is, we will consider

$$\sup_{\|x^*\|\le 1} \| < u(t,x,f), A^*x^* > \|_1$$

rather than $\|Au(t,x,f)\|_1$.

Definition 4.5(from [3]). A function $u \in C([0,\infty),X)$ is a *weak solution* of (4.1) if, $\forall x^* \in \mathcal{D}(A^*)$, the map $t \mapsto < u(t), x^* >$ is absolutely continuous and

$$\frac{d}{dt} < u(t), x^* > = < u(t), A^*x^* > + < f(t), x^* >,$$

for almost all $t \ge 0$.

Proposition 4.6([3]). *Suppose A is closed and densely defined, $T > 0$. Then the following are equivalent.*

(a) *$\forall x \in X, f \in C([0,T],X), \exists$ a unique weak solution of (4.1) on $[0,T]$.*

(b) *A generates a strongly continuous semigroup.*

$u(t,x,f)$ is then given by (4.2).

The operator A generating a strongly continuous semigroup that is uniformly weakly of bounded variation(see Definition 2.4) turns out to be equivalent to the map $f \mapsto Au(t,x,f)$ being a bounded map from L^1 into weak L^1; by the latter we mean the set of all measurable $g : [0,\infty) \to X$ such that

$$\sup_{\|x^*\|\le 1} \| < g, x^* > \|_1 < \infty.$$

Theorem 4.7. *The following are equivalent.*

(a) *A generates a strongly continuous semigroup that is uniformly weakly of bounded variation.*

(b) *(4.1) has a unique weak solution, $\forall x \in X, f \in L^1([0,\infty),X)$ and $\exists$ a constant M such that*

$$\| < u(t,x,f), A^*x^* > \|_1 \le M\|x^*\| (\|x\| + \|f\|_1),$$

$\forall x^* \in \mathcal{D}(A^*), x \in X, f \in L^1([0,\infty),X)$.

Proof: (b) $\to$ (a) is obvious. For (a) $\to$ (b), it is sufficient to assume that $x = 0$.

$$\begin{aligned}
\| < u(t,0,f), A^*x^* > \|_1 &= \int_0^\infty \left| \int_0^t < e^{(t-s)A} f(s), A^*x^* > ds \right| dt \\
&\le \int_0^\infty \int_s^\infty | < e^{(t-s)A} f(s), A^*x^* > | \, dt\, ds \\
&= \int_0^\infty \int_0^\infty | < e^{rA} f(s), A^*x^* > | \, dr\, ds \\
&\le \int_0^\infty M\|f(s)\| \|x^*\| \, ds,
\end{aligned}$$

where M is as in Definition 2.4. ∎

Theorem 2.5 gives us the following.

Corollary 4.8. *Suppose $0 < \theta < \frac{\pi}{2}$ and $-A$ has an $H^\infty(S_\theta)$ functional calculus. Then $\exists$ a constant M such that*

$$\| < u(t,x,f), A^*x^* > \|_1 \leq M\|x^*\| \left(\|x\| + \|f\|_1\right),$$

$\forall x^* \in \mathcal{D}(A^*), x \in X, f \in L^1([0,\infty), X)$.

Combined with Corollary 3.5, this gives us a condition on the range of f that guarantees L^1 behaviour.

Corollary 4.9. *Suppose $0 \in \rho(A)$, A generates a bounded strongly continuous holomorphic semigroup and $\exists r > 0$ such that $s \mapsto A^r f(s) \in L^1([0,\infty), X)$. Then $\forall \alpha < r, \exists$ a constant M_α such that*

$$\| < A^\alpha u(t,0,f), A^*x^* > \|_1 \leq M_\alpha \|A^r f\|_1 \|x^*\|,$$

$\forall x^* \in \mathcal{D}(A^*)$.

Proof: It is clear that $u(t,0,f) \in \mathcal{D}(A^r)$, with

$$A^r u(t,0,f) = u(t,0,A^r f).$$

Fix $\alpha < r$. By Corollary 3.5, $-A$ has an $A^{\alpha-r}$-regularized $H^\infty(S_\theta)$ functional calculus, for some $0 < \theta < \frac{\pi}{2}$. The same arguments as in Theorem 2.5 and 4.7 show that $\exists$ a constant M_α such that

$$\|A^{\alpha-r} < u(t,x,g), A^*x^* > \|_1 \leq M_\alpha \|x^*\| \left(\|x\| + \|g\|_1\right),$$

$\forall x^* \in \mathcal{D}(A^*), x \in X, g \in L^1([0,\infty), X)$.

Thus,

$$\| < A^\alpha u(t,0,f), A^*x^* > \|_1 = \|A^{\alpha-r} < u(t,0,A^r f), A^*x^* > \|_1 \leq M_\alpha \|A^r f\|_1 \|x^*\|,$$

$\forall x^* \in \mathcal{D}(A^*)$. ∎

REFERENCES

[1] M. Balabane and H. Emamirad, *Smooth distribution semigroup and Schrödinger equation in $L^p(\mathbf{R}^n)$*, J. Math. Anal. Appl. 70 (1979), 61–71.

[2] M. Balabane and H. Emamirad, *L^p estimates for Schrödinger evolution equations*, Trans. Amer. Math. Soc. 292 (1985), 357–373.

[3] J. M. Ball, *Strongly continuous semigroups, weak solutions, and the variation of constants formula*, Proc. Amer. Math. Soc. 63 (1977), 370–373.

[4] K. Boyadzhiev and R. deLaubenfels, *H^∞-functional calculus for perturbations of generators of holomorphic semigroups*, Hous. J. Math., 17 (1991), 131–147.

[5] K. Boyadzhiev and R. deLaubenfels, *Boundary values of holomorphic semigroups*, Proc. Amer. Math. Soc., to appear.

[6] K. Boyadzhiev and R. deLaubenfels, *Semigroups and resolvents of bounded variation, imaginary powers and H^∞ functional calculus*, preprint.

[7] K. Boyadzhiev and R. deLaubenfels, *Bounded imaginary powers and H^∞ functional calculus on subspaces*, preprint.

[8] D. C. Champeney, "A Handbook of Fourier Transforms," Cambridge University Press, Cambridge, 1987.

[9] I. Cioranescu and L. Zsido, *Analytic generators for one-parameter groups*, Tohoku Math. J. 28 (1976), 327–362.

[10] G. Da Prato, *Semigruppi regolarizzabili*, Ricerche Mat. 15 (1966), 223–248.

[11] E. B. Davies, "One-Parameter Semigroups," Academic Press, London, 1980.

[12] E. B. Davies, *Kernel estimates for functions of second order elliptic operators,* Quart. J. Math. Oxford (2) 39 (1988), 37–46.
[13] E. B. Davies and M. M. Pang, *The Cauchy problem and a generalization of the Hille-Yosida theorem,* Proc. London Math. Soc. 55 (1987), 181–208.
[14] R. deLaubenfels, *C-semigroups and the Cauchy problem,* J. Func. Anal., to appear.
[15] R. deLaubenfels, *Entire solutions of the abstract Cauchy problem,* Semigroup Forum 42 (1991), 83–105.
[16] R. deLaubenfels, *C-semigroups and strongly continuous semigroups,* Israel J. Math., to appear.
[17] R. deLaubenfels, *Unbounded holomorphic functional calculus and abstract Cauchy problems for operators with polynomially bounded resolvent,* J. Func. Anal., to appear.
[18] G. Dore and A. Venni, *On the closedness of the sum of two closed operators,* Math. Z. 196 (1987), 189–201.
[19] G. Dore and A. Venni, *Some results about complex powers of closed operators,* J. Math. Anal. Appl. 149 (1990), 124–136.
[20] N. Dunford and J. T. Schwartz, "Linear Operators," Part I, Interscience, New York, 1958.
[21] X. T. Duong, *H^∞ functional calculus of elliptic operators with C^∞ coefficients on L^p spaces of smooth domains,* J. Austral. Math. Soc. (Ser. A) 48 (1990), 113–123.
[22] X. T. Duong, *H_∞ functional calculus of second order elliptic PDE on L^p spaces,* Miniconference on Operators in Analysis, Proc. of the Center for Math. Analysis, ANU, Canberra, 24 (1989), 91–102.
[23] H. O. Fattorini, "The Abstract Cauchy Problem," Addison Wesley, Reading, Massachusetts, 1983.
[24] M. J. Fisher, *Imaginary powers of the indefinite integral,* Amer. J. Math. 93 (1971), 317–328.
[25] J. A. Goldstein, "Semigroups of Operators and Applications," Oxford, New York, 1985.
[26] M. Hieber, *Integrated semigroups and differential operators on L^p spaces,* preprint (1991).
[27] E. Hille and R. S. Phillips, "Functional Analysis and Semigroups," Colloq. Publ. Amer. Math. Soc. (1957).
[28] R. J. Hughes, *Semigroups of unbounded linear operators in Banach space,* Trans. Amer. Math. Soc. 230 (1977), 113–145.
[29] R. J. Hughes and S. Kantorovitz, *Boundary values of holomorphic semigroups of unbounded operators and similarity of certain perturbations,* J. Func. Anal. 29 (1978), 253–273.
[30] G. K. Kalisch, *On fractional integrals of purely imaginary order in L^p,* Proc. Amer. Math. Soc. 18 (1967), 136–139.
[31] H. Kober, *On a theorem of Schur and on fractional integrals of purely imaginary order,* Trans. Amer. Math. Soc. 50 (1941), 160–174.
[32] H. Komatsu, *Fractional powers of operators,* Pac. J. Math. 19 (1966), 285–346.
[33] E. R. Love, *Fractional derivatives of imaginary order,* J. London Math. Soc. (2) 3 (1971), 241–259.
[34] T. M. MacRobert, "Functions of a Complex Variable," Macmillan, London, 1962.
[35] E. Marschall, *On the analytic generator of a group of operators,* Indiana U. Math. J. 35 (1986), 289–309.
[36] A. McIntosh, *Operators which have an H^∞ functional calculus,* Miniconference on Operator Theory and PDE, Proc. of the Center for Math. Analysis, ANU, Canberra, 14 (1986), 210–231.
[37] A. McIntosh and A. Yagi, *Operators of type ω without a bounded H^∞- functional calculus,* Miniconference on Operators in Analysis 1989, Proc. of the Center for Math. Analysis, ANU, Canberra, 24 (1989).
[38] R. Nagel, "One-parameter Semigroups of Positive Operators," Lecture Notes in Mathematics 1184, Springer, Berlin, 1986.
[39] M. M. Pang, *Resolvent estimates for Schrödinger operators in $L^p(\mathbf{R}^n)$ and the theory of exponentially bounded C-semigroups,* Semigroup Forum 41 (1990), 97–114.
[40] A. Pazy, "Semigroups of Linear Operators and Applications to Partial Differential Equations," Springer, New York, 1983.

[41] J. Prüss and H. Sohr, *On operators with bounded imaginary powers in Banach spaces,* Math. Z. 203 (1990), 429–452.
[42] M. Reed and B. Simon, "Methods of Modern Mathematical Physics II," Academic Press, New York, 1975.
[43] W. Ricker, *Spectral properties of the Laplace operator in* $L^p(\mathbf{R})$, Osaka J. Math. 25 (1988), 399–410.
[44] B. Ross (ed.), "Fractional Calculus and its Applications," Lecture Notes in Math. 457 (1975), Springer-Verlag.
[45] B. Simon, *Schrödinger semigroups,* Bull. Amer. Math. Soc. 7 (1982), 447–526.
[46] S. Sjöstrand, *On the Riesz means of the solutions of the Schrödinger equation,* Annali Scuola Norm Sup. di Pisa 24 (1970), 331–348.
[47] C. C. Travis, *Differentiability of weak solutions to an abstract inhomogeneous differential equation,* Proc. Amer. Math. Soc. 82 (1981), 425–429.
[48] J. A. van Casteren, "Generators of strongly continuous semigroups," Research Notes in Math. 115, Pitman, 1985.
[49] A. Yagi, *Coincidence entre des espaces d'interpolation et des domaines de puissances fractionnaires d'operateurs,* C. R. Acad. Sci. Paris (Ser. I), 299 (1984), 173–176.
[50] A. Yagi, *Applications of the purely imaginary powers of operators in Hilbert spaces,* J. Func. Anal. 73 (1987), 216–231.

Stability of Linear Evolutionary Systems with Applications to Viscoelasticity

JAN PRÜSS Fachbereich 17 Mathematik, Universität-GH Paderborn, Warburger Str. 100, 4790 Paderborn, Germany

0. Introduction

During the last two decades the mathematical theory of viscoelasticity has undergone a rapid development. In particular, in the linear theory, where the formulation of the general stress-strain relations is quite simple, many problems concerning wellposedness, wave propagation properties, regularity, and asymptotic behaviour have been settled under various conditions. This is in particular true for the case of homogeneous, isotropic, and synchronous or incompressible materials, where the theory is by now fairly complete. However, this is not quite the case for nonsynchronous or even noinisotropic media, in particular regarding the asymptotic behaviour.

In the linear case, there are basically two different approaches to this kind of problems. The first involves semigroup theory; the dynamics are reformulated as an evolution equation in a space of history or forcing functions and then the well established theory of C_0-semigroups is applied.

In the second approach the problem is treated as an abstract Volterra equation with unbounded operator-valued kernels on the line or the halfline, and this equation is then treated by means of techniques which extend those for classical, i.e. finite-dimensional Volterra equations. These methods are based on results from operator theory, transform theory, complex analysis, and harmonic analysis.

It is not possible here to give full account to the literature, we only mention a few papers representing the state of the art. Adali [1], Carr and Hannsgen [4,5], Clément and Prüss [8], Clément and DaPrato [7], Da Prato and Lunardi [9], Desch and Grimmer [11,10] Hannsgen and Wheeler [13], Miller and Wheeler [18,19], Navarro [20], Prüss [25], Tanabe [27,28]. For a much more complete list of references see the forthcoming monograph Prüss [25].

In our opinion, the second approach in general yields better results since no history space is involved, and therefore in this paper we follow it in the case of hyperbolic equations of variational type.

$$(w, v(t)) + \int_0^t \alpha(t-\tau; w, v(\tau))d\tau = < w, f(t) >, \quad t > 0, \quad w \in V, \tag{0.1}$$

Here $V \hookrightarrow H$ are Hilbert spaces, α is a sesquilinear form on V, and $f : \mathbb{R}+ \to V^*$ is continuous. For problems in viscoelasticity this framework is quite natural, as we shall see in Section 1, and physics even gives an indication of the properties the form α should have to obtain energy type estimates. By means of these inequalities, we show by Laplace transform methods that (0.1) is wellposed. Theorem 1, concerned with wellposedness, extends existing results considerably since it does not rely on perturbation results. One exception is the paper Desch and Grimmer [11], where wellposedness and regularity of linear viscoelasticity is proved in a quite special history space setting, under the main assumption that the stress relaxation kernel is completely monotonic; see Section 1 for a definition of the latter. Our approach is strong enough to obtain analogues for the wellknown behaviour of homogeneous, isotropic, and synchronous or incompressible materials. Based on the properties of the Fourier transform in Hilbert spaces, we obtain sufficient conditions for stability, which are direct generalizations of the conditions for the scalar case; see Theorem 4.

This paper is organized as follows. In Section 1 we recall the basic equations of linear viscoelasticity; this is mainly for the motivation of the positivity and coerciveness properties of the relaxation tensor. Also some notation is fixed and the special case of homogeneous isotropic materials is discussed. Section 2 contains the variational formulation of linear viscoelasticity as well as some basic definitions. Thereafter coercive forms are introduced and the basic energy inequalities are proved. We have also included a Laplace transform characterization of coercivity which is in general much simpler to check than the original definition of coercivity. Section 4 contains the first main result of this paper, Theorem 1, which shows existence and t-regularity of the resolvent as well as some of its further boundedness and regularity properties. In Section 5 we take up the study of stability properties, in particular L^p- stability and strong integrability of the resolvent and its primitive and derivative. A discussion of the well understood equations of scalar type which in viscoelasticity correspond to homogeneous, isotropic, and synchronous or incompressible materials is included. The second main result, Theorem 4, which concerns integrability properties of the resolvent for hyperbolic solids is then stated and proved in Section 6. We briefly mention a companion result for hyperbolic fluids, and conclude the paper by translating the assumptions to properties of the stress relaxation tensor of linear nonisotropic viscoelasticity.

1. Linear Nonisotropic Viscoelasticity

Consider a 3-dimensional body which is represented by an open set $\Omega \subset \mathbb{R}^3$ with boundary $\partial\Omega$ of class C^1. Points in Ω (i.e. material points) will be denoted by $x, y, \ldots$. Associated with this body there is a strictly positive function $\rho_0 \in C(\overline{\Omega})$ called the *density of mass*. Acting forces will deform the body, and the material point x will be displaced to its new position $x + u(t,x)$ at time t; the vector field $u(t,x)$ is called the displacement field, or briefly *displacement*. The *velocity* of the material point $x \in \Omega$ at time t is then given by $v(t,x) = \dot{u}(t,x)$, where the dot indicates partial derivative with respect to t. The linearized *strain* in the body due to a deformation is defined by

$$\mathcal{E}(t,x) = \frac{1}{2}(\nabla u(t,x) + (\nabla u(t,x))^T) \quad , \quad t \in \mathbb{R}\,,\ x \in \Omega, \tag{1.1}$$

i.e. $\mathcal{E}(t,x)$ is the symmetric part of the *displacement gradient* ∇u.

A given strain-history of the body causes *stress* in a way to be specified, expressing the properties of the material the body is made of. The stress tensor will be denoted by $\mathcal{S}(t,x)$; both, $\mathcal{E}(t,x)$ and $\mathcal{S}(t,x)$ are symmetric. Let $g(t,x)$ be an external body force field like gravity. Then balance of momentum in the body becomes

$$\rho_0(x)\ddot{u}(t,x) = \operatorname{div}\mathcal{S}(t,x) + \rho_0(x)g(t,x) \quad , \quad t \in \mathbb{R}\,,\ x \in \Omega. \tag{1.2}$$

(1.2) has to be supplemented by boundary conditions; these are basically either 'prescribed displacement' or 'prescribed normal stress (traction)' at the suface of $\partial\Omega$ of the body. Let $\partial\Omega = \Gamma_d \cup \Gamma_s$, where Γ_d, Γ_s are closed, $\overline{\mathring{\Gamma}_s} = \Gamma_s$, $\overline{\mathring{\Gamma}_d} = \Gamma_d$ and such that $\mathring{\Gamma}_d \cap \mathring{\Gamma}_s = \emptyset$; let $n(x)$ denote the outer normal at $x \in \partial\Omega$. The boundary conditions then can be stated as follows.

$$\begin{aligned} u(t,x) &= u_d(t,x) \quad , \quad t \in \mathbb{R}\,,\ x \in \mathring{\Gamma}_d, \\ \mathcal{S}(t,x)n(x) &= g_s(t,x) \quad , \quad t \in \mathbb{R}\,,\ x \in \mathring{\Gamma}_s. \end{aligned} \tag{1.3}$$

In the sequel we always assume $\mathring{\Gamma}_d \neq \emptyset$, and $u_d \equiv 0$, i.e. the body is clamped at a part of its surface.

Taking the inner product of (1.2) with $\dot{u}$ and integrating over Ω and after an integration by parts then over $[0,t]$, we formally obtain the *energy equality*

$$\begin{aligned} \int_\Omega |\dot{u}(t,x)|^2 \rho_0(x)dx + \int_0^t \int_\Omega \mathcal{S}(\tau,x) : \dot{\mathcal{E}}(\tau,x)dxd\tau &= \int_\Omega |\dot{u}(0,x)|^2 \rho_0(x)dx \\ + \int_0^t \int_\Omega g(\tau,x)\cdot\dot{u}(\tau,x)\rho_0(x)dxd\tau &+ \int_0^t \int_{\Gamma_s} g_s(\tau,x)\cdot\dot{u}(\tau,x)dxd\tau. \end{aligned} \tag{1.4}$$

If the body is stress-free and at rest up to time $t = 0$, the total kinetic energy of the body at time t cannot exceed its initial value plus the work done by the acting body and surface forces, the inequality

$$\int_0^t \int_\Omega \mathcal{S}(\tau,x) : \dot{\mathcal{E}}(\tau,x)dxd\tau \geq 0 \tag{1.5}$$

must hold for all values of $t > 0$, and for any choice of initial values and forces.

To complete the system, an equation has to be added which relates the stress $\mathcal{S}(t,x)$ to u and its derivatives; such relations are known as *constitutive laws*. Here we concentrate on linear materials only. Since the stress should only depend on the history of the strain, the general constitutive law is given by

$$\mathcal{S}(t,x) = \int_0^\infty d\mathcal{A}(\tau,x)\dot{\mathcal{E}}(t-\tau,x) \quad , \quad t \in \mathbb{R}\,,\ x \in \Omega \tag{1.6}$$

where the *stress relaxation tensor* $\mathcal{A} : \mathbb{R}_+ \times \Omega \to \mathcal{B}(Sym\{3\})$ is locally of bounded variation w.r.t. $t \in \mathbb{R}_+$; here $Sym\{N\}$ denotes the space of N-dimensional real symmetric matrices. In components the latter means

$$\mathcal{A}_{ijkl}(t,x) = \mathcal{A}_{jikl}(t,x) = \mathcal{A}_{ijlk}(t,x) \quad , \quad t \in \mathbb{R}_+\,,\ x \in \Omega, \tag{1.7}$$

for all $i, j, k, l \in \{1,2,3\}$. The component functions $\mathcal{A}_{ijkl}(t,x)$ are called the *stress relaxation moduli* of the material.

Let us consider some special cases which have been studied extensively in classical continuums mechanics.

(i) *Ideally elastic solids*

Classical elasticity postulates the stress-strain relations

$$\mathcal{S}(t,x) = \mathcal{A}^\infty(x)\mathcal{E}(t,x) \quad , \quad t \in \mathbb{R} \, , \, x \in \Omega,$$

where $\mathcal{A}^\infty \in L^\infty(\Omega; \mathcal{B}(Sym\{3\}))$, and $\mathcal{A}^\infty(x)$ positive definite, uniformly on Ω. This corresponds to the stress relaxation tensor

$$\mathcal{A}(t,x) = t\mathcal{A}^\infty(x) \quad , \quad t > 0 \, , \, x \in \Omega.$$

(ii) *Ideally viscous fluids*

For such fluids the stress strain relations are

$$\mathcal{S}(t,x) = \mathcal{A}^0(x)\dot{\mathcal{E}}(t,x) \quad , \quad t \in \mathbb{R} \, , \, x \in \Omega,$$

with $\mathcal{A}^0 \in L^\infty(\Omega; \mathcal{B}(Sym\{3\}))$, and $\mathcal{A}^0(x)$ positive definite, uniformly on Ω. Here we obtain

$$\mathcal{A}(t,x) = \mathcal{A}^0(x) \quad , \quad t > 0 \, , \, x \in \Omega.$$

(iii) *Kelvin-Voigt solids*

Such materials are governed by constitutive laws of the form

$$\mathcal{S}(t,x) = \mathcal{A}^0(x)\dot{\mathcal{E}}(t,x) + \mathcal{A}^\infty(x)\mathcal{E}(t,x) \quad , \quad t \in \mathbb{R} \, , \, x \in \Omega,$$

with $\mathcal{A}^0, \mathcal{A}^\infty \in L^\infty(\Omega; \mathcal{B}(Sym\{3\}))$, and $\mathcal{A}^\infty(x)$, $\mathcal{A}^0(x)$ positive definite, uniformly on Ω. This gives

$$\mathcal{A}(t,x) = \mathcal{A}^0(x) + t\mathcal{A}^\infty(x) \quad , \quad t > 0 \, , \, x \in \Omega.$$

(iv) *Maxwell fluids*

This class of materials is characterized by the stress-strain relations

$$\mathcal{S}(t,x) = \int_0^\infty \mu e^{-\mu s}\mathcal{A}^1(x)\dot{\mathcal{E}}(t-s,x)ds \quad , \quad t \in \mathbb{R} \, , \, x \in \Omega,$$

with $\mathcal{A}^1 \in L^\infty(\Omega; \mathcal{B}(Sym\{3\}))$, $\mathcal{A}^1(x)$ positive definite, uniformly on Ω, and $\mu > 0$. In this case we have

$$\mathcal{A}(t,x) = (1 - e^{-\mu t})\mathcal{A}^1(x) \quad , \quad t > 0 \, , \, x \in \Omega.$$

A material is called *homogeneous* if ρ_0 and $\mathcal{A}$ do not depend on the material points $x \in \Omega$. It is called *isotropic* if the consitutive laws are invariant under the group of rotations. It can be shown that the general isotropic stress relaxation tensor is given by

$$\mathcal{A}_{ijkl}(t,x) = \frac{1}{3}(3b(t,x) - 2a(t,x))\delta_{ij}\delta_{kl} + a(t,x)(\delta_{ik}\delta_{jl} + \delta_{il}\delta_{jk}).$$

This results into the constitutive law

$$\mathcal{S}(t,x) = 2\int_0^\infty da(\tau,x)\dot{\mathcal{E}}(t-\tau,x) + \frac{1}{3}\mathcal{I}\int_0^\infty (3db(\tau,x) - 2da(\tau,x))\mathrm{tr}\,\dot{\mathcal{E}}(t-\tau,x). \tag{1.8}$$

The kernel b describes the behaviour of the material under compression, while a determines its response in shear; therefore, db is called *compression modulus* and *da shear modulus.* In general, a and b are independent functions, however, if $b(t,x) = \beta a(t,x)$ for some constant $\beta > 0$ then the material is called *synchronous.* The functions a and b are generally believed to be *creep functions*, i.e. positive, nondecreasing, and concave. Therefore $a(t)$ admits the decomposition

$$a(t) = a_0 + a_\infty t + \int_0^t a_1(\tau)d\tau\ , \quad t > 0, \tag{1.9}$$

where a_0, $a_\infty \geq 0$, and $a_1(t)$ is nonnegative and nonincreasing, with $\lim_{t\to\infty} a_1(t) = 0$; similarly for $b(t)$.

Summarizing, we obtain the equations for linear viscoelasticity.

$$\rho_0(x)\ddot{u}(t,x) = \operatorname{div}(\int_0^\infty d\mathcal{A}(\tau,x)\nabla\dot{u}(t-\tau,x)) + \rho_0(x)g(t,x) \quad , \quad t \in \mathbb{R}\ , \ x \in \Omega. \tag{1.10}$$

For homogenenous and isotropic materials (1.10) can be simplified to

$$\ddot{u}(t,x) = \int_0^\infty da(\tau)\Delta\dot{u}(t-\tau,x) + \int_0^\infty (db(\tau) + \frac{1}{3}da(\tau))\nabla\nabla \circ \dot{u}(t-\tau,x) + g(t,x), \tag{1.11}$$

where we have put $\rho_0(x) \equiv \rho_0 = 1$ for simplicity. These equations of course have to be supplemented by the boundary conditions (1.3).

Inequality (1.5) leads to the following restriction on the stress relaxation tensor which will be called *dissipation inequality* in the sequel.

$$\int_0^T \int_0^t d\mathcal{A}(\tau,x) : F(t-\tau) : F(t)dt \geq 0\ , \quad \text{for all } T > 0,\ F \in C(\mathbb{R}+; Sym\{3\}). \tag{1.12}$$

In other words, the matrix-valued measure $d\mathcal{A}$ is of *positive type*; this property will be crucial in later developments. A special case of this are *completely monotonic* relaxation tensors, i.e. $d\mathcal{A}$ is of the form

$$d\mathcal{A}(t,x) = \mathcal{A}^0(x)\delta_0 + \mathcal{A}^\infty(x)dt + (\int_0^\infty e^{-t\xi}\check{\mathcal{A}}(\xi,x)d\alpha(\xi))dt\ , \quad t > 0,\ x \in \Omega, \tag{1.13}$$

with $\mathcal{A}^0, \mathcal{A}^\infty \in L^\infty(\Omega; \mathcal{B}(Sym\{3\}))$ positive semidefinite, $\check{\mathcal{A}} : \mathbb{R}_+ \to L^\infty(\Omega; \mathcal{B}(Sym\{3\}))$ Borel-measurable, bounded, and pointwise positive semidefinite, and $\alpha : \mathbb{R}_+ \to \mathbb{R}_+$ nondecreasing, $\alpha(0) = \alpha(0+) = 0$, and $\int_0^\infty (1+\xi)^{-1}d\alpha(\xi) < \infty$. For the homogeneous and isotropic case the latter means that the kernels a and b are *Bernstein functions*, i.e. $a, b \in C^\infty(0,\infty)$ positive and with completely monotonic derivatives. Observe that the Examples (i)∼(iv) mentioned above are completely monotonic.

For more background information about linear viscoelasticity consult Bland [3], Christensen [6], Leitman and Fisher [16], or Pipkin [23].

2. A Variational Formulation

We now want to rewrite (1.12) in variational form, assuming as before ρ_0 continuous and strictly positive on $\overline{\Omega}$, $u_d \equiv 0$, $\mathring{\Gamma}_d \neq \emptyset$, $\partial\Omega$ of class C^1, and in addition Ω bounded. According to the discussion in Section 1, we restrict our attention to relaxation kernels $\mathcal{A} \in BV_{loc}(\mathbb{R}_+; L^\infty(\Omega; \mathcal{B}(Sym\{3\})))$, which are of positive type and of subexponential growth, i.e.

$$\int_0^\infty e^{-\varepsilon t}|d\mathcal{A}(t,\cdot)|_{L^\infty} < \infty\ , \quad \text{for each } \varepsilon > 0.$$

Observe that this is equivalent to the symmetry property (1.7) and existence of a nondecreasing function α_0 of subexponential growth such that

$$|\mathcal{A}(t,x) - \mathcal{A}(s,x)| \leq \alpha_0(t) - \alpha_0(s)\,, \quad \text{for all } t > s \geq 0 \text{ and for a.a. } x \in \Omega.$$

As a convention we let $\mathcal{A}(0,x) = 0$ for all $x \in \Omega$ as well as $\alpha_0(0) = 0$. These assumptions will be taken for granted in the remainder of this paper.

Consider the Hilbert space $H = L^2(\Omega;\mathbf{R}^3)$ equipped with the inner product

$$(v_1, v_2) = \int_\Omega \rho_0(x) v_1(x) \cdot \overline{v_2(x)} dx\,,$$

and norm $|w| = (w,w)^{1/2}$. Let $V = W^{1,2}_{\Gamma_d}(\Omega;\mathbf{R}^3)$ denote the subspace of $W^{1,2}(\Omega;\mathbf{R}^3)$ of functions vanishing on Γ_d in the sense of traces. As inner product in V we take the usual one

$$((v_1,v_2)) = \int_\Omega \nabla v_1(x) : \overline{\nabla v_2(x)} dx + \int_\Omega v_1(x) \cdot \overline{v_2(x)} dx\,, \quad \text{for } v_1, v_2 \in V.$$

The norm in V will be denoted by $||\cdot||$.

Let V^* denote the anti-dual of V, $< v, v* >$ the natural pairing between $v \in V$ and $v^* \in V^*$, and $||\cdot||_*$ the norm in V^*. Via the identification $< v,w >= (v,w)$ for $v \in V$, $w \in H$, we then have the usual dense embeddings $V \overset{d}{\hookrightarrow} H \overset{d}{\hookrightarrow} V^*$.

Define bounded sesquilinear forms on V by means of

$$\alpha(t; v_1, v_2) = \int_\Omega \nabla v_1 : \overline{\mathcal{A}(t,x)\nabla v_2(x)} dx\,, \quad t \geq 0, v_1, v_2 \in V. \tag{2.1}$$

Then there follows the variational formulation of (1.2), (1.3), (1.6) by an integration by parts.

$$(w, u(t)) + \int_0^t \alpha(t-s, w, u(s))ds =< w, f(t) >\,, \quad t \geq 0,\ w \in V. \tag{2.2}$$

Here $f(t) \in V^*$ contains $g(t,x)$ and $g_s(t,x)$, as well as the history of $v(t,x)$. Since $\mathcal{A}(t,\cdot) \in L^\infty(\Omega; Sym\{3\})$ for each $t \geq 0$, the sesquilinear forms $\alpha(t;\cdot,\cdot)$ are bounded, hence by the Riesz representation theorem, there is a family of bounded linear operators $\{A(t)\}_{t\geq 0} \subset \mathcal{B}(V,V^*)$ such that

$$\alpha(t; w, v) =< w, A(t)v > \quad \text{for all} \quad v, w \in V,\ t \geq 0.$$

Since $\mathcal{A} \in BV_{loc}(\mathbf{R}_+; L^\infty(\Omega; \mathcal{B}(Sym\{3\})))$, there follows $A \in BV_{loc}(\mathbf{R}_+; \mathcal{B}(V,V^*))$. The variational formulation of (1.2), (1.3), (1.6) in V^* becomes now

$$v(t) + \int_0^t A(t-\tau)v(\tau)d\tau = f(t)\,, \quad t \geq 0. \tag{2.3}$$

If the material is homogeneous and isotropic with shear modulus da and compression modulus db, then α is of the following form

$$\alpha(t; w, v) = 2a(t) \int_\Omega E_w(x) : \overline{E_v(x)} dx + (b(t) - 2a(t)/3) \int_\Omega \text{div } w(x) \overline{\text{div } v(x)} dx, \tag{2.4}$$

for all $t \geq 0$, $v, w \in V$; here E_v means the symmetric part of the gradient of E_v, i.e. $E_v(x) = (\nabla v(x) + (\nabla v(x))^T)/2$.

The natural definition of strong and mild solutions of (2.2) or equivalently (2.3) is as follows.

Definition 1 *Let $f \in C(\mathbb{R}_+; V^*)$. A function $v \in C(\mathbb{R}_+; V)$ is called a strong solution of (2.2) or (2.3) if (2.2) holds for every $t \geq 0$ and $w \in V$. $v \in C(\mathbb{R}_+; V^*)$ is called a mild solution of (2.2) or (2.3) if there are $f_n \in C(\mathbb{R}_+; V^*)$ and strong solutions $v_n \in C(\mathbb{R}_+; V)$ of (2.2), with f replaced by f_n, such that $f_n(t) \to f(t)$ and $v_n(t) \to v(t)$ in V^*, uniformly on compact intervals of $\mathbb{R}_+$.*

The most important concept for (2.3) is the notion of the resolvent.

Definition 2 *Let $c \in L^1_{loc}(\mathbb{R}_+)$. A family of linear operators $\{S(t)\}_{t \geq 0} \subset \mathcal{B}(V) \cap \mathcal{B}(V^*)$ is called a resolvent for (2.3) if $S(t)$ is strongly continuous in V and in V^*, $S(0) = I$, and the resolvent equations hold.*

$$S(t)v + \int_0^t A(t-\tau)S(\tau)v = v\,, \quad \textit{for all } t \geq 0,\ v \in V; \tag{2.5}$$

$$S(t)v + \int_0^t S(\tau)A(t-\tau)v = v\,, \quad \textit{for all } t \geq 0,\ v \in V. \tag{2.6}$$

A resolvent $S(t)$ for (2.3) is called c-regular if $c * Sv$ is strongly continuous in V, for each $v \in V^*$.

Without going into details, note that in case a resolvent $S(t)$ for (2.3) exists then it is necessarily unique, is also strongly continuous in H, and the mild solution of (2.3) is given by the variation of parameters formula

$$v(t) = \frac{d}{dt}\int_0^t S(t-\tau)f(\tau)d\tau\,, \quad t \geq 0. \tag{2.7}$$

These facts are known even in a much more general context; see e.g. Prüss [25]. Therefore we restrict our attention to the resolvent $S(t)$.

By the symmetry properties (1.7), the dissipation inequality (1.12) for the relaxation kernel is easily seen to translate to the following property of the form α.

$$2Re\int_0^T\int_0^t d\alpha(\tau; v(t), v(t-\tau))dt \geq 0\,, \quad \text{for all } T > 0 \text{ and } v \in C(\mathbb{R}_+; V). \tag{2.8}$$

Such forms will be called *positive* in the sequel. However, positivity alone does not seem to be strong enough to establish even wellposedness, i.e. existence of a resolvent; for this a stronger notion of positivity is needed.

3. Coercive Forms and Energy Inequalities.

We now strengthen positivity of the form α to coerciveness. It will turn out that this concept is appropriate for hyperbolic solids.

Definition 3 *A form $\alpha : \mathbb{R}_+ \times V \times V \to \mathbb{C}$ as above is called coercive if there is a constant $\gamma > 0$, such that*

$$2\ Re\int_0^T\Big(\int_0^t d\alpha(s, v(t), v(t-s))\Big)dt \geq \gamma\Big\|\int_0^T v(t)dt\Big\|^2 \tag{3.1}$$

for all $v \in C(\mathbb{R}_+;V)$ *and* $T > 0$.

For ideally elastic solids the form α is given by

$$\alpha(t;w,v) = t\int_\Omega E_w(x) : \overline{\mathcal{A}^\infty : E_v(x)}dx;$$

if $\mathcal{A}^\infty(x) \in \mathcal{B}(Sym\{3\})$ is positive definite, uniformly on Ω, then Korn's inequality (see Leis [15] or Marsden and Hughes [17]) implies coerciveness of α. This remark also applies to homogeneous isotropic materials with $a_\infty, b_\infty > 0$.

Suppose $v \in C(\mathbb{R}_+;V)$ is a strong solution of (2.2) and let $f \in W^{1,1}_{loc}(\mathbb{R}_+;V^*)$, $f(0) \in H$. Differentiating (2.2), letting $w = v(t)$ and integrating again we obtain

$$(|v(t)|^2 - |v(0)|^2)/2 + \int_0^t\int_0^s d\alpha(\tau;v(s),v(s-\tau))ds = \int_0^t < v(s),\dot{f}(s) > ds\ , t > 0.$$

Taking real parts in this equation and using coerciveness of α, i.e. (3.1) there results the inequality

$$|v(t)|^2 + \gamma||\int_0^t v(\tau)d\tau||^2 \le |f(0)|^2 + 2Re\int_0^t < v(\tau),\dot{f}(\tau) > d\tau\ , \quad t > 0. \tag{3.2}$$

This is the basic energy inequality for (2.2) in the case of coercive forms. If α is only positive, i.e. $\gamma = 0$, (3.2) is still valid; however, we then do not obtain bounds on any quantity related to the solution $v(t)$ in V. It turns out that (3.2) implies estimates for mild solutions as well.

Proposition 1 *Suppose* $v \in C(\mathbb{R}_+;V^*)$ *is a mild solution of (2.2). Then*
(i) $f \in W^{1,1}_{loc}(\mathbb{R}_+;H)$ *implies* $v \in C(\mathbb{R}_+;H)$, $1 * v \in C(\mathbb{R}_+;V)$, *and*

$$|v(t)|^2 + \gamma||\int_0^t v(\tau)d\tau||^2 \le (\mathrm{Var}_H\, f|_0^t)^2\ , \quad t \ge 0; \tag{3.3}$$

(ii) $f \in W^{2,1}_{loc}(\mathbb{R}_+;V^*)$, $f(0) = 0$, *imply* $v \in C(\mathbb{R}_+;H)$, $1 * v \in C(\mathbb{R}_+;V)$, *and for each* $\delta \in (0,\gamma)$

$$|v(t)|^2 + (\gamma-\delta)||\int_0^t v(\tau)d\tau||^2 \le (\delta^{-1/2} + (\gamma-\delta)^{-1/2})^2 \cdot (\mathrm{Var}_{V^*}\, \dot{f}|_0^t)^2\ , \quad t \ge 0. \tag{3.4}$$

Proof: Since $v \in C(\mathbb{R}_+;V^*)$ is by assumption a mild solution of (2.2), there are $(f_n) \subset C(\mathbb{R}_+;V^*)$ and strong solutions $(v_n) \subset C(\mathbb{R}_+;V^*)$ of (2.2) with f replaced by f_n, such that $f_n \to f$ and $v_n \to v$ in $C(\mathbb{R}_+;V^*)$ as $n \to \infty$. Choose mollifiers $\rho_\varepsilon \in C_0^\infty(0;\varepsilon)$ with $\rho_\varepsilon \ge 0$ and $\int_{-\infty}^\infty \rho_\varepsilon(\tau)d\tau = 1$, and define $f_{n\varepsilon} = f_n * \rho_\varepsilon$, $v_{n\varepsilon} = v_n * \rho_\varepsilon$, $f_\varepsilon = f * \rho_\varepsilon$, $v_\varepsilon = v * \rho_\varepsilon$. Then $f_{n\varepsilon} \in C^1(\mathbb{R}_+;V^*)$, $v_{n\varepsilon} \in C^1(\mathbb{R}_+;V)$ and we have from (3.2)

$$|v_{n\varepsilon}(t)|^2 + \gamma||\int_0^t v_{n\varepsilon}(\tau)d\tau||^2 \le 2Re\int_0^t < v_{n\varepsilon}(\tau),\dot{f}_{n\varepsilon}(\tau) > d\tau\ , \quad t > 0\ , \tag{3.5}$$

since $f_{n\varepsilon}(0) = 0$.
(i) Let $\psi(t) = 2\int_0^t |\dot{f}_{n\varepsilon}(\tau)||v_{n\varepsilon}(\tau)|d\tau$ for $t > 0$; then

$$\frac{d}{dt}\sqrt{\psi(t)} = \frac{\dot{\psi}(t)}{2\sqrt{\psi(t)}} = \frac{2|\dot{f}_{n\varepsilon}(t)||v_{n\varepsilon}(t)|}{2\sqrt{\psi(t)}} \le |\dot{f}_{n\varepsilon}(t)|$$

for each $t \in \mathbb{R}_+$, such that $\psi(t) > 0$. Therefore we conclude

$$(|v_{n\varepsilon}(t)|^2 + \gamma || \int_0^t v_{n\varepsilon}(\tau)d\tau||^2)^{1/2} \leq \sqrt{\psi(t)} \leq \int_0^t |\dot{f}_{n\varepsilon}(\tau)|d\tau\,, \quad t \in \mathbb{R}_+.$$

Letting first $n \to \infty$ we obtain

$$|v_\varepsilon(t)|^2 + \gamma || \int_0^t v_\varepsilon(\tau)d\tau||^2 \leq (\int_0^t |\dot{f}_\varepsilon(\tau)|d\tau)^2 \leq (\int_0^t (\rho_\varepsilon * |df|)(\tau)d\tau)^2$$

and then with $\varepsilon \to 0+$, Inequality (3.3) follows.
(ii) For the proof of (ii), let $w(t) = 1 * v(t)$, $w_n(t) = 1 * v_n(t)$, and $\delta \in (0, \gamma)$; we integrate by parts in (3.5), rearrange, and estimate.

$$\begin{aligned}
|v_{n\varepsilon}(t)|^2 + \gamma||w_{n\varepsilon}(t)||^2 &\leq 2Re \int_0^t < v_{n\varepsilon}(\tau), \dot{f}_{n\varepsilon}(\tau) > d\tau \\
&= 2Re < w_{n\varepsilon}(t), \dot{f}_{n\varepsilon}(t) > -2Re \int_0^t < w_{n\varepsilon}(\tau), \ddot{f}(\tau) > d\tau \\
&\leq \delta||w_{n\varepsilon}(t)||^2 + \delta^{-1}||\dot{f}_{n\varepsilon}(t)||_*^2 + 2\int_0^t ||w_{n\varepsilon}(\tau)|| \cdot ||\ddot{f}_{n\varepsilon}(\tau)||_* d\tau\,.
\end{aligned}$$

Setting

$$\psi(t) = \delta^{-1}||\dot{f}_{n\varepsilon}(t)||_*^2 + 2\int_0^t ||w_{n\varepsilon}(\tau)|| \cdot ||\ddot{f}_{n\varepsilon}(\tau)||_* d\tau,$$

we obtain

$$\begin{aligned}
\dot{\psi}(t) &\leq 2\delta^{-1}||\ddot{f}_{n\varepsilon}(t)||_* \cdot ||\dot{f}_{n\varepsilon}(t)||_* + 2||w_{n\varepsilon}(t)|| \cdot ||\ddot{f}_{n\varepsilon}(t)||_* \\
&\leq 2||\ddot{f}_{n\varepsilon}(t)||_* \cdot (\delta^{-1/2} + (\gamma - \delta)^{-1/2})\sqrt{\psi(t)},
\end{aligned}$$

hence we conclude as in (i)

$$(|v_{n\varepsilon}(t)|^2 + (\gamma - \delta)|| \int_0^t v_{n\varepsilon}(\tau)d\tau||^2)^{1/2} \leq \sqrt{\psi(t)} \leq (\delta^{-1/2} + (\gamma - \delta)^{-1/2}) \int_0^t ||\ddot{f}_{n\varepsilon}(\tau)||_* d\tau, \; t \geq 0.$$

Letting first $n \to \infty$ and then $\varepsilon \to 0$, assertion (ii) follows. □

In practice (3.1) is difficult to check. However, since α_0 is assumed to be of subexponential growth, coerciveness can be characterized in terms of Laplace transforms; cp. Nohel and Shea [21] for the scalar case, i.e. for kernels of positive type. For the sake of completeness, we include an indication of proof for the vector-valued case.

Proposition 2 *Let $\alpha(t; \cdot, \cdot)$ be a sesquilinear form on V such that*

$$|\alpha(t; w, v) - \alpha(s; w, v)| \leq (\alpha_0(t) - \alpha_0(s))||w|| \cdot ||v||\,, \quad \textit{for all } t > s \geq 0\,, v, w \in V, \tag{3.6}$$

where α_0 is nondecreasing and of subexponential growth. Then α satisfies (3.1) iff

$$Re\ \widehat{d\alpha}(\lambda; v, v) \geq \gamma Re(1/\lambda)||v||^2\,, \quad \textit{for each} \quad v \in V, \textit{ and } Re\ \lambda > 0. \tag{3.7}$$

Proof: Replacing $\alpha(t; \cdot, \cdot)$ by $\alpha(t; \cdot, \cdot) - (\gamma t/2)((\cdot, \cdot))$, it is sufficient to verify the equivalence of (3.1) and (3.7) for $\gamma = 0$.

($\Rightarrow$) Suppose α is of positive type, and let $v \in V$, $\lambda \in \mathbb{C}_+$ be given. Choosing $v(t) = e^{-\lambda t}v$, $t > 0$, (3.1) yields

$$2 \text{ Re} \int_0^T \int_0^t d\alpha(\tau; e^{-\lambda t}v, e^{-\lambda(t-\tau)}v)dt \geq 0 , \quad \text{for each} \quad T > 0.$$

Since $\alpha_0(t)$ is of subexponential growth, the limit as $T \to \infty$ of the left hand side exists; from this we obtain for $T \to \infty$

$$\begin{aligned} 0 &\leq 2\text{Re} \int_0^\infty \int_0^t d\alpha(\tau; e^{-\lambda t}v, e^{-\lambda(t-\tau)}v)dt = 2\text{Re} \int_0^\infty (\int_\tau^\infty e^{-(\lambda+\bar{\lambda})t}dt)e^{\bar{\lambda}\tau}d\alpha(\tau; v, v) \\ &= \frac{1}{Re\lambda}\text{Re} \int_0^\infty e^{-(\lambda+\bar{\lambda})\tau}e^{\bar{\lambda}\tau}d\alpha(\tau; v, v) = \frac{1}{Re\lambda}\text{Re } \widehat{d\alpha}(\lambda; v, v), \end{aligned}$$

and so (3.7) with $\gamma = 0$ follows.
($\Leftarrow$) Conversely, assume (3.7) holds and let $v \in C(\mathbb{R}_+; V)$ be given. Redefining $v(t) = 0$ for $t \notin [0, T]$ and using the Fourier inversion formula and Fubini's theorem, we obtain for any $\mu > 0$

$$\begin{aligned} &2\pi \text{ Re} \int_0^T \int_0^t e^{-\mu\tau}d\alpha(\tau; v(t), v(t-\tau))dt = \text{Re} \int_{-\infty}^\infty < \tilde{\bar{v}}(-t), \int_0^t e^{-\mu t}dA(\tau)v(t-\tau) > dt \\ &= \text{Re} \int_{-\infty}^\infty < \tilde{v}(\rho), \widehat{dA}(\mu + i\rho)\tilde{v}(\rho) > d\rho = \text{Re} \int_{-\infty}^\infty \widehat{d\alpha}(\mu - i\rho; \tilde{v}(\rho), \tilde{v}(\rho))d\rho \geq 0 , \end{aligned}$$

and with $\mu \to 0+$, (3.1) with $\gamma = 0$ follows. □

4. Resolvents for Hyperbolic Solids

We are now in position to state and prove our first main result.

Theorem 1 *Suppose* $\alpha : \mathbb{R}_+ \times V \times V \to \mathbb{C}$ *satisfies*
(V1) $\alpha(t; \cdot, \cdot)$ *is a bounded sesquilinear form on* V*, for each* $t \geq 0$*, and* $\alpha(0; \cdot, \cdot) = 0$*;*
(V2) $\alpha(\cdot; u, v) \in W^{1,\infty}_{loc}(\mathbb{R}_+)$ *for each* $u, v \in V$*, and*

$$|\dot{\alpha}(t; u, v) - \dot{\alpha}(s; u, v)| \leq (\alpha_1(t) - \alpha_1(s))||u|| \, ||v|| \quad , \quad u, v \in V \; , \; t \geq s \geq 0,$$

where $\alpha_1(t)$ *is nondecreasing and of subexponential growth, w.l.o.g.* $\alpha_1(0) = 0$*;*
(V3) α *is coercive with coercivity constant* $\gamma > 0$*.*
Then (2.2) admits a t-regular resolvent $S(t)$*. Moreover, with* $R = 1 * S$ *and* $T = t * S$ *we have the following regularity properties*
(a) $\{S(t)\}_{t\geq 0} \subset \mathcal{B}(V) \cap \mathcal{B}(V^*) \cap \mathcal{B}(H)$ *is strongly continuous in* V^**,* H*, and* V*, and*

$$|S(t)|_{\mathcal{B}(V)} \, , |S(t)|_{\mathcal{B}(V^*)} \leq 1 + 2\gamma^{-1}\alpha_1(t) \, , \quad |S(t)|_{\mathcal{B}(H)} \leq 1 \, , \quad \text{for all } t \geq 0;$$

(b) $\{R(t)\}_{t\geq 0} \subset \mathcal{B}(V^*, H) \cap \mathcal{B}(H, V)$ *and* $\{T(t)\}_{t\geq 0} \subset \mathcal{B}(V^*, V)$ *are strongly continuous, and*

$$|R(t)|_{\mathcal{B}(V^*,H)} \, , |R(t)|_{\mathcal{B}(H,V)} \leq \gamma^{-1/2} \, , \quad |T(t)|_{\mathcal{B}(V^*,V)} \leq 2\gamma^{-1} \quad , \text{for all } t \geq 0;$$

(c) $\{S(t)\}_{t\geq 0} \subset \mathcal{B}(V, H) \cap \mathcal{B}(H, V^*)$ *is strongly continuously differentiable,* $\{S(t)\}_{t\geq 0} \subset \mathcal{B}(V, V^*)$ *even twice a.e., and for a.a.* $t \geq 0$ *we have*

$$|\dot{S}(t)|_{\mathcal{B}(V,H)} \, , |\dot{S}(t)|_{\mathcal{B}(H,V^*)} \leq \gamma^{-1/2}\alpha_1(t) \, , \quad |\ddot{S}(t)|_{\mathcal{B}(V;V^*)} \leq \alpha_1(t)(1 + 2\gamma^{-1}\alpha_1(t)).$$

Before we turn to the proof of Theorem 1 several remarks are in order.

Assumption (V2) means $\dot{A} \in BV_{loc}(\mathbb{R}_+; \mathcal{B}(V, V^*))$ even in $BV_{loc}(\mathbb{R}_+; \mathcal{B}(V, V^*))$ if α_1 is bounded. Therefore (2.3) is equivalent to the equation of second order

$$\ddot{v}(t) + \int_0^t d\dot{A}(\tau)v(t-\tau) = g(t)\ , \quad t \geq 0, \tag{4.1}$$
$$v(0) = v_0\ , \quad \dot{v}(0) = v_1.$$

For the solution of (4.1) we have the following variation of parameters formula.

$$v(t) = S(t)v_0 + R(t)v_1 + \int_0^t R(t-\tau)g(\tau)d\tau\ , \quad t \geq 0. \tag{4.2}$$

Thus the resolvent $S(t)$ corresponds to the cosine family, $R(t)$ to the sine family of second order differential equations. The following corollary describes the solvability behavior of (4.1) implied by Theorem 1.

Corollary 1 *Let the assumptions of Theorem 1 be satisfied, let* $v_0\ , v_1 \in V^*$, $g \in L^1_{loc}(\mathbb{R}_+; V^*)$, *and let* $v(t)$ *be given by (4.2). Then*
(i) $v_0 \in V$, $v_1 \in H$, $g \in C(\mathbb{R}_+; H)$ *imply* $v \in C(\mathbb{R}_+; V)$, $\dot{v} \in C(\mathbb{R}_+; H)$, $\ddot{v} - \dot{A}(\cdot)v_0 \in C(\mathbb{R}_+; V^*)$, *and* $v(t)$ *is a strong solution of (4.1);*
(ii) $v_0 \in V$, $v_1 \in H$, $g \in W^{1,1}(\mathbb{R}_+; V^*)$ *imply* $v \in C(\mathbb{R}_+; V)$, $\dot{v} \in C(\mathbb{R}_+; H)$, $\ddot{v} - \dot{A}(\cdot)v_0 \in C(\mathbb{R}_+; V^*)$, *and* $v(t)$ *is a strong solution of (4.1);*
(iii) $v_0 \in H$, $v_1 \in V^*$, $g \in C(\mathbb{R}_+; V^*)$ *imply* $v \in C(\mathbb{R}_+; H)$, $\dot{v} \in C(\mathbb{R}_+; V^*)$, *and* $v(t)$ *is a mild solution of (4.1).*

In the scalar case $\alpha(t; w, v) = a(t)\alpha_\infty(w, v)$, where $a(t)$ is of the form (1.9), the assumptions of Theorem 1 are equivalent to $a_0 = 0$, $a_1(0+) < \infty$, $a_\infty > 0$, and α_∞ a bounded sesquilinear form on V which is coercive. This is known as the case of hyperbolic solids. For hyperbolic fluids see Corollary 2 below.

Considering Examples (i)~(iv) of Section 1 again, observe that (i) and (iv) are hyperbolic, (ii) and (iii) parabolic; (i) and (iii) represent solids, (ii) and (iv) fluids. Thus ideally elastic solids are a special case of Theorem 1; the latter corresponds to the form $\alpha(t; u, v) = t\alpha_\infty(u, v)$, for which (4.1) becomes the Cauchy problem for second order differential equations.

Observe that (4.1) or (2.3) do not have a main part and of course are not of scalar type. Therefore Theorem 1 it is not covered by existing results. Its proof is based on the energy estimates derived in Section 3 and Laplace transform techniques. We use, however, a perturbation result for equations with main part A^0, a generator of an analytic C_0-semigroup, to transfer the energy estimates from time domain to frequency domain.

<u>*Proof:*</u> Introduce sesquilinear forms α_n by means of

$$\dot{\alpha}_n(t; u, v) = \hat{\dot{\alpha}}(n; u, v) + \sum_{k=1}^{\infty} \frac{n^{2k}}{k!} \hat{\dot{\alpha}}^{(k)}(n; u, v)(-1)^k \frac{t^{k-1}}{(k-1)!} e^{-nt},$$

for $t \geq 0$, $u, v \in V$, $n \in \mathbb{N}$; these forms α_n are bounded and sesquilinear. Define operators A^0_n, $A_n(t)$, $B_n(t)$ according to

$$\begin{aligned} &< u, A^0_n v >= \hat{\dot{\alpha}}(n; u, v)\ , \quad \text{for all } u, v \in V \\ &< u, \dot{A}_n(t)v >= \dot{\alpha}_n(t; u, v)\ , \quad \text{for all } t \geq 0,\ u, v \in V \\ &B_n(t) = \dot{A}_n(t) - A^0_n; \end{aligned} \tag{4.3}$$

we obtain this way approximating equations for (2.3) in V^*.

$$\dot{v}(t) = A_n^0 v(t) + \int_0^t B_n(\tau)v(t-\tau)d\tau + h(t)\,, \quad t>0\,, v(0)=v_0. \tag{4.4}$$

By coerciveness of α, the operators A_n^0 are regularly accretive hence generates bounded analytic C_0-semigroups in V, H, V^*; see e.g. Tanabe [26]. On the other hand, $B_n \in C^1(\mathbb{R}_+;\mathcal{B}(V,V^*))$ and B_n as well as $\dot{B}_n$ are of subexponential growth, as simple estimates show. Therefore by a wellknown perturbation result, (4.4) admits a resolvent $S_n(t)$ in V^*; cp. e.g. Grimmer and Prüss [12]. Since

$$\hat{\alpha}_n(\lambda;u,v) = \hat{\alpha}(\frac{\lambda n}{\lambda+n};u,v)\,, \qquad \mathrm{Re}\,\lambda>0\,,\ n\in\mathbb{N}\,,\ u,v\in V,$$

α_n is coercive with the same constant γ of coercivity. Thus the energy estimates (3.3) and (3.4) are also valid for (4.4) and so with $f(t)\equiv x\in H$ resp. $f(t)\equiv tv^*\in V^*$ they imply

$$|S_n(t)|_{\mathcal{B}(H)}\le 1, \quad |R_n(t)|_{\mathcal{B}(H,V)}, \quad |R_n(t)|_{\mathcal{B}(V^*,H)}\le\gamma^{-1/2}, \quad |T_n(t)|_{\mathcal{B}(V^*,V)}\le 2\gamma^{-1}, \quad t>0,$$

where $R_n = 1 * S_n$, $T_n = t * S_n$. These inequalities yield the following estimates in the frequency domain.

$$|(\frac{d}{d\lambda})^k\hat{S}_n(\lambda)|_{\mathcal{B}(H)}\le k!\lambda^{-(k+1)}, \quad |(\frac{d}{d\lambda})^k[\hat{S}_n(\lambda)/\lambda]|_{\mathcal{B}(V^*,H)}\le\gamma^{-1/2}k!\lambda^{-(k+1)},$$
$$|(\frac{d}{d\lambda})^k[\hat{S}_n(\lambda)/\lambda]|_{\mathcal{B}(H,V^*)}\le\gamma^{-1/2}k!\lambda^{-(k+1)}, \quad |(\frac{d}{d\lambda})^k[\hat{S}_n(\lambda)/\lambda^2]|_{\mathcal{B}(V^*,V)}\le 2\gamma^{-1}k!\lambda^{-(k+1)},$$

for all $n\ge 1$, $k\ge 0$, $\lambda>0$. Passing to the limit as $n\to\infty$, the same estimates are valid for $H(\lambda)=(\lambda-\hat{A}(\lambda))^{-1}$; note that $\{H(\lambda)\}_{Re\,\lambda>0}\subset\mathcal{B}(V^*,V)$ exists and is holomorphic, as standard coerciveness arguments show, and that $\hat{S}_n^{(k)}(\lambda)\to H^{(k)}(\lambda)$ in $\mathcal{B}(V^*,V)$ uniformly on compact subsets of $\mathbb{C}_+$, and for all k. The identities

$$H(\lambda) = I/\lambda + \widehat{d\dot{A}}(\lambda)H(\lambda)/\lambda^2 = I/\lambda + [H(\lambda)/\lambda^2]\widehat{d\dot{A}}(\lambda) \quad , \quad \lambda>0,$$

by (V2) imply in addition

$$|(\frac{d}{d\lambda})^k H(\lambda)|_{\mathcal{B}(V^*)}, |(\frac{d}{d\lambda})^k H(\lambda)|_{\mathcal{B}(V)}\le(1+2\gamma^{-1}\widehat{d\alpha_1}\varepsilon))k!(\lambda-\varepsilon)^{-k-1} \quad , \quad \lambda>\varepsilon,$$

for each $\varepsilon>0$. This follows from the estimates for $H(\lambda)/\lambda^2$ and from

$$\sum_{k=0}^{\infty}(\lambda-\varepsilon)^k|\widehat{d\dot{A}}^{(k)}(\lambda)|/k! \le \int_0^\infty e^{-\varepsilon t}|d\dot{A}(t)| \le \widehat{d\alpha_1}(\varepsilon)\,, \varepsilon>0.$$

By the vector-valued version of Widder's characterization of the Laplace transform of L^∞-functions (cp. Arendt [2]) and observing the reflexivity of V, H, V^*, for each $v^*\in V^*$ there are functions $S(t)v^*\in V^*$, $R(t)v^*\in H$, $T(t)v^*\in V$ defined almost everywhere, such that $|R(t)v^*|\le\gamma^{-1/2}$, $||T(t)v^*||\le 2/\gamma$, and $||S(t)v^*||_*\le(1+2\gamma^{-1}\widehat{d\alpha_1}(\varepsilon))e^{\varepsilon t}$ a.e. satisfying $\widehat{Sv^*}(\lambda)=H(\lambda)v^*$, $\widehat{Rv^*}(\lambda)=H(\lambda)v^*/\lambda$, $\widehat{Tv^*}(\lambda)=H(\lambda)v^*/\lambda^2$. Similarly by uniqueness of the Laplace transform for each $h\in H$, we have $S(t)h\in H$, $R(t)h\in V$, $|S(t)h|\le 1$, $||R(t)h||\le\gamma^{-1/2}$ a.e., and for each $v\in V$ also $S(t)v\in V$ with $||S(t)v||\le(1+2\gamma^{-1}\widehat{d\alpha_1}(\varepsilon))e^{\varepsilon t}$

a.e.. Since $\dot{T}(t)v^* = R(t)v^*$, $T(t)v^*$ is Lipschitz in V for $v^* = h \in H$, hence by density of H in V^*, $T(t) \subset \mathcal{B}(V^*, V)$ is strongly continuous. Similarly, one obtains also the strong continuity properties of $R(t)$ asserted in Theorem 1 from the boundedness of $S(t)$ in H and V. The first resolvent equation (2.6) and its once differentiated version then yield strong continuity of $S(t)$ in V^* and of $\dot{S}(t)$ from H to V^*, as well as the corresponding estimates claimed in Theorem 1. The remaining properties follow analogously from the second resolvent equation (2.7) and from (2.6) twice differentiated. □

Another coercivity concept different from Definition 3 is based on an inequality of the form

$$\text{Re}\ \widehat{d\alpha}(\lambda, v, v) \geq \gamma \text{Re}\ \frac{1}{\lambda+\eta}||v||^2 \quad , \quad v \in V\ ,\ \text{Re}\ \lambda > 0, \tag{4.5}$$

where γ and η are positive constants; compare with (3.7). Forms satisfying (4.5) will be called *η-coercive* in the sequel. This concept is appropriate for hyperbolic fluids; in fact, Example (iv) of Section 1, i.e. the Maxwell fluid is μ-coercive. In the scalar case $\alpha(t; w, v) = a(t)\alpha_\infty(w, v)$, where $a(t)$ is of the form (1.9), the assumptions of Corollary 2 below are equivalent to $a_0 = 0$, $a_1(0+) < \infty$, $a_\infty \geq 0$, a_1 strongly positive in the sense of Nohel and Shea [21], and α_∞ a bounded sesquilinear form on V which is coercive.

By the same methods as in the proof of Theorem 1 the following result for η-coercive forms is obtained.

Corollary 2 *Suppose $\alpha : \mathbb{R}_+ \times V \times V \to \mathbb{C}$ is η-coercive for some $\eta > 0$, and satisfies (V1), (V2) of Theorem 1. Then (2.2) admits a t-regular resolvent $S(t)$. Moreover, with $R_\eta(t) = (e^{-\eta t} * S)(t)$, $T_\eta(t) = (e^{-\eta t} * R_\eta)(t)$ we have the estimates*

$$|S(t)|_{\mathcal{B}(H)} \leq 1\ ;\ |R_\eta(t)|_{\mathcal{B}(V^*,H)}\ ,\ |R_\eta(t)|_{\mathcal{B}(H,V)} \leq \gamma^{-1/2}\ ;\ |T_\eta(t)|_{\mathcal{B}(V^*,V)} \leq 2\gamma^{-1},$$

and

$$||R_\eta(\cdot)x||_2 \leq (2\eta\gamma)^{-1/2}|x|\ ;\ ||T_\eta(\cdot)x||_2 \leq 2(2\eta\gamma^2)^{-1/2}||x||_*.$$

In the situation of Corollary 2 one can also obtain bounds for the remaining quantities, i.e. $S(t)$ in $\mathcal{B}(V)$ and $\mathcal{B}(V^*)$, $\dot{S}(t)$ in $\mathcal{B}(V, H)$ and $\mathcal{B}(H, V^*)$, as well as $\ddot{S}(t)$ in $\mathcal{B}(V, V^*)$, to the result that similar estimates as in Theorem 1 are valid; in particular, all of these operator families are bounded on $\mathbb{R}_+$ if α_1 is bounded.

Concerning the proof of Corollary 2 we only mention the basic energy inequality corresponding to (3.2) which is implied by η-coercivity.

$$|v(t)|^2 + \gamma||w(t)||^2 + 2\eta\gamma \int_0^t ||w(s)||^2 ds \leq |f(0)|^2 + 2\text{Re} \int_0^t < v(s), \dot{f}(s) > ds\ , \quad t \geq 0, \tag{4.6}$$

where $w(t) = e^{-\eta t} * v(t)$.

5. Stability on the Halfline

We now turn attention to the asymptotic behaviour of the solutions of (2.2) or (2.3) as $t \to \infty$ in the case of solids as well as of fluids. For this purpose we first summarize several versions of the variation of parameters formula (2.8).

$$v(t) = f(t) + \int_0^t \dot{S}(t-\tau)f(\tau)d\tau\ , \quad t \geq 0; \tag{5.1}$$

$$v(t) = S(t)f(0) + \int_0^t S(t-\tau)\dot{f}(\tau)d\tau\,, \quad t \geq 0; \tag{5.2}$$

$$v(t) = S(t)f(0) + R(t)\dot{f}(0) + \int_0^t R(t-\tau)\ddot{f}(\tau)d\tau\,, \quad t \geq 0; \tag{5.3}$$

$$v(t) = S(t)f(0) + R(t)\dot{f}(0) + T(t)\ddot{f}(0) + \int_0^t T(t-\tau)\dddot{f}(\tau)d\tau\,, \quad t \geq 0. \tag{5.4}$$

The strong continuity properties of S, R, and T obtained in Theorem 1 and Corollary 2 then yield regularity properties of $v(t)$, according to those of $f(t)$; Corollary 1 is an example for this. Similarly, asymptotic properties of S, R, and T in combination with those of f imply a certain asymptotic behaviour of the solution v. For example, suppose $S \in L^1(\mathbb{R}_+; \mathcal{B}(H))$, then it is easily seen that (5.2) implies $v \in L^p(\mathbb{R}_+; H)$, whenever $\dot{f}$ has this property and $f(0) \in H$; note that $|S(t)| \leq 1$ holds. On the other hand, suppose the solution v of (2.3) belongs to $L^1(\mathbb{R}_+; H)$ whenever $f(t) \equiv h \in H$; then $S(\cdot)h \in L^1(\mathbb{R}_+; H)$, for each $h \in H$. This shows that integrability properties of S, R, and T are important. There are several different notions of integrability; cp. Prüss [14].

Definition 4 *Let X and Z be Banach spaces, and $\{W(t)\}_{t\geq 0} \subset \mathcal{B}(X,Z)$ be a strongly measurable family of operators, i.e. $W(\cdot)x$ is Bochner-measurable in Z, for each $x \in X$. Then $W(t)$ is called*
(i) strongly integrable (from X to Z), if $W(\cdot)x \in L^1(\mathbb{R}_+; Z)$ for each $x \in X$;
(ii) integrable (from X to Z) if there is $\varphi \in L^1(\mathbb{R}_+)$ such that $|W(t)| \leq \varphi(t)$ a.e. on $\mathbb{R}_+$;
(iii) uniformly integrable (from X to Z), if $W(\cdot) \in L^1(\mathbb{R}_+; \mathcal{B}(X,Z))$.

Obviously, every uniformly integrable operator family $W(t)$ is integrable, however not conversely, unless $W(\cdot)$ is Bochner-measurable in $\mathcal{B}(X,Z)$. Similarly, every integrable family $W(t)$ is also strongly integrable, but the converse is not true, in general.

Suppose $W(t)$ is integrable from X to Z, and let $g \in L^p(\mathbb{R}_+; X)$, where $1 \leq p \leq \infty$; then approximating g by simple functions, it is not difficult to see that $W * g$ belongs to $L^p(\mathbb{R}_+; Z)$, and that the estimate

$$|W * g|_{Z,p} \leq |\varphi|_1 |g|_{X,p}\,, \quad g \in L^p(\mathbb{R}_+; X),$$

is satisfied; here $|\cdot|_{X,p}$ indicates the norm in $L^p(\mathbb{R}_+; X)$. Similarly the subspaces $C_b(\mathbb{R}_+; X)$, $C_{ub}(R_+; X)$, and $C_0(\mathbb{R}_+; X)$ of $L^\infty(\mathbb{R}_+; X)$ are mapped into the corresponding subspaces of $L^\infty(\mathbb{R}_+; Z)$.

If $W(t)$ is only strongly integrable from X to Z the situation is more delicate. Since the map $x \mapsto W(\cdot)x$ from X to $L^1(\mathbb{R}_+; Z)$ is closed linear, hence bounded, thanks to the closed graph theorem, there is a constant $C > 0$, such that

$$|W(\cdot)x|_{Z,1} \leq C|x|_X\,, \quad x \in X.$$

For a Riemannian step function $g = \sum_{i=0}^n \chi_{[t_i,t_{i+1})} x_i$, where $0 \leq t_0 < \ldots < t_n < \infty$, this inequality yields with $|g|_{X,1} = \sum_{i=0}^n |x_i|_X (t_{i+1} - t_i)$

$$|W * g|_{Z,1} \leq C|g|_{X,1}. \tag{5.5}$$

Since such step functions are dense in L^1, (5.5) remains valid for each $g \in L^1(\mathbb{R}_+; X)$. Approximating $W(t)x$ by $\rho_n * Wx$, where ρ_n denotes a Dirac sequence, we see that (5.5) is even equivalent to the strong integrability of W from X to Z.

Unfortunately, the argument just given does not work for $p \neq 1$, strong integrability alone does not seem to be sufficient for (5.5) in case $p \neq 1$. However, if the family $\{W^*(t)\}_{t\geq 0} \subset \mathcal{B}(Z^*, X^*)$ is also strongly integrable, we may obtain L^∞-estimates as follows. Let $g \in L^\infty(\mathbb{R}_+; X)$, and $z^* \in Z^*$. Then

$$\begin{aligned} < W * g(t), z^* >_Z &= \int_0^t < g(t-\tau), W^*(\tau)z^* >_X d\tau \leq \int_0^t |g(t-\tau)|_X |W^*(\tau)z^*|_{X^*} d\tau \\ &\leq C^*|z^*|_{Z^*}|g|_{X,\infty}, \end{aligned}$$

for all $t \geq 0$. Taking the supremum over the unit ball in Z^*, this inequality implies

$$|W * g|_{Z,\infty} \leq C^*|g|_{X,\infty}. \tag{5.6}$$

Conversely, if W^*z^* is locally integrable on $\mathbb{R}_+$ for each $z^* \in Z^*$ and (5.6) holds, then W^* must necessarily be strongly integrable.

Finally, if both, (5.5) and (5.6), hold then the vector-valued Riesz-Thorin interpolation theorem implies that the convolution with W maps also $L^p(\mathbb{R}_+; X)$ into $L^p(\mathbb{R}_+; Z)$, for each $p \in [1, \infty]$. It is also easy to show that convolution with W leaves the spaces $C_b(\mathbb{R}_+; X)$ of bounded continuous functions, $C_{ub}(\mathbb{R}_+; X)$ of bounded uniformly continuous functions, and $C_0(\mathbb{R}_+; X)$ of continuous functions converging to zero as $t \to \infty$ invariant. We summarize these observations in the following proposition.

Proposition 3 *Let X and Z be Banach spaces, and let $\mathcal{F}$ denote any of the symbols L^p, $p \in [1, \infty]$, C_b, C_{ub}, and C_0. Assume either of the following conditions.*
(i) W is integrable from X to Z;
(ii) W and W^* are strongly integrable from X to Z resp. from Z^* to X^*.
Then $g \in \mathcal{F}(\mathbb{R}_+; X)$ implies $W * g \in \mathcal{F}(\mathbb{R}_+; Z)$, and there is a constant $C > 0$ such that $|W * g|_{Z,\mathcal{F}} \leq C|g|_{X,\mathcal{F}}$.

Clearly, Proposition 3 yields results on the solvability behavior of (2.3) on the halfline from integrability of S, R, and T, resp. strong integrability of these quantities and their duals; the statement of these results we leave to the reader.

After these preparations let us now consider the integrability properties of the resolvent S for (2.3) in the hyperbolic case. To obtain an indication of what is to be expected for nonisotropic viscoelasticity, let us recall first some wellknown results for the homogeneous, isotropic, and synchronous or incompressible case. Thus consider the equation of scalar type

$$v(t) + \int_0^t a(t-\tau)Av(\tau)d\tau = f(t), \quad t \geq 0, \tag{5.7}$$

in the Hilbert space H, where A is a positive semidefinite operator in H, and the kernel $a(t)$ is of the form

$$a(t) = a_0 + a_\infty t + \int_0^t a_1(s)ds, \quad t \geq 0, \tag{5.8}$$

with a_0 , $a_\infty \geq 0$, and $a_1 \in L^1_{loc}(\mathbb{R}_+)$ nonnegative, nonincreasing, and of positive type, i.e. Re $\hat{a}_1(\lambda) \geq 0$ for all Re $\lambda > 0$. Here one should think of da being the shear modulus of the material in question and A the Stokes operator in the incompressible case or the classical elasticity operator for synchronous compressible media. The first part of the following result is due to Carr and Hannsgen [4], while the second part is taken from Prüss [14].

Theorem 2 *In addition to the assumptions stated above, assume either of the following.*
(i) a_1 *and* $-\dot{a}_1$ *are convex;*
(ii) $\log a_1$ *is convex.*
Then the resolvent $S(t)$ *is integrable in* H *iff* $0 \in \rho(A)$ *and* $a(t) \not\equiv a_\infty t$. *In this case*

$$\int_0^\infty S(t)dt = \begin{cases} 0 & \text{if } a_\infty > 0 \text{ or } a_1 \notin L^1(\mathbb{R}_+) \\ (a_0 + \int_0^\infty a_1(t)dt)^{-1}A^{-1} & \text{if } a_\infty = 0 \text{ and } a_1 \in L^1(\mathbb{R}_+). \end{cases} \tag{5.9}$$

For viscoelasticity this means that $S(t)$ is integrable in H iff the material is not ideally elastic and A is invertible; the latter is the case, e.g. if the underlying domain Ω is bounded.

Observe the sharp difference between a *fluid*, i.e. $a_\infty = 0$ and $a_1 \in L^1(\mathbb{R}_+)$, or equivalently $a \in BV(\mathbb{R}_+)$, and a *solid*, i.e. $a_\infty > 0$. For a fluid one cannot expect integrability of $R(t)$ since $\lim_{t\to\infty} R(t) = \int_0^\infty S(t) \neq 0$, while for a solid integrability of $R(t)$ in H follows from integrability of $tS(t)$, which was proved in the paper of Hannsgen and Wheeler [13]. Observe that one can never expect $T(t) = t * S(t)$ to be strongly integrable, since this would imply boundedness of $\widehat{T}(\lambda) = (I + \widehat{a}(\lambda)A)^{-1}/\lambda^3 = (\lambda^2 + (a_0\lambda + a_\infty + \lambda\widehat{a}_1(\lambda))A)^{-1}/\lambda$ as $\lambda \to 0+$, which is impossible.

Concerning uniform integrability, note that in Prüss [14] the equivalence "$S(t)$ is uniformly integrable in H iff $0 \in \rho(A)$, and $a_0 > 0$ or $-a_1(0+) = \infty$" is obtained if $\log a_1$ is convex. This characterization shows that uniform integrability is not the right concept for hyperbolic problems.

Integrability of $\dot{S}$ and $\ddot{S}$ are in the hyperbolic case $a_0 = 0$ and $a_1(0+) < \infty$ easy consequences of that of S and R; differentiate the resolvent equation to see this. For the general case we refer to Carr and Hannsgen [5], Hannsgen and Wheeler [13], Noren [22], and Prüss [24].

For kernels a_1 which are convex but do not satisfy (i) or (ii) of Theorem 2, there is still a characterization of strong integrability of the resolvent in terms of frequency domain conditions; see Prüss [25] for the proof.

Theorem 3 *Let the general assumptions on* A *and* $a(t)$ *stated behind (5.7) hold, and let* a_1 *be convex. Then for each positive definite operator* A *the resolvent* $S(t)$ *for (5.7) is strongly integrable in* H *iff the following conditions are satisfied.*
(i) $Re\, \widehat{da}(i\rho) > 0$ *for all* $\rho \in \mathbb{R}, \rho \neq 0$;
(ii) $\overline{\lim}_{|\rho|\to\infty}$ $-Im\, \widehat{da}(i\rho)/\rho Re\, \widehat{da}(i\rho) < \infty$.

Conditions (i) and (ii) of Theorem 3 have nice interpretations: (i) means $\widehat{a}(\lambda) \notin (-\infty, 0)$ which is necessary for $\widehat{S}(\lambda) = (\lambda + \widehat{da}(\lambda)A)^{-1}$ to exist on $\mathbb{C}_+$, while (ii) implies its boundedness as $|\lambda| \to \infty$.

6. Strong Integrability of Resolvents

Consider now the problems of variational type as introduced in Section 2 in the *hyperbolic* case, i.e. the form α is subject to assumptions (V1) and (V2) of Theorem 1. It seems that in this situation integrability of, say, $S(t)$ in H is quite difficult to prove since this requires pointwise estimates of $|S(t)|_{\mathcal{B}(H)}$. However the circumstances are much better concerning strong integrability of S, R, $\dot{S}$, and $\ddot{S}$, since Parseval's theorem is available due to the Hilbert space setting.

Since Theorem 3 refers to the special case $\alpha(t; u, v) = a(t)(A^{1/2}u, A^{1/2}v)$, it becomes apparent that for strong integrability of $S(t)$ for (2.3) frequency domain conditions reflecting

(i) and (ii) of Theorem 3 will be needed, as well as some regularity assumptions on the form $\alpha(t;u,v)$ which replace convexity of $a_1(t)$. We first concentrate on the case of *solids* which corresponds to $\dot{A}(t) \to A_\infty \neq 0$ as $t \to \infty$.

Theorem 4 *Let $\alpha : \mathbb{R}_+ \times V \times V \to \mathbb{C}$ satisfy (V1) and (V2) of Theorem 2 with $\alpha_1(t)$ bounded, and let α be η-coercive for some $\eta > 0$. In addition assume*
(V4) $\alpha(\cdot;u,v) \in W^{2,1}_{loc}(\mathbb{R}_+)$, for each $u, v \in V$, and

$$|\ddot{\alpha}(t;u,v) - \ddot{\alpha}(s;u,v)| \le (\alpha_2(s) - \alpha_2(t))||u||\,||v||\,, \quad \textit{for all } t > s > 0,\ u,v \in V,$$

where $\alpha_2(t)$ is nonincreasing and such that $-\int_0^\infty t d\alpha_2(t) < \infty$;
(V5) $\overline{\lim}_{|\rho|\to\infty} \mid Im\ \widehat{\alpha}(i\rho,u,u)/[\rho\ Re\ \widehat{\alpha}(i\rho,u,u)]| \le \phi_\infty < \infty$ for each $u \in V$;
(S) For each $u \in V$, $\alpha_\infty(u,u) = \lim_{t\to\infty} \dot{\alpha}(t,u,u)$ exists, and, for some $\gamma_\infty > 0$,

$$Re\ \alpha_\infty(u,u) \ge \gamma_\infty ||u||^2\,, \quad \textit{for all } u \in V.$$

Then the resolvent $S(t)$ for (2.3) and its integral $R(t) = (1 * S)(t)$ satisfy
(i) $v \in V \Rightarrow S(\cdot)x \in L^1(\mathbb{R}_+;V)$, $\dot{S}(\cdot)v \in L^1(\mathbb{R}_+;H)$, $\ddot{S}(\cdot)v \in L^1(\mathbb{R}_+;V^*)$;
(ii) $x \in H \Rightarrow S(\cdot)x \in L^1(\mathbb{R}_+;H)$, $\dot{S}(\cdot)x \in L^1(\mathbb{R}_+;V^*)$, $R(\cdot)x \in L^1(\mathbb{R}_+;V)$;
(iii) $u \in V^* \Rightarrow S(\cdot)u \in L^1(\mathbb{R}_+;V^*)$, $R(\cdot)u \in L^1(\mathbb{R}_+;H)$, $e^{-\eta t} * R(\cdot)u \in L^1(\mathbb{R}_+;V)$.

Note that $\widehat{\alpha}(\lambda;u,v) = \widehat{d\alpha}(\lambda;u,v)/\lambda$ is welldefined on $\mathbb{C}_+ \setminus \{0\}$ since α_1 is bounded by assumption; therefore (V5) makes sense.

η-coercivity together with (S), (V2), and boundedness of α_1 imply coercivity of α, hence Theorem 1 applies, and consequently the exponent 1 in (i), (ii), (iii) of Theorem 4 can be replaced by ∞, hence by any $p \in [1,\infty]$, via interpolation. Since the adjoint form $\alpha^*(t,u,v) = \overline{\alpha(t;v,u)}$ also satisfies the assumptions of Theorem 4, the same properties are true for the the adjoints S^* and R^*; therefore we may apply Proposition 3(ii) to obtain the corresponding solvability properties of (2.3) on the halfline.

Proof: (a) We first show how that $H(\lambda)$ is bounded in $\mathcal{B}(V^*)$, $\mathcal{B}(H)$, $\mathcal{B}(V)$, $H(\lambda)/\lambda$ in $\mathcal{B}(V^*,H)$, $\mathcal{B}(H,V)$, and $H(\lambda)/(\lambda(\lambda+\eta))$ in $\mathcal{B}(V^*,V)$ on $\mathbb{C}_+$, which are necessary for (i), (ii), (iii) of Theorem 4 to be valid.

Taking the Laplace transform of (2.2) we obtain with $f \equiv v \in V$

$$(w, H(\lambda)v) + \widehat{\alpha}(\bar{\lambda}; w, H(\lambda)v) = \lambda^{-1} < w, v >\,, \quad \text{for all } v, w \in V,\ \text{Re}\ \lambda > 0;$$

For $w = H(\lambda)v$ this identity implies

$$\lambda|H(\lambda)x|^2 + \widehat{\alpha}(\bar{\lambda}, H(\lambda)x, H(\lambda)x) =< H(\lambda)x, x >, \tag{6.1}$$

valid for Re $\lambda > 0$ and $x \in V^*$ since V is dense in V^*. Taking real parts in (6.1), by η-coercivity we obtain

$$\eta\gamma||H(\lambda)x|| \le ((\eta+\sigma)^2 + \rho^2)||x||_* \quad , \quad \text{Re}\ \lambda > 0\ ,\ x \in V^*, \tag{6.2}$$

where $\lambda = \sigma + i\rho$. We have

$$-\lambda^2 \widehat{\dot{A}}'(\lambda) = \lambda^2 \widehat{t\dot{A}}(\lambda) = \widehat{td\ddot{A}}(\lambda) + 2\widehat{d\dot{A}}(\lambda) - \dot{A}(0), \tag{6.3}$$

hence $|\lambda^2 \widehat{A}'(\lambda)|_{\mathcal{B}(V,V^*)} \leq M$ since $-\int_0^\infty t d\alpha_2(t) < \infty$ and $\alpha_1(t)$ is bounded. With

$$H'(\lambda) = -H(\lambda)(I + \widehat{A}'(\lambda))H(\lambda) \quad , \quad \text{Re}\,\lambda > 0, \tag{6.4}$$

(6.2) and (6.3) then imply the existence of $H(i\rho) = \lim_{\lambda \to i\rho} H(\lambda)$ in $\mathcal{B}(V^*, V)$ for each $\rho \neq 0$. η-coerciveness and (6.1) for $\lambda = i\rho$ yield

$$\eta\gamma ||H(i\rho)(\eta + i\rho)^{-1}x||^2 \leq |< H(i\rho)x, x >| \quad , \quad \rho \neq 0 \, , \, x \in V^*, \tag{6.5}$$

while the decomposition of (6.1) into real and imaginary parts for $\lambda = i\rho$ with (V5) implies

$$|\rho||H(i\rho)x|^2 \leq c(1 + |\rho|)| < H(i\rho)x, x > | \quad , \quad \rho \neq 0 \, , \, x \in V^*. \tag{6.6}$$

Finally (S) gives

$$||H(i\rho)x|| \leq c|\rho| \, ||x||_* \quad , \quad \rho \neq 0 \, , \, x \in V^*; \tag{6.7}$$

in particular $H(0) = \lim_{\lambda \to 0} H(\lambda) = 0$. From (6.5), (6.6), (6.7) one easily obtains the boundedness of $H(i\rho)$ in $\mathcal{B}(H)$, of $H(i\rho)/(i\rho(\eta+i\rho))$ in $\mathcal{B}(V^*, V)$, and of $H(i\rho)/i\rho$ in $\mathcal{B}(H, V)$, and $\mathcal{B}(V^*, H)$. By means of the Laplace transformed resolvent equations and boundedness of α_1, the latter imply boundedness of $H(i\rho)$ in $\mathcal{B}(V)$, $\mathcal{B}(V^*)$, $H(i\rho)i\rho$ in $\mathcal{B}(H, V^*)$, and $\mathcal{B}(V, H)$, and of $H(i\rho)(i\rho)^2$ in $\mathcal{B}(V, V^*)$. By the maximum principle all of these bounds hold also on $\overline{\mathbb{C}}_+$.

(b) Next we consider the identity

$$H(i\rho)x = H(i\rho + \eta)x + \eta H(i\rho)H(i\rho + \eta)x + H(i\rho)(\widehat{A}(i\rho + \eta) - \widehat{A}(i\rho))H(i\rho + \eta)x \tag{6.8}$$

where $x \in V^*$, $\rho \in \mathbb{R}$. (V4) and

$$(i\rho + \eta)^2(\widehat{A}(i\rho) - \widehat{A}(i\rho + \eta)) = \eta^2 \widehat{A}(i\rho) + \eta \dot{A}(0) + 2\eta \widehat{\dot{A}}(i\rho) + \widehat{d\ddot{A}}(i\rho) - \widehat{d\ddot{A}}(i\rho + \eta),$$

imply

$$|(i\rho + \eta)^2(\widehat{A}(i\rho) - \widehat{A}(i\rho + \eta))|_{\mathcal{B}(V,V^*)} \leq C(1 + 1/|\rho|) \quad , \quad \rho \neq 0,$$

for $-\int_0^\infty t d\alpha_2(t) < \infty$ and α_1 is bounded. Since e.g. $e^{-\eta t}S(t)x \in L^2(\mathbb{R}_+; V)$ for $x \in V$, by Parseval's theorem, we obtain $H(i\rho + \eta)x \in L^2(\mathbb{R}_+; V)$, hence $H(i\rho)x \in L^2(\mathbb{R}_+; V)$ by (6.8) and (a), and so by Parseval's theorem also $S(t)x \in L^2(\mathbb{R}_+; V)$. Similar arguments yield the assertions of Theorem 4 with exponent 2 instead of 1.

(c) To obtain e.g. $tS(t)x \in L^2(\mathbb{R}_+; V)$ consider (6.4) for $\lambda = i\rho$. Then for $x \in V$ one deduces from (6.4)

$$||H'(i\rho)x|| \leq |H(i\rho)|_{\mathcal{B}(V)}||H(i\rho)x|| + |H(i\rho)/\rho^2|_{\mathcal{B}(V^*,V)}M||H(i\rho)x||,$$

hence $H(i\rho)x \in L^2(\mathbb{R}_+; V)$, by the boundedness properties of $H(i\rho)$, and by (6.7). Parseval's theorem gives $tS(t)x \in L^2(\mathbb{R}_+; V)$ for each $x \in V$, which together with (b) yields strong integrability of $S(t)$ in V. The remaining assertions follow by similar arguments. □

The same type of arguments lead to the following result, in which the case of a *fluid*, i.e. $\int_0^\infty |\dot{A}(t)|_{\mathcal{B}(V,V^*)}dt < \infty$, is considered.

Corollary 3 *Let the assumptions of Theorem 4 be satisfied, with the exception that (S) is replaced by*
(F) $|\dot{\alpha}(t;u,v)| \leq \alpha_3(t)||u||\,||v||$, $t > 0$, $u, v \in V$, *where* $\alpha_3 \in L^1(\mathbb{R}_+)$.
Then the assertions of Theorem 4 for $S(t)$, $\dot{S}(t)$, $\ddot{S}(t)$ *remain valid.*

Concluding we return to the beginning of this paper, to viscoelasticity. let $\mathcal{A}(t,x)$ denote the stress relaxation tensor and α the form on $V = W^{1,2}_{\Gamma_d}(\Omega;\mathbb{R}^3)$ defined by (2.1). Then clearly (V1) holds, provided $\mathcal{A}(,x) \equiv 0$, and $\mathcal{A}(t,\cdot)$ is a.e. bounded for each $t \in \mathbb{R}_+$. Assumption (V2) is implied by $\mathcal{A}_{ijkl}(\cdot,x) \in W^{1,1}(\mathbb{R}_+)$, and

$$\text{ess sup}_{x\in\Omega}|\dot{\mathcal{A}}_{ijkl}(t,x) - \dot{\mathcal{A}}_{ijkl}(s,x)| \leq \alpha_1(t) - \alpha_1(s)\,, \quad \text{for all } t > s \geq 0, \text{ and } i,j,k,l,$$

where α_1 is nondecreasing, while (V4) follows from $\mathcal{A}_{ijkl}(\cdot,x) \in W^{2,1}(\mathbb{R}_+)$, and

$$\text{ess sup}_{x\in\Omega}|\ddot{\mathcal{A}}_{ijkl}(t,x) - \ddot{\mathcal{A}}_{ijkl}(s,x)| \leq \alpha_2(s) - \alpha_2(t)\,, \quad \text{for all } t > s \geq 0, \text{ and } i,j,k,l,$$

where α_2 is nonincreasing and such that $\int_0^\infty t d\alpha_2(t) > -\infty$. η-coercivity and coercivity, i.e. 0-coercivity, are inherited to α by the pointwise estimate

$$\text{Re}\,\hat{\mathcal{A}}(\lambda) : F : \bar{F} \geq \gamma \text{Re}\frac{1}{\lambda+\eta}|F|^2 \quad \text{for all Re}\,\lambda > 0, \text{ and all symmetric } F \in \mathbb{C}^{3\times 3},$$

thanks to Korn's inequality. The frequency domain condition (V5) is implied e.g. by

$$|\text{Im}\,\hat{\mathcal{A}}(i\rho) : F : \bar{F}| \leq \phi_\infty \text{Re}\,|\rho|[|\hat{\mathcal{A}}(i\rho) : F : \bar{F}|]\,, \quad |\rho| \geq R, \text{ for all symmetric } F \in \mathbb{C}^{3\times 3}.$$

The material is a solid, i.e. (S) holds, if

$$\dot{\mathcal{A}}_{ijkl}(t,x) \to \mathcal{A}^\infty_{ijkl}(x)\,, \quad \text{as } t \to \infty, \text{ for a.a. } x \in \Omega, \text{and } i,j,k,l \in \{1,2,3\},$$

while it is a fluid if

$$\text{ess sup}_{x\in\Omega}|\dot{\mathcal{A}}(t,x)| \leq \alpha_3(t)\,, \quad t > 0,$$

where $\alpha_3 \in L^1(\mathbb{R}+)$.

In the special case of homogeneous, isotropic materials which are incompressible but nonsynchronous, these conditions follow from $a_0 = b_0 = 0$, $a_1(0+)$, $b_1(0+) < \infty$ (hyperbolic case), a_1 , b_1 convex, strongly positive, and subject to assumption (ii) of Theorem 3; (S) is equivalent to a_∞ , $b_\infty > 0$, while (F) means $a_\infty = b_\infty = 0$, and a_1 , $b_1 \in L^1(\mathbb{R}_+)$.

References

[1] S. Adali. Existence and asymptotic stability of solutions of an abstract integrodifferential equation with applications to viscoelasticity. *SIAM J. Math. Anal.*, 9:185–206, 1978.

[2] W. Arendt. Vector Laplace transforms and Cauchy problems. *Israel J. Math.*, 59:327–352, 1987.

[3] D.R. Bland. *The Theory of Linear Viscoelasticity*, volume 10 of *Pure and Applied Mathematics*. Pergamon Press, New York, 1960.

[4] R.W. Carr and K.B. Hannsgen. A nonhomogeneous integrodifferential equation in Hilbert space. *SIAM J. Math. Anal.*, 10:961–984, 1979.

[5] R.W. Carr and K.B. Hannsgen. Resolvent formulas for a Volterra equation in Hilbert space. *SIAM J. Math. Anal.*, 13:453–483, 1982.

[6] R.M. Christensen. *Theory of Viscoelasticity.* Academic Press, New York, second edition, 1982.

[7] Ph. Clément and G. Da Prato. Existence and regularity results for an integral equation with infinite delay in a Banach space. *Integral Eqns. Operator Theory*, 11:480–500, 1988.

[8] Ph. Clément and J. Prüss. Completely positive measures and Feller semigroups. *Math. Ann.*, 287:73–105, 1990.

[9] G. Da Prato and A. Lunardi. Solvability on the real line of a class of linear Volterra integrodifferential equations of parabolic type. *Ann. Mat. Pura Appl.*, 55:67–118, 1988.

[10] W. Desch and R. Grimmer. Singular relaxation moduli and smoothing in three dimensional viscoelasticity. *Trans. Amer. Math. Soc.*, 314:381–404, 1989.

[11] W. Desch and R. Grimmer. Smoothing properties of linear Volterra integrodifferential equations. *SIAM J. Math. Anal.*, 20:116–132, 1989.

[12] R. Grimmer and J. Prüss. On linear Volterra equations in Banach spaces. *J. Comp. Appl. Math.*, 11:189–205, 1985.

[13] K.B. Hannsgen and R.L. Wheeler. Viscoelastic and boundary feedback damping : Precise energy decay rates when creep modus are dominant. *J. Integral Eq. and Appl.*, 2:495–527, 1990.

[14] J.Prüss. Positivity and regularity of hyperbolic Volterra equations in Banach spaces. *Math. Ann.*, 279:317–344, 1987.

[15] R. Leis. *Initial Boundary Value Problems in Mathematical Physics.* B. G. Teubner, Stuttgart, 1986.

[16] M.J. Leitman and G.M.C. Fisher. *The linear theory of viscoelasticity*, volume 3, pages 1–123. Springer Verlag, Berlin, Heidelberg, New York, 1983.

[17] J.E. Marsden and T.J.R. Hughes. *Mathematical Foundations of Elasticity.* Prentice-Hall, Englewood Cliffs, New Jersey, 1983.

[18] R. Miller and R.L. Wheeler. Asymtotic behavior for a linear Volterra integral equation in Hilbert space. *J. Diff. Equations*, 23:270–284, 1977.

[19] R. Miller and R.L. Wheeler. Wellposedness and stability of linear Volterra integrodifferential equations in abstract spaces. *Funk. Ekv.*, 21:279–305, 1978.

[20] C. Navarro. Asymptotic stability in linear thermovisco-elasticity. *J. Math. Anal. Appl.*, 65:399–431, 1978.

[21] J.A. Nohel and D.F. Shea. Frequency domain methods for Volterra equations. *Advances Math.*, 22:278–304, 1976.

[22] R.D. Noren. Uniform l^1-behavior of the solution of a Volterra equation with parameter. Preprint.

[23] A.C. Pipkin. *Lectures on Viscoelasticity Theory*, volume 7 of *Appl. Math. Sci.* Springer, Berlin, Heidelberg, New York, 1972.

[24] J. Prüss. Regularity and integrability of resolvents of linear Volterra equations. In M. Iannelli G.Da Prato, editor, *Volterra Integrodifferential Equations in Banach Spaces and Applications*, pages 339–367, Harlow, Essex, 1989. Lonman Scientific and Technical.

[25] J. Prüss. *Linear Evolutionary Integral Equations and Applications.* Birkhäuser, Basel, 1993. to appear.

[26] H. Tanabe. *Equations of Evolution*, volume 6 of *Monographs and Studies in Mathematics.* Pitman, London, 1979.

[27] H. Tanabe. Volterra intergrodifferential equations of parabolic type of higher order in t. *J.Fac. Sci. Univ. Tokyo*, 34:111–125, 1987.

[28] H. Tanabe. On fundamental solutions of differential equation with time delay in Banach space. *Proc. Japan Acad.*, 64:131–134, 1988.

Generation of Analytic Semigroups by Variational Operators with L^∞ Coefficients

VINCENZO VESPRI Australian National University, Centre for Mathematics and its Applications, Canberra, ACT 2601 *

Summary

In the last few years many papers were devoted to study linear and nonlinear evolution equations by using the semigroup approach. The first step consists in proving that an operator generates an analytic semigroup in a suitable function space. The aim of this note is to prove such a property for an elliptic operator with L^∞ coefficients. Moreover some interpolation spaces are characterised.

1 Definitions

Let Ω be a bounded set of $\mathbf{R}^n$ with Lipschitz continuous boundary. A function u belongs to the Morrey space $L^{2,\mu}(\Omega)$, $0 \leq \mu \leq n$, if $u \in L^2(\Omega)$ and

$$\|u\|^2_{L^{2,\mu}(\Omega)} \quad = \quad \sup_{x\in\Omega\,,\,r>0} r^{-\mu} \int_{B(x,r)\cap\Omega} u^2 dx \quad < \infty.$$

For more details about these function spaces, see for example [3] and [6]. We say that $u \in H^{1,2}_\mu(\Omega)$ if $u \in H^{1,2}_0(\Omega)$ and for each $i = 1, \cdots, n$ $D_i u \in L^{2,\mu}(\Omega)$. Let $H^{-1,p}(\Omega)$, $1 < p < \infty$, be the dual space of $H^{1,p'}_0(\Omega)$. We say that a functional f belongs to $H^{-1,\infty}(\Omega)$ ($C^{-1,\alpha}(\Omega)$, $H^{-1,2}_\mu(\Omega)$ resp.) if f belongs to $H^{-1,2}(\Omega)$ and admits a distributional representation $f = f_0 +$

*Permanent address: Università di Pavia, Dipartimento di Matematica, via Strada Nuova 65, Pavia, Italy

$\sum_{i=1}^{n} D_i f_i$,where $f_0, \cdots, f_n$ belong to $L^\infty(\Omega)$ ($C^{0,\alpha}(\bar{\Omega})$, $L^{2,\mu}(\Omega)$, resp.). $H^{-1,\infty}(\Omega)$ ($C^{-1,\alpha}(\Omega)$, $H_\mu^{-1,2}(\Omega)$, resp.) is a Banach space with the norm

$$\|f\|_{H^{-1,\infty}(\Omega)} \equiv \inf_{f=f_0+\sum_{i=1}^{n} D_i f_i} \left(\sum_{i=0}^{n} \|f_i\|_{L^\infty(\Omega)}\right).$$

Analogously

$$\|f\|_{C^{-1,\alpha}(\Omega)} \equiv \inf_{f=f_0+\sum_{i=1}^{n} D_i f_i} \left(\sum_{i=0}^{n} \|f_i\|_{C^{0,\alpha}(\bar{\Omega})}\right)$$

$$\|f\|_{H_\mu^{-1,2}(\Omega)} \equiv \inf_{f=f_0+\sum_{i=1}^{n} D_i f_i} \left(\sum_{i=0}^{n} \|f_i\|_{L^{2,\mu}(\Omega)}\right).$$

Let $a_{i,j}, a_i, b_i, c$, $i,j = 1, \cdots, n$ belong to $L^\infty(\Omega)$. Assume that there is $k > 0$ such that for each $x \in \bar{\Omega}$ for each $\xi \in \mathbf{R}^n$

$$\sum_{i,j=1}^{n} a_{i,j}\xi_i\xi_j \geq k|\xi|^2.$$

Set

$$Eu = \sum_{i,j=1}^{n} D_i(a_{i,j}D_j u) + \sum_{i=1}^{n} D_i(a_i u) + \sum_{i=1}^{n} b_i D_i u + cu.$$

In the sequel, we will show that the realisations of E in suitable Banach spaces generate analytic semigroups. Let X be a Banach space. We recall that an operator $A : D(A) \subseteq X \to X$ generates an analytic semigroup, if there are $\omega \in \mathbf{R}, M > 0$ and $\frac{\pi}{2} < \theta < \pi$ such that for each $\lambda \in \mathbf{C}$ with $|\arg(\lambda - \omega)| \leq \theta$ the following inequality holds:

$$\|(A - \lambda I)^{-1}\|_{L(X)} \leq \frac{M}{|\lambda|}.$$

Note that we do not require $\overline{D(A)} = X$ because in this framework the abstract semigroup theory can be applied [10] .

2 Generation results in $L^p(\Omega)$

In this section, an essential tool is a regularity result due to DeGiorgi [5] (for the formulation we need, see for instance [7] and [6]).

Theorem 2.1
Let $u \in H_0^{1,2}(\Omega)$ be a weak solution of the equation

$$Eu = f_0 + \sum_{i=1}^{n} D_i f_i \qquad \text{in } \Omega \tag{1}$$

where $f_i \in L^{2,\mu}(\Omega)$,$i = 1, \cdots, n$, $f_0 \in L^{2,\mu-2}(\Omega)$ .. Then there exists a $\mu_0 > n-2$ such that for each $0 \leq \mu \leq \mu_0$, the function u belongs to $H_\mu^{1,2}(\Omega)$. Moreover

$$\|u\|_{H_\mu^{1,2}(\Omega)} \leq k_\mu \left(\sum_{i=1}^{n} \|f_i\|_{L^{2,\mu}(\Omega)} + \|f_0\|_{L^{2,\mu-2}(\Omega)}\right) \tag{2}$$

where k_μ *is a constant independent of* u *and* f_i.

The next result can be proved via standard variational inequalities.

Theorem 2.2
There is a $\gamma \geq 0$ *such that for each* $\lambda \in \mathbf{C}$ *with* $Re\lambda \geq \gamma$ *and for each* $f_i \in L^2(\Omega)$ $i = 0, 1, \cdots, n$ *there is a unique solution of*

$$(\lambda - E)u = f_0 + \sum_{i=1}^{n} D_i f_i \qquad \text{in } \Omega. \tag{3}$$

Moreover there is a constant k_0, *independent of* λ, u *and* f_i, *such that*

$$\|u\|_{H^{1,2}(\Omega)} + |\lambda|^{\frac{1}{2}} \|u\|_{L^2(\Omega)} \leq k_0 \left(\sum_{i=1}^{n} \|f_i\|_{L^2(\Omega)} + |\lambda|^{-\frac{1}{2}} \|f_0\|_{L^2(\Omega)} \right) \tag{4}$$

The next result is an essential step in order to extend estimate (4) .

Lemma 2.3
Assume that (4) *holds for a* $\nu \in [0, n[$,*i.e. for each* $f_i \in L^{2,\nu}(\Omega)$ $i = 0, 1, \cdots, n$ *there is a constant* k_ν,*independent of* λ, u *and* f_i, *such that*

$$\|u\|_{H^{1,2}_\nu(\Omega)} + |\lambda|^{\frac{1}{2}} \|u\|_{L^{2,\nu}(\Omega)} \leq k_\nu \left(\sum_{i=1}^{n} \|f_i\|_{L^{2,\nu}(\Omega)} + |\lambda|^{-\frac{1}{2}} \|f_0\|_{L^{2,\nu}(\Omega)} \right). \tag{5}$$

Then,
for each $\mu \in [\nu, \nu + 2]$ *if* $\nu < n - 2$
for each $\mu \in [\nu, n[$ *if* $\nu = n - 2$
for each $\mu \in [\nu, n]$ *if* $\nu > n - 2$
there is a constant k_μ,*independent of* λ, u *and* f_i,*such that*

$$|\lambda|^{\frac{1}{2}} \|u\|_{L^{2,\mu}(\Omega)} \leq k_\mu \left(\sum_{i=1}^{n} \|f_i\|_{L^{2,\mu}(\Omega)} + |\lambda|^{-\frac{1}{2}} \|f_0\|_{L^{2,\mu}(\Omega)} \right). \tag{6}$$

PROOF
Here we follow a technique introduced by Stewart [11] in a different framework.
Let x_0 belong to Ω and let θ_r be a cutoff function such that

$$\begin{cases} 0 \leq \theta_r \leq 1 & \theta_r = 1 \text{ in } B(x_0, r) \\ \theta_r = 0 \text{ outside } B(x_0, 2r) & \left(r \sum_{i=1}^{n} |D_i \theta_r| + r^2 \sum_{i,j=1}^{n} |D_{i,j} \theta_r| \right) \leq 8n^2. \end{cases} \tag{7}$$

Let $\Omega(x_0, r) = \Omega \cap B(x_0, r)$.
The function $\theta_r u$ solves the equation

$$\begin{cases} (\lambda - E)\theta_r u = -\sum_{i,j=1}^{n} D_i(a_{ij}u) D_j \theta_r - \sum_{i,j=1}^{n} a_{ij} D_j u D_i \theta_r - \sum_{i,j=1}^{n} a_{ij} u D_{ij} \theta_r + \\ -\sum_{i=1}^{n} a_i u D_i \theta_r - \sum_{i=1}^{n} b_i u D_i \theta_r - \sum_{i=1}^{n} D_i(f_i \theta_r) + \sum_{i=1}^{n} f_i D_i \theta_r + f_0 \theta_r = \\ = g_0 + \sum_{i=1}^{n} D_i g_i. \end{cases} \tag{8}$$

Therefore, by (5)

$$\begin{cases} \|u\|_{H^{1,2}_\nu(\Omega(x_0,r))} + |\lambda|^{\frac{1}{2}}\|u\|_{L^{2,\nu}(\Omega(x_0,r))} \le k_\nu(\sum_{i=1}^n \|f_i\|_{L^{2,\nu}(\Omega(x_0,2r))}+ \\ +|\lambda|^{-\frac{1}{2}}(r^{-1}\sum_{i=1}^n \|f_i\|_{L^{2,\nu}(\Omega(x_0,2r))} + \|f_0\|_{L^{2,\nu}(\Omega(x_0,2r))}) + r^{-1}\|u\|_{L^{2,\nu}(\Omega(x_0,2r))}+ \\ +|\lambda|^{-\frac{1}{2}}r^{-2}\|u\|_{L^{2,\nu}(\Omega(x_0,2r))} + |\lambda|^{-\frac{1}{2}}r^{-2}\|u\|_{H^{1,2}_\nu(\Omega(x_0,2r))}). \end{cases} \tag{9}$$

Now, we recall that in such a framework one can prove (see [9] and [4])

$$\begin{cases} \|u\|_{H^{1,2}_\nu(\Omega(x_0,2r))} \le K(r^{-1}\|u\|_{L^{2,\nu}(\Omega(x_0,4r))} + \sum_{i=1}^n \|f_i\|_{L^{2,\nu}(\Omega(x_0,4r))}+ \\ +|\lambda|^{-\frac{1}{2}}\|f_0\|_{L^{2,\nu}(\Omega(x_0,4r))} + |\lambda|^{-\frac{1}{2}}r^{-1}\sum_{i=1}^n \|f_i\|_{L^{2,\nu}(\Omega(x_0,4r))}). \end{cases} \tag{10}$$

where K is a constant independent of λ, u and f_i.

Putting togheter (9) and (10), one gets

$$\begin{cases} \|u\|_{H^{1,2}_\nu(\Omega(x_0,r))} + |\lambda|^{\frac{1}{2}}\|u\|_{L^{2,\nu}(\Omega(x_0,r))} \le \\ \le k_\nu K(\sum_{i=1}^n \|f_i\|_{L^{2,\nu}(\Omega(x_0,4r))} + |\lambda|^{-\frac{1}{2}}(r^{-1}\sum_{i=1}^n \|f_i\|_{L^{2,\nu}(\Omega(x_0,4r))} + \|f_0\|_{L^{2,\nu}(\Omega(x_0,4r))})+ \\ +r^{-1}\|u\|_{L^{2,\nu}(\Omega(x_0,4r))} + +|\lambda|^{-\frac{1}{2}}r^{-2}\|u\|_{L^{2,\nu}(\Omega(x_0,4r))}). \end{cases} \tag{11}$$

Recalling that $\|u\|_{L^{2,\nu}(\Omega(x_0,r))} \le r^{\frac{\mu-\nu}{2}}\|u\|_{L^{2,\mu}(\Omega(x_0,r))}$, it holds

$$\begin{cases} \|u\|_{H^{1,2}_\nu(\Omega(x_0,r))} + |\lambda|^{\frac{1}{2}}\|u\|_{L^{2,\nu}(\Omega(x_0,r))} \le \\ \le k_\nu K r^{\frac{\mu-\nu}{2}}(\sum_{i=1}^n \|f_i\|_{L^{2,\mu}(\Omega(x_0,4r))} + |\lambda|^{-\frac{1}{2}}(r^{-1}\sum_{i=1}^n \|f_i\|_{L^{2,\mu}(\Omega(x_0,4r))}+ \\ +\|f_0\|_{L^{2,\mu}(\Omega(x_0,4r))}) + r^{-1}\|u\|_{L^{2,\mu}(\Omega(x_0,4r))} + +|\lambda|^{-\frac{1}{2}}r^{-2}\|u\|_{L^{2,\mu}(\Omega(x_0,4r))}). \end{cases} \tag{12}$$

Now, by a well-known interpolation result (see e.g. [2]), for each $\varepsilon > 0$ there exists $c(\varepsilon) > 0$ such that

$$\|u\|_{L^{2,\mu}(\Omega(x_0,r))} \le \varepsilon r^{\frac{\nu-\mu}{2}}\|u\|_{H^{1,2}_\nu(\Omega(x_0,r))} + c(\varepsilon)r^{-1+\frac{\nu-\mu}{2}}\|u\|_{L^{2,\nu}(\Omega(x_0,r))}. \tag{13}$$

By choosing $r = \frac{4c(\varepsilon)}{\varepsilon|\lambda|^{\frac{1}{2}}}$ and combining (12) with (13), we get

$$\begin{cases} |\lambda|^{\frac{1}{2}}\|u\|_{L^{2,\mu}(\Omega(x_0,r))} \le 2\varepsilon k_\nu K|\lambda|^{\frac{1}{2}}\|u\|_{L^{2,\mu}(\Omega(x_0,4r))}+ \\ +2c(\varepsilon)k_\nu K\left(\sum_{i=1}^n \|f_i\|_{L^{2,\mu}(\Omega(x_0,4r))} + |\lambda|^{-\frac{1}{2}}\|f_0\|_{L^{2,\mu}(\Omega(x_0,4r))})\right). \end{cases} \tag{14}$$

Let $\{x_i\}_{i=1}^N$ be a net of points in $\mathbf{R}^n$ such that $\bigcup_{i=1}^N B(x_i,r) \supseteq \bar{\Omega}$ and for each $i = 1, \cdots, N$, $B(x_i, 4r)$ has at most $(n+1)8^n$ nonempty intersections with the other spheres.

By centering (14) on x_i and adding it for $i = 1, \cdots, N$ one may deduce

$$\begin{cases} |\lambda|^{\frac{1}{2}}\|u\|_{L^{2,\mu}(\Omega)} \le c_\mu\varepsilon(n+1)8^n|\lambda|^{\frac{1}{2}}\|u\|_{L^{2,\mu}(\Omega)}+ \\ +c_\mu c(\varepsilon)\varepsilon^{-1}\left(\sum_{i=1}^n \|f_i\|_{L^{2,\mu}(\Omega)} + |\lambda|^{-\frac{1}{2}}\|f_0\|_{L^{2,\mu}(\Omega)})\right). \end{cases} \tag{15}$$

The statement follows just choosing $\varepsilon = (2c_\mu(n+1)8^n)^{-1}$.

Remark 2.4
An easy consequence of the previous results is that for each $0 < \mu \leq n$

$$|\lambda|^{\frac{\mu-\nu}{4}} \sup_{x_0 \in \bar{\Omega}} \|u\|_{H^{1,2}_\nu(\Omega(x_0,|\lambda|^{-\frac{1}{2}}))} \leq c_\mu \left(\sum_{i=1}^{n} \|f_i\|_{L^{2,\mu}(\Omega)} + |\lambda|^{-\frac{1}{2}} \|f_0\|_{L^{2,\mu}(\Omega)} \right). \tag{16}$$

Lemma 2.5
Let $0 \leq \mu \leq \mu_0$. Under the same assumptions of Lemma 2.2

$$\|u\|_{H^{1,2}_\mu(\Omega)} \leq c_\mu \left(\sum_{i=1}^{n} \|f_i\|_{L^{2,\mu}(\Omega)} + |\lambda|^{-\frac{1}{2}} \|f_0\|_{L^{2,\mu}(\Omega)}. \right) \tag{17}$$

PROOF
Let θ_r be the cutoff function defined in (7). $\theta_r u$ is a solution of the equation

$$E(\theta_r u) = \lambda \theta_r u + g_0 + \sum_{i=1}^{n} D_i g_i.$$

(g_i are defined in (8)). By choosing r as in lemma 2.2 and by using DeGiorgi estimate, one gets

$$\begin{cases} \|u\|_{H^{1,2}_\mu(\Omega(x_0,|\lambda|^{-\frac{1}{2}}))} \leq Kk_\mu(|\lambda| \|u\|_{L^{2,\mu-2}(\Omega(x_0,4|\lambda|^{-\frac{1}{2}}))} + |\lambda|^{\frac{1}{2}} \|u\|_{L^{2,\mu}(\Omega(x_0,4|\lambda|^{-\frac{1}{2}}))} + \\ + \left(\sum_{i=1}^n \|f_i\|_{L^{2,\mu}(\Omega(x_0,4|\lambda|^{-\frac{1}{2}}))} + |\lambda|^{\frac{1}{2}} \sum_{i=1}^n \|f_i\|_{L^{2,\mu-2}(\Omega(x_0,4|\lambda|^{-\frac{1}{2}}))} \right) + \\ + \|f_0\|_{L^{2,\mu-2}(\Omega(x_0,4|\lambda|^{-\frac{1}{2}}))}). \end{cases} \tag{18}$$

Hence

$$\begin{cases} \|u\|_{H^{1,2}_\mu(\Omega(x_0,|\lambda|^{-\frac{1}{2}}))} \leq c_\mu(|\lambda|^{\frac{1}{2}} \|u\|_{L^{2,\mu}(\Omega(x_0,4|\lambda|^{-\frac{1}{2}}))} + \\ + \sum_{i=1}^n \|f_i\|_{L^{2,\mu}(\Omega(x_0,4|\lambda|^{-\frac{1}{2}}))} + |\lambda|^{-\frac{1}{2}} \|f_0\|_{L^{2,\mu}(\Omega(x_0,4|\lambda|^{-\frac{1}{2}}))}). \end{cases} \tag{19}$$

Hence (17) follows by (19) via the same covering argument of lemma 2.2.

By putting together statements 2.2-2.5, one obtains

Theorem 2.6
There is a $\omega > 0$ such that for each $0 \leq \mu \leq n$, for each $\lambda \in \mathbf{C}$ such that $Re\lambda > \omega$, for each $f_i \in L^p(\Omega)$, $i = 0, 1, \cdots, n$, there is a unique solution $u \in H^{1,2}_{\min(\mu,\mu_0)}(\Omega)$ of the equation

$$(\lambda - E)u = f_0 + \sum_{i=1}^{n} D_i f_i.$$

Moreover there is a constant c_μ, independent of λ, u, f_i such that *if* $0 \leq \mu \leq \mu_0$

$$\|u\|_{H^{1,2}_{\mu}(\Omega)} + |\lambda|^{\frac{1}{2}}\|u\|_{L^{2,\mu}(\Omega)} \leq c_{\mu}\left(\sum_{i=1}^{n}\|f_i\|_{L^{2,\mu}(\Omega)} + |\lambda|^{-\frac{1}{2}}\|f_0\|_{L^{2,\mu}(\Omega)}.\right) \tag{20}$$

if $\mu_0 < \mu \leq n$

$$\begin{cases} |\lambda|^{\frac{n-\mu_0}{4}} \sup_{x_0 \in \bar{\Omega}} \|u\|_{H^{1,2}_{\mu_0}(\Omega(x_0,|\lambda|^{-\frac{1}{2}}))} + |\lambda|^{\frac{1}{2}}\|u\|_{L^{2,\mu}(\Omega)} \leq \\ \leq c_{\mu}\left(\sum_{i=1}^{n}\|f_i\|_{L^{2,\mu}(\Omega)} + |\lambda|^{-\frac{1}{2}}\|f_0\|_{L^{2,\mu}(\Omega)}\right). \end{cases} \tag{21}$$

Let $\mu \in [0, n]$ and let us define the operator $\bar{E}_{\mu}$

$$\begin{cases} \bar{E}_{\mu} : D(\bar{E}_{\mu}) = \{u \in H^{1,2}_{\min(\mu,\mu_o)(\Omega)} : Eu \in L^{2,\mu}(\Omega)\} \to L^{2,\mu}(\Omega) \\ (\bar{E}_{\mu})u = Eu \qquad \text{for each } u \in D(\bar{E}_{\mu}). \end{cases}$$

Inequalities (20) and (21) imply

Theorem 2.7
The operator $\bar{E}_{\mu}$ *generates an analytic semigroup in* $L^{2,\mu}(\Omega)$.

Remark 2.8
Recalling that $L^{2,n}(\Omega) \equiv L^{\infty}(\Omega)$, the previous result implies a generation result in $L^{\infty}(\Omega)$. On the other hand, Theorem 2.2 implies a generation result in $L^2(\Omega)$. Hence the following generation theorem is an easy consequence of the Riesz-Thorin interpolation theorem.

Theorem 2.9
Let $p \in [2, \infty]$ *and let us define the operator* $\bar{E}_p$

$$\begin{cases} \bar{E}_p : D(\bar{E}_p) = \{u \in H^{1,2}_0 \cap L^p(\Omega) : Eu \in L^p(\Omega)\} \to L^p(\Omega) \\ (\bar{E}_p)u = Eu \qquad \textit{for each } u \in D(\bar{E}_p). \end{cases}$$

The operator $\bar{E}_p$ *generates an analytic semigroup in* $L^p(\Omega)$.

Remark 2.10
By using standard duality techniques, it is possible to prove that the above generation result is true for each $1 \leq p \leq \infty$.

3 Generation results in $H^{-1,p}(\Omega)$

Also in this section we prove a result that will imply the generation result:

Theorem 3.1
There is a $\omega > 0$ *such that for each* $2 \leq p \leq \infty$, *for each* $\lambda \in \mathbf{C}$ *such that* $Re\lambda > \omega$, *for each* $f_i \in L^p(\Omega), i = 0, 1, \cdots, n$, *there is a unique solution* $u \in H^{1,2}_0(\Omega)$ *of the equation*

$$(\lambda - E)u = f_0 + \sum_{i=1}^{n} D_i f_i.$$

Moreover there is a constant c_p, *independent of* λ, u, f_i *such that*
if $2 \leq p < \infty$

$$|\lambda|^{\frac{1}{2}} \|u\|_{L^p(\Omega)} + |\lambda| \|u\|_{H^{-1,p}(\Omega)} \leq c_p \left(\sum_{i=0}^{n} \|f_i\|_{H^{-1,p}(\Omega)}. \right) \tag{22}$$

if $p = \infty$

$$\begin{cases} |\lambda|^{\frac{n-\mu_0}{4}} \sup_{x_0 \in \bar{\Omega}} \|u\|_{H^{1,2}_{\mu_0}(\Omega(x_0, |\lambda|^{-\frac{1}{2}}))} + |\lambda|^{\frac{1}{2}} \|u\|_{L^p(\Omega)} + \\ + |\lambda| \|u\|_{H^{-1,p}(\Omega)} \leq c_p \left(\sum_{i=0}^{n} \|f_i\|_{H^{-1,p}(\Omega)} \right). \end{cases} \tag{23}$$

PROOF

If $p = \infty$, in [9] it is proved that (22) can be deduced by (21) for solutions of variational equations with continuous coefficients. The same proof holds in the case of equations with discontinuous coefficients.

Lastly, if $2 \leq p \leq \infty$ the statement follows by the previous ones via Riesz-Thorin interpolation theorem.

Let $p \in [2, \infty]$ and let us define the operator E_p

$$\begin{cases} E_p : D(E_p) = \{u \in H_0^{1,2} \cap L^p(\Omega) : Eu \in H^{-1,p}(\Omega)\} \rightarrow H^{-1,p}(\Omega) \\ E_p u = Eu \quad \text{for each } u \in D(E_p). \end{cases}$$

Inequalities (22) and (23) imply

Theorem 3.2
The operator E_p *generates an analytic semigroup in* $H^{-1,p}(\Omega)$ *for each* $2 \leq p \leq \infty$.

4 Characterisation of some interpolation spaces

An essential tool in order to apply the semigroup theory to evolution problems is the characterisation of some interpolation spaces. This section is devoted to this aim.

Theorem 4.1

$$D_{E_\infty}(\theta, \infty) = \begin{cases} C_0^{0,2\theta-1}(\bar{\Omega}) & \text{if } \frac{1}{2} < \theta < \frac{\mu_0+4-n}{4} \\ C^{-1,2\theta}(\Omega) & \text{if } 0 < \theta < \frac{1}{2} \end{cases} \tag{24}$$

where $D_{E_\infty}(\theta, \infty)$ *are the interpolation spaces introduced by Lions-Peetre* [8].

Here we follow an approach introduced by Acquistapace-Terreni in [1] to which we refer for more details.

Let

$$u(t, x) = t^{-1}[R(t^{-1}, E)f](x) \text{ where } R(z, E) = (z - E_\infty)^{-1} \text{ and } f \in D_{E_\infty}(\theta, \infty).$$

By the results of the previous sections it is possible to prove (see [1]):

$$t^{-\frac{1}{2}}\|u(t,\cdot)\|_{L^\infty(\Omega)} + t^{-\frac{n-\mu_0}{4}}\sup_{x_0\in\bar{\Omega}}\|u(t,\cdot)\|_{H^{1,2}_{\mu_0}(\Omega(x_0,t^{-\frac{1}{2}}))} \leq c_1 t^{-1+\theta}\|f\|_{D_{E_\infty}(\theta,\infty)}. \tag{25}$$

$$t^{-\frac{1}{2}}\|u'(t,\cdot)\|_{L^\infty(\Omega)} + t^{-\frac{n-\mu_0}{4}}\sup_{x_0\in\bar{\Omega}}\|u'(t,\cdot)\|_{H^{1,2}_{\mu_0}(\Omega(x_0,t^{-\frac{1}{2}}))} \leq c_2 t^{-2+\theta}\|f\|_{D_{E_\infty}(\theta,\infty)}. \tag{26}$$

The following two lemmas are an essential step in proving theorem 4.1:

Lemma 4.2
Let $\frac{1}{2} < \theta < 1$, *then*

$$\|f(\cdot) - u(t,\cdot)\|_{C^0(\bar{\Omega})} \leq c_3 t^{\theta-\frac{1}{2}}\|f\|_{D_{E_\infty}(\theta,\infty)}.$$

PROOF

$$\|f(\cdot) - u(t,\cdot)\|_{C^0(\bar{\Omega})} \leq \int_0^t \|u'(t,\cdot)\|_{C^0(\bar{\Omega})} ds \leq$$

$$\leq c_4 \int_0^t s^{\theta-\frac{3}{2}}\|f\|_{D_{E_\infty}(\theta,\infty)} ds \leq c_3 t^{\theta-\frac{1}{2}}\|f\|_{D_{E_\infty}(\theta,\infty)}.$$

Lemma 4.3
For each $0 < \alpha < \frac{\mu_0+2-n}{2}$, *and for each* $\frac{1+\alpha}{2} \leq \theta < \frac{\mu_0+4-n}{4}$

$$\sup_{x_0\in\bar{\Omega}}[u(t,\cdot)]_{C^{0,\alpha}(\overline{\Omega(x,t^{-\frac{1}{2}})})} \leq c_\alpha\|f\|_{D_{E_\infty}(\theta,\infty)}.$$

PROOF
Note that

$$\|u(t,\cdot)\|_{H^{1,2}_{\mu_0}(\Omega(x_0,t^{-\frac{1}{2}}))} \leq \int_t^1 \|u'(s,\cdot)\|_{H^{1,2}_{\mu_0}(\Omega(x_0,t^{-\frac{1}{2}}))} ds + \|u(1,\cdot)\|_{H^{1,2}_{\mu_0}(\Omega(x_0,t^{-\frac{1}{2}}))}.$$

Hence by (25) and (26)

$$\sup_{x_0\in\bar{\Omega}}\|u(t,\cdot)\|_{H^{1,2}_{\mu_0}(\Omega(x_0,t^{-\frac{1}{2}}))} \leq c_{\mu_0} t^{-1+\theta+\frac{n-\mu_0}{4}}\|f\|_{D_{E_\infty}(\theta,\infty)}.$$

Let ν be equal to $n-2+2\alpha$. Hence

$$[u(t,\cdot)]_{C^{0,\alpha}(\overline{\Omega(x,t^{-\frac{1}{2}})})} \leq c_5\|u(t,\cdot)\|_{H^{1,2}_{\nu}(\Omega(x_0,t^{-\frac{1}{2}}))} \leq$$

$$\leq c_6\|u(t,\cdot)\|_{H^{1,2}_{\mu_0}(\Omega(x_0,t^{-\frac{1}{2}}))} t^{\frac{\mu_0-\nu}{4}}$$

Therefore

$$[u(t,\cdot)]_{C^{0,\alpha}(\overline{\Omega(x,t^{-\frac{1}{2}})})} \leq c_7 t^{-1+\theta+\frac{n-\mu_0}{4}+\frac{\mu_0-\nu}{4}}\|f\|_{D_{E_\infty}(\theta,\infty)} =$$

$$= c_\alpha t^{\theta-1+\frac{1-\alpha}{2}}\|f\|_{D_{E_\infty}(\theta,\infty)} \leq c_\alpha\|f\|_{D_{E_\infty}(\theta,\infty)}.$$

for each $t \in (0,1)$ and $\frac{\mu_0+4-n}{4} > \theta \geq \frac{1+\alpha}{2}$.

Now we are able to prove theorem 4.1:

PROOF
Consider first the case $\frac{1}{2} < \theta < \frac{\mu_0+4-n}{4}$.
The inclusion $D_{E_\infty}(\theta, \infty) \subseteq C_0^{0,2\theta-1}(\bar{\Omega})$ can be proved exactly as in [9].
Let us show the reverse inclusion, i.e.

$$D_{E_\infty}(\theta, \infty) \supseteq C_0^{0,2\theta-1}(\bar{\Omega}).$$

Let $f \in D_{E_\infty}(\theta, \infty)$. Then let x, y belong to $\bar{\Omega}$ and let $t = |x-y|^2$. Then

$$|f(x) - f(y)| \le 2\|f(\cdot) - u(t,\cdot)\|_{L^\infty(\Omega)} + [u(t,\cdot)]_{C^{0,2\theta-1}(\overline{\Omega(x,t^{-\frac{1}{2}})})}|x-y|^{2\theta-1}.$$

Hence, by lemmata 4.2-4.3

$$|f(x) - f(y)| \le c_8|x-y|^{2\theta-1}\|f\|_{D_{E_\infty}(\theta,\infty)}.$$

and the previous inequality implies the inclusion.

In the case $0 < \theta < \frac{1}{2}$ the statement follows by the characterisation of the interpolation spaces obtained in [9] where some interpolation spaces between $H^{-1,\infty}(\Omega)$ and $C_0^{0,\alpha}(\bar{\Omega})$ are studied.

References

[1] P.ACQUISTAPACE-B.TERRENI *Hölder classes with boundary conditions as interpolation spaces.* Math. Z. **195** (1987) 451-471.

[2] R.A.ADAMS *Sobolev spaces.* Academic Press. New York (1975).

[3] S.CAMPANATO *Sistemi ellittici in forma divergenza regolarità all'interno.* Quaderni Scuola Normale Superiore. Pisa (1980).

[4] P.CANNARSA-B.TERRENI-V.VESPRI *Analytic semigroups generated by nonvariational elliptic systems of second order under Dirichlet boundary conditions* . J. Math. Anal. Appl. **112** (1985) 56-103.

[5] E. DE GIORGI *Sulla differenziabilitá e analiticitá delle estremali degli integrali multipli regolari.* Mem. Acad. Sci. Torino **3** (1957) 25-43.

[6] M.GIAQUINTA *Multiple integrals in the calculus of variation and nonlinear elliptic systems.* Annals of Mathematics studies. Princeton (1983).

[7] O.A. LADYZHENSKAJA - N.N.URAL'TSEVA *Linear and quasilinear elliptic equations.* Academic Press, New York (1968).

[8] J.L.LIONS-J.PEETRE *Sur une classes d'espaces d'interpolation.* I. H. E. S. Publ. Math. **19** (1964) 5-68.

[9] A.LUNARDI-V.VESPRI *Hölder regularity for variational nonhomogeneous boundary problems.* J. Diff. Equs. **94** (1991) 1-40.

[10] E. SINESTRARI *On the abstract Cauchy problem of parabolic type in spaces of continuous functions.* J. Math. Anal. Appl. **107** (1985) 16-66.

[11] H.B.STEWART *Generation of analytic semigroups by strongly elliptic operators under general boundary conditions.* T. A. M. S. **259** (1980) 299-310.

Asynchronous Exponential Growth in Differential Equations with Homogeneous Nonlinearities

G. F. WEBB* Department of Mathematics, Vanderbilt University, Nashville TN 37235, U.S.A.

1. Introduction. The objective of this paper is to investigate the asymptotic behavior of solutions of abstract semilinear differential equations with homogeneous nonlinearities. The equation we consider has the form

$$z_x'(t) = Az_x(t) + F(z_x(t)),\ t \geq 0,\ z_x(0) = x \tag{1.1}$$

We suppose that A is the infinitesimal generator of a strongly continuous semigroup of positive linear operators in a Banach lattice X and F is a nonlinear operator in X_+ satisfying $F(cx) = cF(x)$, $x \in X_+$, $c \geq 0$. If $z_x(t) \equiv x$ is an equilibrium solution of (1.1), then so is $z_{cx}(t) \equiv cx$ for any $c > 0$. Consequently, there is no possibility of an attracting nontrivial equilibrium. There is, however, the possibility of another type of asymptotic behavior which is defined as follows: The solutions of (1.1) have *asynchronous exponential growth* on $U \subset X_+$ provided there is a constant $\lambda_1 \in \mathbb{R}$ (called the *intrinsic growth constant*) such that $Qx := \lim_{t\to\infty} e^{-\lambda_1 t} z_x(t)$ exists, Qx is nonzero for all $x \in U$, and $R(Q)$ lies in a one-dimensional subspace of X.

The method we use to show this behavior is a linearization technique. We suppose there exists a solution to the nonlinear eigenvalue problem $Ax_1 + F(x_1) = \lambda_1 x_1$. It follows that $e^{-\lambda_1 t} z_{cx_1}(t) \equiv cx_1$ for $c > 0$ and thus that asynchronous exponential growth holds on $U_1 = \{cx_1 : c > 0\}$. Can the set U_1 be enlarged? The answer is yes if the linear problem

$$w_x'(t) = Aw_x(t) + F'(x_1)w_x(t),\ t \geq 0,\ w_x(0) = x \tag{1.2}$$

has asynchronous exponential growth on a subspace of X. Our theorem is typical of linearization results in that it is local. It says, roughly speaking, that if the linear problem (1.2) has asynchronous exponential growth on a subspace of X, then the nonlinear problem (1.1) has asynchronous exponential growth on a certain kind of neighborhood of U_1.

The term asynchronous exponential growth is used by cell biologists to describe the behavior of growing cell populations. In this context it means that the cell population

*Supported in part by the National Science Foundation under Grant No. DMS-9001790

loses any initial synchronization of its age or size structure as it grows at an exponential rate. That is, after multiplication by an exponential factor in time, the solutions converge to a nonzero one-dimensional image of the initial higher dimensional state space. This type of behavior is common in models of population growth. It occurs in linear models or in nonlinear models when nonlinearities are not strong enough to force convergence to an attracting equilibrium. It thus describe population growth in early stages before resource limitations become dominant.

There are many examples of linear population models with asynchronous exponential growth and some of these are found in our references. In [11] and [12] sufficient conditions for asynchronous exponential growth were established and illustrated with nonlinear models of tumor cell population growth. These conditions did not involve homogeneous nonlinearities, but instead required the nonlinear problem to be asymptotic to a linear problem with asynchronous exponential growth. An example of asynchronous exponential growth with homogeneous nonlinearity is found in an epidemic model of Diekmann and Kretzschmar [6]. In fact homogeneous nonlinearities are natural in population problems in which vital rates may depend on proportions of the population in population subclasses determined by compartments or structure. In Section 3 we provide illustrative examples. In Section 2 below we prove our main theorem.

2. Main Result. Let X be a Banach lattice with positive cone X_+. We assume the following hypotheses:

(H.1) A is the infinitesimal generator of a strongly continuous semigroup of positive bounded linear operators $T(t)$, $t \geq 0$ in X;

(H.2) F is a (nonlinear) operator from X_+ to X_+, F is Lipschitz continuous on each bounded set of X_+, and $\lim_{h \to 0^+} \min_{y \in X_+} \frac{1}{h}\|x + hF(x) - y\| = 0$ for each $x \in X_+$;

(H.3) $F(cx) = cF(x)$ for $x \in X_+$, $c \geq 0$;

(H.4) There exists $x_1 \in X_+$ with $\|x_1\| = 1$ and $\lambda_1 \in \mathbb{R}$ such that $Ax_1 + F(x_1) = \lambda_1 x_1$, F is Fréchet differentiable at x_1, $A_1 := A + F'(x_1)$ is the infinitesimal generator of a strongly continuous semigroup of bounded linear operators $T_1(t)$, $t \geq 0$ in X, and there exists a nonzero rank one projection P in X such that $\lim_{t\to\infty} e^{-\lambda_1 t}T_1(t) = P$.

Consider the abstract semilinear equation:

$$z_x(t) = T(t)x + \int_0^t T(t-s)F(z_x(s))ds, \ t \geq 0 \tag{2.1}$$

A solution of (2.1) is called a *mild solution* of (1.1) and the hypotheses (H.1) and (H.2) insure the existence of a unique solution in X_+ of (2.1) for each $x \in X_+$ on some maximal interval of existence $[0, t_x)$ with either $t_x = \infty$ or $\lim_{t \to t_x^-} \|z_x(t)\| = \infty$ (see [15], Theorem 2.1, p. 335 and Proposition 4.1, p. 194 and Proposition 4.16, p. 194 in [21]).

Theorem 1. Let (H.1) - (H.4) hold. There exists $\delta > 0$ such that if $x \in U := \{x \in X_+ - \{0\} : \|(I-P)x\|/\|Px\| < \delta\}$, then $t_x = \infty$, $Qx := \lim_{t\to\infty} e^{-\lambda_1 t} z_x(t)$ exists, $Qx \in R(P)$, and $Qx \neq 0$.

Proof. Let $S(t)x := e^{-\lambda_1 t} z_x(t)$, $0 \le t < t_x$, $x \in X_+$. For $h \in X_+$ and $c > 0$ define $o_c(h) := F(cx_1+h) - F(cx_1) - F'(x_1)h$. Since F is Fréchet differentiable at x_1, there exists a continuous increasing function $b : \mathbb{R}_+ \longrightarrow \mathbb{R}_+$ such that $b(0) = 0$ and $\|o_1(h)\| \le b(\|h\|)\|h\|$ for $h \in X_+$. Since F is homogeneous, $o_c(h) = co_1(\frac{h}{c})$ for $h \in X_+$ and $c > 0$. Further, $\|F(x_1) - F'(x_1)x_1\| = \|\frac{1}{c}o_1(cx_1)\| \le b(\|cx_1\|)$, which implies $F'(x_1)x_1 = F(x_1)$.

Define $\hat{P} := I - P$. By (H.4) $PT_1(t) = e^{\lambda_1 t}P$ for $t \ge 0$ and there exists $M > 1$ and $\tau > 0$ such that $|\hat{P}T_1(t)| \le Me^{(\lambda_1-\tau)t}|\hat{P}|$ for $t \ge 0$ (see [22]). Choose $\delta_1 > 0$ such that $\delta_1 < |\hat{P}|/|P|$ (if $\dim X > 1$, then $|\hat{P}| > 0$) and $b(r) < \tau/2M|\hat{P}|$ if $0 < r < \delta_1$. Define $\delta = \delta_1|\hat{P}|/(M|\hat{P}| + \delta_1|P|)$ (note that $\delta < \delta_1$). Let $x \in U$ such that $\hat{P}x \neq 0$ and let $\hat{t}_x \le t_x \le \infty$ be the largest extended real number such that if $0 \le t < \hat{t}_x$, then

$$\|\hat{P}S(t)x\|/\|PS(t)x\| \le \delta_1, \tag{2.2}$$

and

$$\|\hat{P}S(t)x\| \le Me^{-\tau t/2}\|\hat{P}x\| \tag{2.3}$$

For $0 \le t < \hat{t}_x$ let $PS(t)x = c(t)x_1$, $c(t) \in \mathbb{R}$ (where we have used the fact that $R(P)$ is one-dimensional) and let $\hat{P}S(t)x = h(t)$.

From (2.1) we have for $0 \le t < \hat{t}_x$

$$\begin{aligned} S(t)x &= e^{-\lambda_1 t}T(t)x + \int_0^t e^{-\lambda_1(t-s)}T(t-s)F(S(s)x)ds \\ &= e^{-\lambda_1 t}T_1(t)x + \int_0^t e^{-\lambda_1(t-s)}T_1(t-s)[F(S(s)x) - F'(x_1)S(s)x]ds \end{aligned} \tag{2.4}$$

(see [21], Proposition 4.17, p. 198)

$$\begin{aligned} &= e^{-\lambda_1 t}T_1(t)x + \int_0^t e^{-\lambda_1(t-s)}T_1(t-s)o_{c(s)}(h(s))ds \\ &= e^{-\lambda_1 t}T_1(t)x + \int_0^t e^{-\lambda_1(t-s)}T_1(t-s)c(s)o_1(h(s)/c(s))ds \end{aligned}$$

From (2.4) we obtain for $0 \le t < \hat{t}_x$

$$\|\hat{P}S(t)x\| \le Me^{-\tau t}\|\hat{P}x\| + \int_0^t Me^{-\tau(t-s)}|\hat{P}|b(h(s)/c(s))\|h(s)\|ds \tag{2.5}$$

and

$$\|\hat{P}S(t)x\| < Me^{-\tau t}\|\hat{P}x\| + \int_0^t Me^{-\tau(t-s)}|\hat{P}|(\tau/2M|\hat{P}|)\|\hat{P}S(s)x\|ds$$

which by Gronwall's Inequality implies that for $0 \le t < \hat{t}_x$

$$\|\hat{P}S(t)x\| < M\|\hat{P}x\|e^{-\tau t/2} \tag{2.6}$$

Also, from (2.4) we obtain for $0 \le t < \hat{t}_x$

$$\begin{aligned} \|PS(t)x\| &\ge \|Px\| - \int_0^t |P|(\tau/2M|\hat{P}|)\|\hat{P}S(s)x\|ds \\ &\ge \|Px\| - \int_0^t |P|(\tau/2M|\hat{P}|)M\|\hat{P}x\|e^{-\tau s/2}ds \\ &\ge \|Px\| - |P|\ \|\hat{P}x\|/|\hat{P}| \\ &> 0 \end{aligned} \tag{2.7}$$

Then, (2.6) and (2.7) yield for $0 \le t < \hat{t}_x$

$$\|\hat{P}S(t)x\|/\|PS(t)x\| \le M\|\hat{P}x\|\ |\hat{P}|/(\|Px\|\ |\hat{P}| - |P|\ \|\hat{P}x\|) < \delta_1 \tag{2.8}$$

From (2.4) and (2.6) we obtain for $0 \le t < \hat{t}_x$

$$\|PS(t)x\| \le \|Px\| + \int_0^t |P|(\tau/2M|\hat{P}|)M\|\hat{P}x\|e^{-\tau s/2}ds \le \|Px\| + |P|\ \|\hat{P}x\|/|\hat{P}| \tag{2.9}$$

Thus, (2.6), (2.8), and (2.9) imply that $\hat{t}_x = t_x = \infty$,

$$Qx := \lim_{t\to\infty} S(t)x = P[x + \int_0^\infty c(s)o_1(h(s))/c(s))ds]$$

exists, and $Qx \in R(P)$. From (2.6) and (2.7) we see that $Qx \ne 0$. If $\hat{P}x = 0$, then $x = cx_1$ for some $c > 0$ and $S(t)cx_1 = cS(t)x_1 = cx_1$ for $t \ge 0$. ∎

3. Examples. The examples below illustrate the ideas of Theorem 1.

Example 1. Let $X = \mathbb{R}^2$ and consider the system of scalar ordinary differential equations

$$u' = 2u - \frac{uv}{u+v},\quad u(0) = u_0 \ge 0 \tag{3.1}$$

$$v' = u + \frac{uv}{u+v},\quad v(0) = v_0 \ge 0 \tag{3.2}$$

Set $A[u,v] = [2u, u]$ and $F([u,v]) = [\frac{-uv}{v+v}, \frac{uv}{u+v}]$. Notice that for $r+s>0,\ u+v>0$

$$\begin{aligned} &\left|\frac{rs}{r+s} - \frac{uv}{u+v}\right| \\ &\le \left|\frac{s}{r+s}\right|\ |r-u| + \left|\frac{u}{u+v}\right|\ \left|\frac{s}{r+s}\right|\ |u-r|\ + \left|\frac{u}{u+v}\right|\ \left|\frac{r}{r+s}\right|\ |s-v| \\ &\le 2|r-u| + |s-v| \end{aligned} \tag{3.3}$$

Notice also that for $u > 0, v > 0$

$$F'([u,v]) = \frac{1}{(u+v)^2}\begin{bmatrix} -v^2 & -u^2 \\ v^2 & u^2 \end{bmatrix}$$

The nonlinear eigenvalue problem $A\vec{x} + F(\vec{x}) = \lambda\vec{x}$ has solutions $\lambda_1 = \frac{3}{2}$, $\vec{x}_1 = [1,1]$ and $\lambda_2 = 0$, $\vec{x}_2 = [0,1]$.

Consider first $\lambda_1 = \frac{3}{2}$, $\vec{x}_1 = [1,1]$, and

$$A_1 := A + F'([1,1]) = \frac{1}{4}\begin{bmatrix} 7 & -1 \\ 5 & 1 \end{bmatrix}$$

The eigenvalues of A_1 are $\frac{3}{2}$ and $\frac{1}{2}$ and e^{tA_1}, $t \geq 0$ has asynchronous exponential growth with intrinsic growth constant $\frac{3}{2}$ and $\lim_{t\to\infty} e^{-\frac{3}{2}t}e^{tA_1} = P_1$, where $P_1[u,v] = \frac{5u-v}{4}[1,1]$. Hypotheses (H.1)-(H-4) are satisfied and thus Theorem 1 applies. Consider now the other solution of the nonlinear eigenvalue problem $\lambda_2 = 0$, $\vec{x}_2 = [0,1]$. In this case F is not Fréchet differentiable at $[0,1]$ and Theorem 1 does not apply.

In fact the solution of (3.1), (3.2) for $u_0 > 0$ is

$$u(t) = \exp\left[\int_0^t \left(\frac{3 - ce^{-s}}{2 - ce^{-s}}\right) ds\right] u_0, \; v(t) = u(t)[1 - ce^{-t}]$$

with $c = 1 - v_0/u_0$ and for $u_0 = 0$ is $u(t) \equiv 0$, $v(t) \equiv v_0$. Consequently, $\lim_{t\to\infty} e^{-\frac{3}{2}t}[u(t), v(t)] = [u_0, u_0]$ for all $[u_0, v_0] \in \mathbb{R}_+$ and asynchronous exponential growth holds on $\{[u_0, v_0] \in \mathbb{R}_+ : u_0 \neq 0\}$.

Example 2. Let $X = \mathbb{R}^2$ and consider the system of scalar ordinary differential equations

$$u' = 2v - \frac{uv}{u+v}, \; u(0) = u_0 \geq 0 \tag{3.4}$$

$$v' = -v + \frac{uv}{u+v}, \; v(0) = v_0 \geq 0 \tag{3.5}$$

Let $A[u,v] = [2v, -v]$ and let F be as in Example 1. The nonlinear eigenvalue problem $A\vec{x}_1 + F(\vec{x}_1) = \lambda_1\vec{x}_1$ has only the solution $\lambda_1 = 0$, $\vec{x}_1 = [1,0]$. Since F is not Fréchet differentiable at $[1,0]$, Theorem 1 does not apply. The solution of (3.4), (3.5) is

$$[u(t), v(t)] = \left(\frac{u_0 + v_0}{u_0 + v_0 + 2tv_0}\right)^{1/2} [u_0 + 2tv_0, v_0]$$

Obviously, $\lim_{t\to\infty} t^{-1/2}u(t) = (2(u_0 + v_0)v_0)^{1/2}$ and $\lim_{t\to\infty} t^{1/2}v(t) = ((u_0 + v_0)v_0/2)^{1/2}$, so that asynchronous exponential growth does not hold on any subset of $\mathbb{R}^2$.

Example 3. Let $X = L^1(0,\infty;\mathbb{R})$ and consider the age-structured population model

$$u_t(a,t) + u_a(a,t) = -\mu u(a,t) - \frac{f(u(\cdot,t))}{g(u(\cdot,t))}u(a,t) \tag{3.6}$$

$$u(0,t) = \beta\int_0^\infty u(a,t)da, \; t > 0 \tag{3.7}$$

$$u(a,0) = \phi(a), \; a > 0 \tag{3.8}$$

Here $u(\cdot,t) \in X$ for $t \geq 0$, μ and β are positive constants, f and g are positive linear functionals on X, and $\phi \in X_+$. We suppose there exists $C > 0$ such that $\|\phi\|/g(\phi) \le C$ for all $\phi \in X_+ - \{0\}$. Examples for f and g are $f(\phi) = \int_{a_1}^{a_2} \phi(a)da$ and $g(\phi) = \int_0^\infty \phi(a)da$, where $0 \le a_1 < a_2 \le \infty$. In this case there is an increased mortality due to crowding which is proportional to the population in the age range $[a_1, a_2]$.

The abstract formulation (1.1) for the problem (3.6)-(3.8) has

$$\begin{aligned} A\phi &:= -\phi' - \mu\phi, \quad \phi \in D(A) \\ D(A) &:= \{\phi \in X : \phi' \in X,\ \phi(0) = \beta \int_0^\infty \phi(a)da\} \\ F(\phi) &:= -f(\phi)\phi/g(\phi),\ \phi \in X_+ - \{0\}, \\ F(0) &= 0. \end{aligned}$$

In [21] it is shown that A is the infinitesimal generator of a strongly continuous semigroup $T(t)$, $t \ge 0$ of positive linear operators in X. An argument similar to (3.3) shows that $\|F(\phi) - F(\psi)\| \le C|f|(2 + C|g|)\|\phi - \psi\|$ for $\phi, \psi \in X_+$. The Fréchet derivative of F at $\phi_1 \in X_+ - \{0\}$ is $F'(\phi_1)\phi := -[h'(\phi_1)\phi]\phi_1 - h(\phi_1)\phi$, $\phi \in X$, where $h(\phi) := f(\phi)/g(\phi)$ and $h'(\phi_1)\phi := (g(\phi_1)f(\phi) - g(\phi)f(\phi_1))/g(\phi_1)^2$.

The only solution of the nonlinear eigenvalue problem $A\phi_1 + F(\phi_1) = \lambda_1\phi_1, \phi_1(0) = 1$, is $\phi_1(a) := e^{-\beta a}$, $\lambda_1 := \beta - \mu - c_1$, where $c_1 := f(\phi_1)/g(\phi_1)$. Consider the linear eigenvalue problem $A_1\phi + F'(\phi_1)\phi = \lambda\phi$, $\phi(0) = 1$, $\lambda \in \mathbb{C}$. Then,

$$\begin{aligned} \phi(a) &= e^{-(\mu + c_1 + \lambda)a} - h'(\phi_1)\phi \int_0^a e^{-(\mu + c_1 + \lambda)(a-b)}\phi_1(b)db \\ 1 &= \beta \int_0^\infty \phi(a)da \end{aligned}$$

(since $\phi \in D(A)$). Thus, $h'(\phi_1)\phi = \beta - \mu - c_1 - \lambda$, and so $\phi = \phi_1$, $\lambda = \lambda_1$ is the only solution of the linear eigenvalue problem as well.

To apply Theorem 1 we must show that the linear semigroup $T_1(t)$, $t \ge 0$ generated by A_1 has asynchronous exponential growth with intrinsic growth constant λ_1. By [22], Proposition 2.3, we need only show that (i) $\omega_1(A_1) < \omega_0(A_1)$ and (ii) λ_1 is a simple pole of $(\lambda I - A_1)^{-1}$ and $N(\lambda_1 I - A_1)$ is one-dimensional. To see (i) define $A_2\phi = A\phi - h(\phi_1)\phi$, $D(A_2) = D(A)$. By [21], Theorem 4.6, $\omega_1(A_2) \le -\mu - h(\phi_1) = -\mu - c_1$. Since $A_1\phi = A_2\phi - [h'(\phi_1)\phi]\phi_1$, A_1 is a compact perturbation of A_2 and by [21], Proposition 4.14, $\omega_1(A_1) = \omega_1(A_2)$. Thus, $\omega_1(A_1) < \lambda_1 = \omega_0(A_1)$ by [21], Proposition 4.13. To show (ii) let $\phi \in N((\lambda_1 I - A_1)^2)$, $\phi(0) = 1$. Then $\psi := A\phi + F'(\phi_1)\phi - \lambda_1\phi$ and $A\psi + F'(\phi_1)\psi - \lambda_1\psi = 0$. We have seen that $\psi(a) = e^{-\beta a}\psi(0)$ and thus $-\phi'(a) - \mu\phi(a) - [h'(\phi_1)\phi]\phi_1(a) - h(\phi_1)\phi(a) - \lambda_1\phi(a) = e^{-\beta a}\psi(0)$. Integrate this last equation to obtain

$$\phi(0) - h'(\phi_1)\phi/\beta - \beta \int_0^\infty \phi(a)da = \psi(0)/\beta$$

Since $\phi \in D(A)$, $\psi(0) = -h'(\phi_1)\phi$. Thus, $\phi' + \beta\phi = 0$ and $\phi(a) = e^{-\beta a}$, which means $\phi = \phi_1$, $h'(\phi_1)\phi = \psi(0) = 0$. Thus, λ_1 is a simple pole of $(\lambda I - A)^{-1}$ and obviously $N(\lambda_1 I - A) = \{ce^{-\beta a} : c \in \mathbb{R}\}$ is one-dimensional. Hence, Theorem 1 can be applied.

In fact, the solutions of (3.6)-(3.8) have asynchronous exponential growth on $X_+ - \{0\}$. In [21], p.188 it is shown that $\lim_{t\to\infty} e^{-(\beta-\mu)t}T(t) = P$, where

$$(P\phi)(a) = \beta e^{-\beta a}\int_0^\infty \phi(b)db$$

and

$$e^{-(\beta-\mu)t}PT(t) = P, \text{ and } |(I-P)T(t)| \le Me^{(\beta-\mu-\tau)t}$$

for some $M \ge 1$ and $\tau > 0$. Since $R(P)$ is one-dimensional, $h(P\phi) = c_1$ is independent of $\phi \in X_+ - \{0\}$. Rewrite the problem as

$$z'_\phi(t) = (A - c_1 I)z_\phi(t) + (h(z_\phi(t)) - c_1 I)z_\phi(t), \; z_\phi(0) = \phi$$

and observe that

$$e^{-\lambda_1 t}z_\phi(t) = \exp\left[\int_0^t (h(z_\phi(s)) - c_1 I)ds\right] e^{-(c_1+\lambda_1)t}T(t)\phi \tag{3.9}$$

Then, (3.9) implies that

$$h(z_\phi(t)) = h(e^{-\lambda_1 t}z_\phi(t)) = h(e^{-(\beta-\mu)t}T(t)\phi) = h(P\phi + (I-P)e^{-(\beta-\mu)t}T(t)\phi).$$

Thus,

$$\|\int_0^t (h(z_\phi(s)) - c_1)ds\| \le \text{const}\int_0^t \|(I-P)e^{-(\beta-\mu)s}T(s)\phi\|ds \le \text{const } M\|\phi\|/\tau$$

which implies $\lim_{t\to\infty} e^{-\lambda_1 t}z_\phi(s) = \exp[\int_0^\infty (h(z_\phi(s) - c_1)ds]P\phi$ for all $\phi \in X_+$.

REFERENCES

1. W. Arendt, A. Grabosch, G. Greiner, U. Groh, H. P. Lotz, U. Moustakis, R. Nagel, F. Neubrander, U. Schlotterbeck, *One-parameter Semigroups of Positive Operators*, R. Nagel, ed., Lecture Notes in Mathematics 1184, Springer-Verlag, Berlin Heidelberg New York Tokyo, 1986.

2. O. Arino and M. Kimmel, *Asymptotic analysis of a cell-cycle model based on unequal division*, SIAM J. Appl. Math. **47**(1987), 128-145.

3. J. Carr, *Applications of Centre Manifold Theory*, Applied Mathematical Sciences, Vol. 35, Springer-Verlag, New York, 1981.

4. Ph. Clément, H. J. A. M. Heijmans, S. Angenent, C. J. van Duijn, and B. de Pagter, *One-Parameter Semigroups*, North Holland, 1987.

5. O. Diekmann, H. Heijmans, and H. Thieme, *On the stability of the cell size distribution* II, Hyperbolic Partial Differential Equations III, Inter. Series in Modern Appl. Math. Computer Science, Vol. 12., M. Witten, ed., Pergamon Press (1986), 491-512.

6. O. Diekmann and M. Kretzschmar, *Patterns in the effects of infections diseases on population growth*, to appear.

7. G. Greiner, *A typical Perron-Frobenius theorem with application to an age-dependent population equation*, Infinite-Dimensional Systems, Proceedings, Retzhof 1983, F. Kappel and W. Schappacher, eds., Lecture Notes in Mathematics, Vol. 1076, Springer-Verlag, Berlin Heidelberg New York Tokyo, 1984.

8. G. Greiner and R. Nagel, *On the stability of strongly continuous semigroups of positive operators on* $L^2(\mu)$, Ann. Scuola Norm. Sup. Pisa, Serie IV-Vol. X, n. 2(1983), 257-262.

9. M. Gyllenberg, *Nonlinear age-dependent population dynamics in continuously propagated bacterial cultures*, Math. Biosci. **62**(1982), 45-74.

10. M. Gyllenberg and G. F. Webb, *Age-size structure in populations with quiescence*, Math. Biosci, 86 (1987), 67-95.

11. M. Gyllenberg and G. F. Webb, *A nonlinear structured population model of tumor growth with quiescence*, J. Math. Biol. **28**(1990), 671-694.

12. M. Gyllenberg and G. F. Webb, *Asynchronous expoential growth of semigroups of nonlinear operators*, to appear in J. Math. Anal. Appl.

13. H. Inaba, *A semigroup approach to the strong ergodic theorem of the multistate stable population process*, Math. Population Studies Vol. 1 (1) (1988), 49-77.

14. P. Jagers, *Balanced exponential growth: What does it mean and when is it there?* Biomathematics and Cell Kinetics, Development in Cell Biology, Vol. **2**, A. Valleron and P. Macdonald, eds., Elsevier/North-Holland Press (1978), 21-29.

15. R. H. Martin, *Nonlinear Operators and Differential Equations in Banach Spaces*, John Wiley and Sons, 1976.

16. Prüss, J., Equilibrium solutions of age-specific population dynamics of several species, *J. Math. Biol.* **11**(1981), 65-84.

17. M. Rotenberg, *Equilibrium and stability in populations whose interactions are age-specific*, J. Theoret. Biol. **54**(1975), 207-224.

18. E. Sinestrari and G. F. Webb, *Hyperbolic systems with non-local boundary conditions with applications to epidemic models*, J. Math. Anal Appl. **121**(1987), 449-464.

19. J. Song, C.-H. Tuan, J.-Y. Yu, *Population Control in China, Theory and Applications*, Praeger, New York, 1985.

20. S. L. Tucker and S. O. Zimmerman, *A nonlinear model of population dynamics containing an arbitrary number of continuous variables*, SIAM J. Appl. Math. **48**, No. 3 (1988), 549-591.

21. G. Webb, *Theory of Nonlinear Age-Dependent Population Dynamics*, Monographs and Textbooks in Pure and Applied Mathematics Series, Vol. 89, Marcel Dekker, New York and Basel, 1985.

22. G. Webb, *An operator-theoretic formulation of asynchronous exponential growth*, Trans. Amer. Math. Soc. 303, No. 2 (1987), 751-763.

23. G. Webb and A. Grabosch, *Asynchronous exponential growth in transition probability models of the cell cycle*, SIAM J. Math. Anal. Vol. 18, No. 4 (1987), 897-907.

Inversion of the Vector-Valued Laplace Transform in $L_p(X)$-Spaces

L. WEIS* Department of Mathematics, Louisiana State University, Baton Rouge, LA 70803, U.S.A.

1. Introduction

Recently, there has been renewed interest in the vector-valued Laplace transform, in the context of (integrated) semigroups and the abstract Cauchy problem (see e.g. [1], [13], [14]). This made it desirable to refine and extend the classical results due to Hille, Miyadera and Zaidman on the inversion theory in Bochner spaces $L_p(X)$ and Pettis spaces $PL_p(X)$.

In Section 3 we identify the precise class of Banach spaces for which the classical inversion theorem holds (in this respect we answer a question of Zaidman) but we also give an 'integrated' representation which holds for general Banach spaces as it was done in the case $p=\infty$ by Arendt ([1], see also [13]). Furthermore, we consider L_p-spaces with densities of power-type which we expect to be useful in the study of Cauchy problems.

Section 4 contains a complex inversion theorem which is derived from the real conditions of Section 3 and inspired by the approach of Rooney [22] and Sova [25]. This leads to a result on the range of the vector-valued Fourier transform. We also formulate a Paley-Wiener theorem for the Pettis norm.

*Permanent address: Mathematical Institute, Kiel, Germany

In Section 5 we give some applications to semi group theory and the abstract Cauchy problem, in particular to characterization theorems of certain singular semi-groups, stability theory and weak solutions of Cauchy-problems.

Section 6 gives an outline of the proof of our main result (detailed proofs will appear in [27], [28] and Section 2 contains preliminaries on functions with bounded p-variation and the Pettis norms.

I am especially indebted to Frank Neubrander who drew my interest to the subject and in many discussions convinced me of his "philosophy" on the L-transform.

2. Preliminaries on functions of bounded p-Variation

For some aspects of the classical theory of L_p-functions (e.g. duality, boundary values of harmonic functions, inversion of the Laplace transform) the Bochner spaces $L_p(X)$ are not sufficient in the vector-valued case but one needs an extension of Bochner p-integrable functions, namely the functions of bounded p-variation introduced in [5] (see also [10] for the related notation vector measures of bounded p-variation).

First we fix some notations.

2.1 Notations: Let I be an interval in $\mathbb{R}$ with interior $\mathring{I}$. Let w: $\mathring{I} \to (0,\infty)$ be continuous and put $\mu_p(E) = \int_E w(t)^p \, dt$.

X is always a Banach space. By $L_p(w,X)$ we denote the Bochner space of all norm-measurable functions $f: I \to X$ with norm

$$\| f \|_{L_p(w,X)} = \left(\int_I (\| f(t) \| \, w(t))^p dt \right)^{1/p}$$

for $1 \leq p < \infty$ and for $p=\infty$ we put

$$\| f \|_{L_\infty(w,X)} = \sup \{ \| f(t) \| \, w(t) : t \in I \}.$$

A norm-measurable function $\phi : \mathring{I} \to X$ is of <u>bounded $L_p(w)$-variation</u> if the supremum

$$\sup\left\{\left(\sum\left(\frac{\|\phi(t_i)-\phi(t_{i-1})\|}{t_i-t_{i-1}}\right)^p \mu_p(t_{i-1}, t_i)\right)^{1/p} : t_0 < t_1 < \ldots < t_n\right\}$$

taken over all choices of $t_i \in \mathring{I}$, is finite. If $p=\infty$ we consider

$$\sup\left\{\frac{\phi(t)-\phi(s)}{t-s}\, w(u) : s \leq u \leq t,\ s<t,\ s,t \in \mathring{I}\right\}$$

$V_p(w,X)$ is the space of all X-valued functions of bounded $L_p(w)$-variation with the above norm, modulo constant functions. If $p=1$ we also assume that the functions are continuous from the right.

<u>2.2. Proposition</u>: ([5]) a) If $\phi : \mathring{I} \to X$ is of bounded $L_p(w)$-variation, then ϕ is of bounded variation (if $p>1$ even absolutely continuous) on every compact subinterval of $\mathring{I}$.
b) $V_p(w,X)$ is a Banach space that contains $L_p(w,X)$ as a subspace via the embedding

$$\rho \in L_p(w,X) \quad \to \quad \phi(t) = \int_{t_0}^{t} \rho(s)ds = \begin{cases} \displaystyle\int_{t_0}^{t} \rho(s)ds & \text{for } t \geq t_0 \\ \displaystyle -\int_{t}^{t_0} \rho(s)ds & \text{for } t \leq t_0 \end{cases}$$

for a fixed $t_0 \in \mathring{I}$.
c) For $1 < p \leq \infty$ we have $V_p(w,X) = L_p(w,X)$ in the sense of b) if and only if X has the Radon-Nikodym property.

Recall that X has the <u>Radon-Nikodym property</u> if and only if every function $\phi : \mathbb{R} \to X$ of bounded variation is differentiable almost everywhere. For some details and references concerning this and the next statement, see [10] p. 115, and [5].

<u>2.3. Proposition</u>. The dual of $L_p(w,X)$ can be identified with $V_q(w^{-1},X^*)$, $\frac{1}{p}+\frac{1}{q}=1$, via the duality

$$(\rho,\phi) = \int_I \rho(t)d\phi(t)$$

where the integral is a Stieltjes integral in the sense of [5].

We will also find the following "weak" notions useful which are inspired by the so-called Pettis norm (see [26] for p=l).

2.3. Definition: The space $PL_p(w,X)$ consists of all norm-measurable $\rho : I \to X$ such that the Pettis-norm

$$\|\rho\|_{PL_p(w)} = \sup\left\{\|x^* \circ \rho\|_{L_p(w)} : x^* \in X^*, \|x^*\| \le 1\right\}$$

is finite. Similarly, $PV_p(w,X)$ is the space of norm-measurable functions $\phi : \mathring{I} \to X$ with finite p-semi-variation

$$\|\phi\|_{PV_p(w)} = \sup\left\{\|x^* \circ \phi\|_{V_p(w)} : x^* \in X^*, \|x^*\| \le 1\right\},$$

modulo constant functions.

From a Banach space point of view, these spaces are rather different from the Bochner spaces. For example, it follows from the Dvoretzky-Rogers theorem that there is always a $\rho \in PL_p(w,X) \setminus L_p(w,X)$ and that $PL_p(w,X)$ is incomplete unless X is finite dimensional. Also, even in a Hilbert space H, one cannot write every function $\phi : \mathbb{R} \to H$ of bounded p-semi-variation as the indefinite integral of a $\rho \in PL_p(H)$. But one can analyse these spaces in terms of operators:

2.4. Proposition: a) The completion of $PL_p(w,X)$, $1 < p \le \infty$, can be identified with the space of all operators $T : L_q(w^{-1}) \to X$, $\frac{1}{p} + \frac{1}{q} = 1$, for which there is a partition $(E_n)_{n \in \mathbb{N}}$ of I such that $\chi_{E_n} T : L_q(w^{-1}) \to X$ is compact for all p.

The embedding $\rho \in PL_p(w,X) \to T \in B(L_q(w^{-1}),X)$ is given by the Pettis integral (see [26])

$$Tf = \int_I f(t)\, \rho(t)dt \quad , \ f \in L_q(w^{-1}) .$$

b) $PV_p(w,X)$ is isometric to $B(L_q(w^{-1}),X)$ and the isometry is given by

$$Tf = \int_I f(t)\, d\phi(t) \quad , \ f \in C_0(I) .$$

for $\phi \in PV_p(w,X)$.

Next we consider boundary values of harmonic functions.

2.5. Definition: By $h_p(X)$ [$H_p(X)$], $1 \le p \le \infty$, we denote the space of all harmonic [analytic] functions $f:\mathbb{C}^+ \to X$, $\mathbb{C}^+ = \{\lambda \in \mathbb{C} : \operatorname{Re}\lambda > 0\}$, with the norm

$$\| f \| = \sup_{\alpha>0}\left(\int_{-\infty}^{\infty} \| f(\alpha+i\beta)\|^p d\beta\right)^{1/p} < \infty$$

Furthermore, $Ph_p(X)$ [$PH_p(X)$], $1 \le p \le \infty$, stands for the space of harmonic [analytic] functions $f:\mathbb{C}^+ \to X$ with the norm

$$\| f \| = \sup\left\{\| x^*\circ f\|_{H_p(\mathbb{R})} : x^* \in X^*, \|x^*\| \le 1\right\} < \infty .$$

The following is a variation of results of O. Blasco ([3], [4] :

2.6. Proposition: Let X be arbitrary Banach space except in the case $PH_1(X)$ where we assume that X does not contain c_0

a) The (improper) Poisson-Stieltjes integral

$$(+) \qquad f(\alpha+i\beta) = \frac{1}{\pi}\int_{-\infty}^{\infty} \frac{\alpha}{\alpha^2+(t-\beta)^2}\, d\phi(t)$$

defines an isometry between $V_p(\mathbb{R},X)$ $[PV_p(\mathbb{R},X)]$ and $h_p(X)$ [$Ph_p(X)$] .

b) The f in (+) belongs to $H_p(X)$ [$PH_p(X)$] if and only if for all $x^* \in X^*$ and $t<0$

$$\int_{-\infty}^{\infty} e^{-its} dx^*\circ\phi(t) = 0 .$$

Remark: It follows from Proposition 2.2. that (+) defines an isometry between $h_p(X)$ and $L_p(X)$ if and only if X has the Radon-Nikodym property. This was shown first in [7] where it is also shown that $H_p(X)$ is isometric to $\{f \in L_p(X), \hat{f}(t) = 0 \text{ for } t<0\}$ if and only if X has the analytic Radon-Nikodym property.

3. Real inversion of the Laplace transform

We now give our main inversion theorem in terms of the classical Widder-Post operators L_k (cf [32]).

3.1. Definition: For a C^∞-function we put for $k \in \mathbb{N}$ and $t>0$

$$L_k[f](t) = (-1)^k \frac{1}{k!}\left(\frac{k}{t}\right)^{k+1} f^{(k)}\!\left(\frac{k}{t}\right).$$

The special form of the L_k's is partially motivated by the fact that if $f(\lambda) = \int_0^\infty e^{-\lambda t}\rho(t)dt$ then

$$L_k[f](t) = \int_0^\infty l_k(t,s)\rho(s)ds \quad \text{where } l_k(t,s) = \frac{1}{k!}\left(\frac{k}{t}\right)^{k+1} s^k e^{-ks/t}.$$

Since $l_k(t,s)$ goes to δ_t for $k \to \infty$ weakly in the sense of measures, this leads to the inversion formula $L_k[f](t) \to \rho(t)$.

In the following we consider the densities

$$w_{\alpha,\beta}(t) = \begin{cases} t^\alpha \text{ for } 0 < t \le 1 \\ t^\beta \text{ for } 1 \le t < \infty. \end{cases}$$

3.2. Theorem: Let X be an arbitrary Banach space (only in the case $\phi \in PV_1(X)$ we assume that X does not contain a subspace isomorphic to c_0) and $f:(0,\infty) \to X$ a C^∞-function with $f(\lambda) \to 0$ for $\lambda \to \infty$.

Let $1 \le p \le \infty$ and $\alpha, \beta < 1 + \frac{1}{q}$, $\frac{1}{q} + \frac{1}{p} = 1$.

a) There is a $\phi \in V_p(w_{\alpha,\beta},X)$ [$PV_p(w_{\alpha,\beta},X)$] with $\|\phi\| \le M$ and

$$(+) \qquad f(\lambda) = \lambda \int_0^\infty e^{-\lambda t}\,\phi(t)\,dt$$

if and only if $L_k[f] \in L_p(w_{\alpha,\beta},X)$ [$PL_p(w_{\alpha,\beta},X)$] for k=1,2,... and

$$\overline{\lim_k}\, \|L_k[f]\| \le M.$$

b) If (+) holds for a $\phi \in PV_p(X)$ with $\phi(t) = \frac{1}{2}[\phi(t^+) + \phi(t^-)]$ then for all $0 < a < b < \infty$

$$\int_a^b L_k[f](t)\,dt \to \phi(b) - \phi(a) \quad , \text{for } k \to \infty.$$

If X has the Radon-Nikodym property one obtains results similar to the scalar valued case:

3.3. Corollary: Let X be a Banach space with the Radon-Nikodym property and $f:(0,\infty)\to X$ a C^∞-function.

Let $1 \le p \le \infty$ and $\alpha < \frac{1}{q}$, $\beta < 1 + \frac{1}{q}$, $\frac{1}{p} + \frac{1}{q} = 1$.

a) There is a $\rho \in L_p(w_{\alpha,\beta},X)$ with $\|\rho\| \le M$ and

$$(++) \qquad f(\lambda) = \int_0^\infty e^{-\lambda t}\, \rho(t)\, dt$$

if and only if $f(\lambda) \to 0$ for $\lambda \to \infty$ and $L_k[f] \in L_p(w_{\alpha,\beta},X)$ for k=1,2,... with

$$\overline{\lim_k} \|L_k[f]\| \le M .$$

(If p=1 we may allow α=0 but have to assume in addition that the $L_k[f]$'s are equi-integrable in $L_p(w_{\alpha,\beta},X)$).

b) If (++) holds for a $\rho \in L_p(w_{\alpha,\beta})$ then for all Lebesgue points t of ρ we have

$$\lim_{k\to\infty} L_k[f](t) = \rho(t).$$

3.4. Remarks: a) The integral in (+) is a Bochner integral. If $\alpha < \frac{1}{q}$ one can rewrite (+) as an improper Stieltjes integral

$$f(\lambda) = \int_0^\infty e^{-\lambda t}\, d\phi\ (t)$$

b) For the existence of a norm-measurable $\phi \in PV_1(X)$ representing an f with $L_k[f] \in PL_1\ (X)$, the condition that X does not contain c_0 is not only sufficient but also necessary. This answers a question left open in Zaidman's paper [37].

c) In Corollary 3.3., the Radon-Nikodym property is not only sufficient but also necessary to obtain a norm-measurable $\phi \in L_p(X)$. In a general Banach space X one can still prove the following weaker representation. If $\overline{\lim_k} \| L_k[f] \|_{L_p} \le M$ there is a separable subspace Y of X* which norms span $\{ L_k[f]\,(t) : t > 0\,,\, k \in \mathbb{N}\ \}$ and a w*-measurable function $\psi : \mathbb{R}^+ \to Y^*$ with

$$\int_0^\infty (\| \psi(t) \|\ w_{\alpha,\beta}(t))^p\, dt \le M^p$$

and (++) holds in the sense of a w*-integral for $x^* \in Y$.

d) If (X,≤) is a Banach lattice then ϕ is increasing if and only if f is completely monotonic in the sense that

$$(-1)^k f^{(k)}(\lambda) \ge 0 \qquad \text{for } k=0,1,2,...$$

This is an extension of the classical theorem of Bernstein (see e.g. [32] and [1]).

e) Exponential densities can be handled in the following way:

$t \in \mathbb{R}^+ \to e^{-at}\rho(t)$, $a \in \mathbb{R}$, belongs to $L_p(w_{\alpha,\beta},X)$, $\alpha < \frac{1}{q}$, $\beta < 1 + \frac{1}{q}$ if and only if

$$f(\lambda) = \int_0^\infty e^{-\lambda t}\,\rho(t)\,dt$$

exists for $\lambda > a$, $f(\lambda) \to 0$ for $\lambda \to \infty$ and for $f_a(\lambda) = f(\lambda + a)$ the sequence $L_k[f_a]$, $k=1,2,\dots$, is bounded in $L_p(w_{\alpha,\beta},X)$.

3.5. Remarks on the history of the theorem.
The scalar valued version with $w_{\alpha,\beta} \equiv 1$ goes back to Widder [32] .The first vector-valued inversion theorems by Hille and Miyadera ([18]) were of the kind (++) for a reflexive Banach space X and $w_{\alpha,\beta} \equiv 1$.Zaidman [36] was the first to recognize that for $p=\infty$ the representation (++) is possible if and only if X has the Radon-Nikodym property. He also gave inversion theorems using the Stieltjes representation in 3.4.a) for the norms $V_1(X)$, $PV_1(X)$ and sequentially complete X. For $V_1(X)$ this restriction was removed by Whitford [31] who used vector measures. Finally,Arendt[1] gave theorem 1 for $p=\infty$ and $w_{\alpha,\beta} \equiv 1$ in its present form.
There are different, but closely related inversion theorems by Rooney [21] and Leviatan [17], using different operators and - in Leviatan's case - vector measures in (+).
To my knowledge there was so far no systematic study of densities $w \not\equiv 1$ for the Widder-Post operators, even in the scalar valued case (for a different inversion operator see [21]). We feel that these densities should be useful in applications to semi-groups and the abstract Cauchy problem.
See Section 6 for a sketch of the proof of 3.2 and 3.3, for details and further extentions, see [27, 28]

4. Complex inversion and the vector-valued Fourier transform

For an analytic function $f: \mathbb{C}^+ \to X$ one is also interested in conditions along vertical lines (complex conditions) that ensure the existence of a Laplace representation. If $\| f(\lambda) \| \le C(1 + |\lambda|)^p$ on $\{\lambda : \mathrm{Re}\,\lambda \ge \alpha\}$ then Cauchy's formula gives

$$f^k(\lambda) = (-1)^k \frac{k!}{2\pi} \int_{-\infty}^{\infty} \frac{f(\alpha+i\beta)}{(\lambda-\alpha-i\beta)^{k+1}}\,d\beta$$

for $\lambda > \alpha$ and $k=p+1, p+2, \dots$, or

$$L_k[f](t) = U_{\alpha,k}[f](t) \quad , \; t < \frac{k}{\alpha}$$

where

$$U_{\alpha,k}[f](t) = \begin{cases} \dfrac{1}{2\pi}\displaystyle\int_{-\infty}^{\infty}\left(1-\dfrac{(\alpha+i\beta)\,t}{k}\right)^{-k-1} f(\alpha+i\beta)d\beta & \text{for } t<\dfrac{k}{\alpha} \\ 0 & \text{otherwise} \end{cases}$$

(see e.g. [25] or [13] for details). From 3.2 we get now

<u>4.1. Theorem</u>: Let X be an arbitrary Banach space and $f:\mathbb{C}_+ \to X$ an analytic function.

For $w=w_{\alpha,\beta}$, $\alpha < \frac{1}{q}$, $\beta < 1 + \frac{1}{q}$, $1 \le p \le \infty$ $\frac{1}{p}+\frac{1}{q} = 1$ the following are equivalent

a) There is a $\phi \in V_p(w,X)$, $\| \phi \| \le M$, with

$$f(\lambda) = \int_0^\infty e^{-\lambda t}\, d\phi(t) \qquad \text{for } \operatorname{Re}\lambda > 0$$

b) We have $f(\lambda) \to 0$ for $\lambda \to \infty$, $f(\lambda)$ is bounded on every $\{\lambda : \operatorname{Re}\lambda \ge a\}$, $a > 0$, and $U_{a,k}[f] \in L_p(w,X)$ for all $k=1,2,\ldots$, and $a>0$ with

$$\overline{\lim_{\substack{k\to\infty \\ a\to 0}}} \| U_{a,k}[f] \|_{L_p(w,X)} \le M .$$

This leads to a result on the range of the vector-valued Fourier transform (see [22] for the scalar-valued case).

<u>4.2. Corollary</u>. Let $g \in L_q(\mathbb{R},X)$ with $2 \le q \le \infty$. There is a $\phi \in V_p(\mathbb{R},X)$, $\frac{1}{p}+\frac{1}{q}=1$, with

$$x^*{\circ}g(t) = \int_{-\infty}^{\infty} e^{-ist}\, dx^*{\circ}\phi(s) \ , \quad \|\phi\|_{V_p} \le M$$

for all $x^* \in X^*$ if and only if for $k=1,2,\ldots$

$$\| U_{0,k}[g] \|_{L_p(\mathbb{R},X)} \le M$$

(In essence one applies 4.1. separately to the functions

$$g^{\pm}(\lambda) = \frac{1}{2\pi}\int_{-\infty}^{\infty} \frac{1}{\lambda \pm i\beta}\, g(\beta)\, d\beta \quad)$$

4.2. derives some additional interest from the fact that the vector-valued Hausdorff-Young inequality holds only for special Banach spaces, namely for Banach spaces of the so-called Fourier type p (see [6]). In particular,Parseval's inequality only holds for Hilbert spaces X (cf. [16]}. 4.1. and 4.2. can also be formulated for the Pettis norm and the semi - variation, but in this case one deals essentially with estimates for scalar-valued functions and therefore the description of $PH_2(X)$ in terms of "boundary values" given in 2.6. allows to formulate a Paley-Wiener type theorem.

4.3. Corollary. The Laplace transform

$$f(\lambda) = \frac{1}{\sqrt{2\pi}}\int_{-\infty}^{\infty} e^{-\lambda t}\, d\phi(t) \quad , \quad \mathrm{Re}\,\lambda > 0$$

defines an isometry of $PV_2(\mathbb{R}^+,X)$ onto $PH_2(X)$.

5. Some applications to semi-groups

In this section, A is always a closed linear operator in a Banach space X with $\overline{D(A)} = X$. We also assume that $\{\lambda : \mathrm{Re}\,\lambda > 0\}$ belongs to the resolvent set of A and denote $R(\lambda) = (\lambda I - A)^{-1}$ for $\mathrm{Re}\,\lambda > 0$. If A generates a bounded semi-group T_t we have

$$R(\lambda)x = \int_0^{\infty} e^{-\lambda t} T_t\, x dx$$

for $x \in X$ and $\lambda > 0$, and for $k \to \infty$

$$L_k\left[R(\lambda)x\right](t) \to T_t x \quad \text{where} \quad L_k\left[R(\lambda)\right](t) = \left(\frac{k}{t}\right)^{k+1} R\left(\frac{t}{k}\right)^{k+1} .$$

We sketch now some applications of the Laplace inversion to typical problems in semi-group theory.

5a. Characterization of generators.

The Hille Yosida condition $\| \lambda^n R(\lambda)^n \| \le M$ says precisely that $\| L_k[R(\lambda)] \|_{L_\infty(w_{1,1})} \le M$ and Arendt [1] has detailed how to derive the Hille Yosida theorem for C_0-semi-groups from Widder's theorem. But many authors also considered semi-groups "of growth α" or "type L_p", which have singularities at $t = 0$ (see [10] 2.5 for a survey on the literature). We give here a variant of a generation result of Zabreiko, Zafievskii and Sobolevskii [33, 34]

We say that a semi-group of bounded, linear operators T_t, $t \geq 0$, on X is of type $L_p(w_{\alpha,\beta})$, $\alpha < \frac{1}{q}$, $\beta < 1 + \frac{1}{q}$, $\frac{1}{p} + \frac{1}{q} = 1$, if the functions $t \in (0, \infty) \to T_t x$ are continuous and belong to $L_p(w_{\alpha,\beta}, X)$, for all $x \in X$.

<u>5.1. Theorem</u>: An operator A as above generates a semi-group T_t of type $L_p(w_{\alpha,\beta})$ if and only if there are continuous functions $0 \leq \rho_x \in L_p(w_{\alpha,\beta})$ for all $x \in X$ with

i) for every $1 < a < \infty$ there is a $C < \infty$ such that $\rho_x(t) \leq C \| x \|$ for all $\frac{1}{a} \leq t \leq a$ and $x \in X$.

ii) $\| R^{(n)}(\lambda) x \| \leq \int_0^\infty e^{-\lambda t} t^n \rho_x(t)\, dt$ for $\lambda > 0$, $n \in \mathbb{N}$ and $x \in X$.

Sketch of proof: Given a semi-group T_t of type $L_p(w_{\alpha,\beta})$ one can check easily that $\rho_x(t) = \| T_t x \|$ satisfies i) and ii).

Conversely, i) and ii) imply that

$$\overline{\lim_k} \| L_k[R(\lambda)]x \|_{L_p(w_{\alpha,\beta})} \leq \| \rho_x \|_{L_p(w_{\alpha,\beta})}$$

and the main theorem gives functions $\phi_x(t)$ of bounded $L_p(w_{\alpha,\beta})$-variation with

$$R(\lambda)\, x = \int e^{-\lambda t}\, d\phi_x(t), \qquad \phi(0) = 0$$

If $x \in D(A)$ one can show as in the case of integrated semi-groups (see e.g. [1]) that the ϕ_x's have continuous derivatives τ_x. An approximation argument using i) shows that the same is true for all $x \in X$ and we obtain T_t as $T_t x = \tau_x(t)$.

<u>5b. Stability theory</u>.

An important question in stability theory is whether the spectral bound

$$s(A) = \sup \{ \operatorname{Re} \lambda : \lambda \in \sigma(A) \}$$

of a generator equals the growth bound w of the semi-group which (by a theorem of Pazy [19], Section 4.4) can be expressed as

$$(*)\qquad w = \inf\left\{\omega : \int_0^\infty \| e^{-\omega t} T_t x \| \, dt < \infty \quad \text{for all} \quad x \in X\right\}.$$

While $s(A) = w$ may fail even in a Hilbert space, we have the following formula for $s(A)$ which together with (*) illustrates the difference between the $L_p(X)$ and $PL_p(X)$-norms.

5.2. Theorem. If A generates a positive semi-group T_t on a Banach lattice X then

$$s(A) = \inf\left\{\omega : \int e^{-\omega t} \; x^*(T_t x) \, dt < \infty \quad \text{for } x \in X_+ \, , \; x^* \in X_+^*\right\}$$

This formula is implicit in [29]1.3 and 2.2 and also in [8], 7.3 and 7.4, where the Pringsheim-Landau theorem is used (see [32], Chap. II, Theorem 5b). Here we give a simple proof based on the inversion theorem. In particular we use the following remarkable reduction of the Widder conditions for p=1 and positive operators.

5.3 Lemma: Let X be a Banach lattice and $R(\lambda) \geq 0$ for $\lambda > 0$. Then

i) $\| L_k[R(\lambda)x] \|_{PL_1(X)} \leq M \quad , \; k = 1,2,\ldots$

is equivalent to the much simpler conditions,

ii) $\lim_{\lambda \to 0} x^*(R(\lambda)x) \leq M \; , \quad \lim_{\lambda \to \infty} x^* R(\lambda)x = 0$

iii) $\lambda^k x^*(R(\lambda)^{k+1}x) \to 0$ for $\lambda \to 0$ and $\lambda \to \infty \; , \quad k = 1,2,\ldots$

for all $x \in X_+ , \; x^* \in X_+^*$.

This follows from the fact that the kth condition

$$\| L_k[R(\lambda)x \|_{PL_1(X)} = \sup_{\|x^*\| \leq 1} k \int s^{k-1} x^*(R(s)^{k+1}x) \, ds$$

in i) can be obtained from the $(k-1)^{th}$ condition by integration by parts and in this way they can be reduced to condition ii). The conditions iii) ensure that the boundary conditions appearing in the integration by parts formula disappear.

Now one can prove theorem 4.2 by reducing the conditions $\int e^{-\omega t} x^*(T_t x) \, dt \leq M$ to

$$\| L_k [R(\omega + \lambda)x] \|_{PL_1(X)} \leq M$$

using the inversion theorem for the $PL_1(w_{0,0},X)$-norm.

By 4.3. this is equivalent to

$$\lim_{\lambda\to 0} x^*(R(\omega+\lambda)x) \le M, \qquad \lim_{\lambda\to\infty} x^*R(\omega+\lambda)x = 0$$

$$\lambda^k x^*(R(\omega+\lambda)^{k+1}x) \to 0 \qquad \text{for } \lambda \to 0 \text{ and } \lambda \to \infty$$

for all k=1,2,..., $x \in X_+$, $x^* \in X_+^*$.

Clearly this holds for some $M < \infty$ as long as $\omega > \sigma(A)$.

5c. L_2-Solutions of the Cauchy-problem.

The following kind of weak solution. was considered e.g. by Beals ([2]): A norm-measurable $u : \mathbb{R}^+ \to X$ is a weak solution of

$$u'(t) = Au(t) \quad , \quad u(0) = x$$

if for $y \in D(A^*)$ the function $t \to \langle u(t), y \rangle$ is equal a.e. to an absolutely continuous function $t \to u(y,t)$ with

$$\frac{d}{dt} u(y,t) = \langle u(t), A^*y \rangle \ \text{a.e.} \quad , \quad u(y,0) = \langle y,x \rangle .$$

Using the complex conditions of Section 4 one can show:

5.4 Theorem: Suppose there are $M < \infty$ and $\lambda_0 \in \mathbb{R}$ such that for $\operatorname{Re}\lambda \ge \lambda_0$

$$\| R(\lambda,A) \| \le M .$$

Then, for every $x \in D(A)$ there is a weak solution u with $\| u \|_{PL_2([0,T],X)} < \infty$ for all $T < \infty$.

In [2] Beals even requires that $\int_0^T \| u(t) \|^2 dt < \infty$ for all $T < \infty$, but has to restrict himself to Hilbert spaces X to obtain such L_2-solutions via the vector-valued Fourier transform.

Note that every resolvent-positive operator A satisfies the assumption of the theorem, even if A does not generate a semi-group.

For a proof of 5.4 together with a systematic study of L_p-solutions, see [30].

6. Sketch of the proof of Theorem 3.2.

We concentrate on the case p=1 which requires some additional measurability considerations. Also put $\alpha=\beta=0$. In this case the implication "$\Rightarrow$" is similar to the scalar-valued case treated in [30].

a) Assume that $\overline{\lim_k} \|L_k[f]\|_{PL_1(X)} \leq M$. Let $C_0[0,\infty)$ denote the continuous functions on $[0,\infty)$ vanishing at ∞ and define operators $T_k : C_0[0,\infty) \to X$ by

$$T_k(g) = \int_0^\infty L_k[f](t)g(t)\,dt \quad , \; g \in C_0[0,\infty)$$

the assumption gives $\overline{\lim_k}\|T_k\| \leq M$ and for $g_\lambda(t) = e^{-\lambda t}$ we have $\lim_k T_k(g_\lambda) = f(\lambda)$ . . This follows from the fact that for $k \to \infty$

$$\int_0^\infty (-1)^k \frac{1}{k!}\left(\frac{k}{t}\right)^{k+1} f^{(k)}\left(\frac{k}{t}\right) e^{-\lambda t}\,dt \to f(\lambda) \; ,$$

which was shown already by Widder ([30], p.303). Since the span of the g_λ , $\lambda>0$ is dense in $C_0[0,\infty)$ we can define an operator $T : C_0[0,\infty) \to X$ by $T(g) = \lim_k T_k(g)$. with $\|T\| \leq M$ and $T(g_\lambda) = f(\lambda)$. It remains to find a representation of T in terms if a Stieltjes integral

$$(+) \qquad T(g) = \int_0^\infty g(t)\,d\phi(t)$$

with respect to a norm-measurable $\phi : \mathbb{R}^+ \to X$ with finite semi-variation. If X does not contain c_0 , a theorem of Pelczynski ([9], VI, theorem 15 , p. 159) implies that $T : C_0[0,\infty) \to X$ is a weakly compact operator. In particular $T^{**} : C_0[0,\infty)^{**} \to X$. If we identify $\chi_{[0,1]}$ with an element of X^{**} we can define the function $\phi : \mathbb{R}^+ \to X$, by $\varphi(t) = T^{**}(\chi_{[0,t]})$. ϕ is weakly measurable with separable range, i.e. ϕ is norm-measurable and also satisfies (+).

b) If X contains a sequence (e_n) equivalent to the unit vector basis of c_0 we can find a C^∞-function $f : \mathbb{R}^+ \to X$ with $\| L_k[f] \|_{PL_1(X)} \le 1$ but without a representation (+) by a norm-measurable $\phi : \mathbb{R}^+ \to X$ with $\phi(0) = 0$.

Indeed, assume first that there is $(a_n) \in U_{l_\infty}$ such that $\sum_{n=1}^{\infty} a_n e_n$ does not belong to X. Put

$$h_n = 2n(\chi_{[1-\frac{1}{n},1]} - \chi_{(1,1+\frac{1}{n}]}) \ , \ n \in \mathbb{N}$$

and define the operator $T : C_0[0,\infty) \to X$ by $Tg = \sum_n (\int g(t)h_n(t)\,dt)\, a_n\, e_n$.

Since (e_n) spans l_∞ in X** we can define $\phi : \mathbb{R}^+ \to X^{**}$ by $\phi(t) = \sum (\int_0^t h_n(t)\,dt)\, a_n e_n$.

Put $f(\lambda) = T(g_\lambda)$. Then f satisfies for all $x^* \in X^*$

$$x^*f(\lambda) = \int_0^\infty e^{-\lambda t} dx^*\circ\phi(t) \quad , \quad \phi(0) = 0$$

In particular $f \in C^\infty((0,\infty),X)$ and

$$\text{(++)} \qquad \|x^*\circ L_k[f]\|_{L_1} \le \|x^*\circ\phi\|_{V_1} \le \|T^*x^*\|_{M[0,\infty]} \le \|x^*\| .$$

But $\phi(1) = \sum \frac{1}{2} a_n e_n \notin X$.

On the other hand, if $\sum a_n e_n \in X$ for all $(a_n) \in l_\infty$ there is an isomorphic embedding $j: C[0,1] \to X$. In this case we define $T : C_0[0,\infty) \to X$ by $Tg = j(g|_{[0,1]})$ and put

$$f(\lambda) = T(g_\lambda) \quad , \quad \phi(t) = T^{**}(\chi_{[0,t]}) .$$

Again (+) and (++) are satisfied but even if ϕ takes values in X it is not norm-measurable. Indeed, for s<t we have

$$\|\phi(t) - \phi(s)\| = \|T^{**}(\chi_{[s,t]})\| \ge C$$

for a constant C>0 since T** is an isomorphic embedding. Hence ϕ does not have essentially separable range.

c) Now assume that

$$(°) \qquad \|L_k[f]\|_{L_1(X)} \le M \qquad \text{for } k = 1,2,\ldots .$$

As in a) we construct operators $T_k : C_0[0,\infty) \to X$ which converge to an operator

$T : C_0[0,\infty) \to X$ with $T(g) = \lim_k T_k(g)$ and $T(g_\lambda) = f(\lambda)$. The assumption ($\circ$) implies that the T_k are absolutely summing (see [9] VI.3, Theorem 3, p. 162), i.e. for all $g_1,...g_n \in C_0[0, \infty)$:

$$\sum_{i=1}^{n} \| T_k g_i \| \leq M \sup_{t>0} \sum_{i=1}^{n} | g_i(t) | .$$

Hence T is also absolutely summing and therefore weakly compact. So in this case we obtain $\| \phi(t) \| = T^{**}(\chi_{[0,t]}) \in X$ without special assumptions on X and we can continue as in a) to show (+).

d) The case $1 < p \leq \infty$ is handled in a similar fashion. By Hölder's inequality we can define operators, $(\frac{1}{p} + \frac{1}{q} = 1,)$

$$T_k : L_q(\mathbb{R}^+) \to X \quad , \quad T_k(g) = \int_0^\infty L_k[f](t)\; g(t)\; dt$$

which again converge in the strong operator topology to a $T : L_q(\mathbb{R}^+) \to X$.. Since $\chi_{[0,t]} \in L_q(\mathbb{R}^+)$ we find the representing function $\phi : \mathbb{R}^+ \to X$ directly by $\phi(t) = T(\chi_{[0,t]})$.

The case of general densities $w_{\alpha,\beta}$ requires some technical changes in the convergence argument $T = \lim T_k$ and the definition of ϕ.

References

1) W. Arendt: Vector valued Laplace transforms and Cauchy problems, J. of Math. 59 (1987), 327-352.

2) R. Beals : On the Abstract Cauchy Problem, J. of Funct. Anal. 10 (1972), 281-299.

3) O. Blasco: Boundary values of vector-valued harmonic functions considered as operators, Studia Math. 86 (1987), 19-33.

4) O. Blasco: Boundary Values of Functions in Vector-valued Hardy Spaces and Geometry of Banach Spaces, J. of Funct. Analysis 78, 346-364 (1968).

5) S. Bochner and A. Taylor: Linear functionals on certain spaces of abstractly-valued functions, Ann. of Math. 39 (1938), 913-944.

6) J. Bourgain: Vector-valued Hausdorff-Young inequalities and applications, Geometric Aspects of Functional Analysis 1986/87, 239-249, Lecture Notes in Mathematics 1317, Springer Verlag 1988.

7) A.V. Bukhvalov and A.A. Danilevich: Boundary values of analytic and harmonic functions with values in Banach spaces, Math. Notes 31, (1982), 104-110.

8) Ph. Clément, H.J.A.M. Heijmans, et al: One parameter semi-groups, CWI Monographs 5, North Holland, Amsterdam 1987.

9) J. Diestel and J. Uhl: Vector measures, American Math. Soc., Providence 1977.

10) N. Dinculeanu : Vector Measure, Pergamon New York 1967

11) H.O. Fattorini: The Cauchy Problem, Addison-Wesley, Reading 1983.

12) J. Goldstein: Semigroups of linear operators and Applications, Oxford University Press, Oxford 1985.

13) B. Hennig and F. Neubrander: On the representation, inversions and approximations of Laplace Transforms in Banach spaces, Preprint.

14) M. Hieber, A. Holderrieth and F. Neubrander: Abstract Petrovski Operators and Systems of Linear Partial Differential Equations in $L_p(\mathbb{R}^n)^N$, Preprint.

15) E. Hille and R.S. Phillips: Functional Analysis and Semi-groups, Amer. Math. Soc. Coll. Publ. 31, Providence, 1957.

16) S. Kwapien: Isomorphic Characterizations of inner product spaces by orthogonal series with vector coefficients, Studia Math. 44 (1972), 583-595.

17) D. Leviatan: Some vector-valued Laplace transforms, Israel J. Math. 16 (1973), 73-86.

18) I. Miyadera: On the representation theorem by the Laplace transformation of vector-valued functions, Tohoku Math. J. 8 (1956), 170-180.

19) A. Pazy: Semi-groups of linear operators and Applications to Partial Differential Equations, Springer Verlag, New York-Berlin, 1983.

20) G. Pisier: Factorization of linear operators and geometry of Banach-spaces, Amer. Math. Soc., Providence, 1986, CBMS-Lecture Series 60

21) P. Rooney: An inversion and representation theory for the Laplace integral of abstractly-valued functions, Can. J. Math. 6(1954), 190-209.

22) P. Rooney: On the representation of functions as Fourier transforms, Can. J. Math. 11 (1959), 168-174.

23) H.P. Rosenthal: On Factors of C[0,1] with non-separable Dual, Israel J. Math. 13 (1972), 361-378.

24) H.H. Schaefer: Banach lattices and positive operators, Springer Verlag, Berlin, 1974.

25) M. Sova: Relation between the real and complex properties of the Laplace transform, Casopis pest. mat. 105 (1980), 111-119.

26) M. Talagrand, Pettis integral and measure theory, Mem. Amer. Math. Soc. Vol. 51, No. 307, Amer. Math. Soc., Providence 1984.

27) E. Teske and L. Weis: On the inversion of the vector-valued Laplace transform, in preparation.

28) P. Vieten and L. Weis: The classical integral transforms and functions of bounded semi-variation, in preparation.

29) J. Voigt: On the abscissa of convergence for the Laplace transform of vector valued measures, Archiv der Mathematik 39 (1982), 455-462.

30) L.W. Weis: The range of an operator on C(X) and its representing measure, Archiv Math. 46, (1986), 171-178.

31) A.K. Whitford: Laplace-Stieltjes transforms of vector valued measures, Math. Casopis Sloven. Akad Vied. 22, 1972, 156-163.

32) D. Widder: The Laplace Transform, Princeton University Press, 1946.

33) P. Zabreiko and V. Zafievskii: On a certain class of semi-groups, Dokl. Akad. Nauk SSSR 189, (1969), 1523-1526.

34) A. Zafievskii: On semi-groups with singularities summable with a power weight at zero, Dokl. Akad. Nauk SSSR 195 (1970), 1408-1411.

35) S. Zaidman: La réprésentation des fonctions vectorielles par des intégrales de Laplace-Stieltjes, Annals of Mathematics 68, 1958 (260-277).

36) S. Zaidman: Sur un theorem de I. Miyadera concernant la réprésentation des fonctions vectorielles par des intégrales de Laplace, Tohoku Math. J. 12 (1960), 47-51.

37) S. Zaidman: La réprésentation des fonctions vectorielles par des integrales de Laplace-Stieltjes II, Tohoku Math. J. 12, (1960), 52-70.

Some Quasilinear Parabolic Problems in Applied Mathematics

ATSUSHI YAGI Department of Mathematics, Himeji Institute of Technology, 2167 Shosha, Himeji, Hyogo 671-22, Japan

1. Introduction.

We study a strongly coupled parabolic system:

$$
\text{(P.S)} \quad \begin{cases}
\dfrac{\partial u}{\partial t} = \operatorname{div}[\nabla(a_1 u + \alpha_{11} u^2 + \alpha_{12} uv) + b_1(\nabla\Phi(x))u] \\
\qquad\qquad + c_1 u - \gamma_{11} u^2 - \gamma_{12} uv \quad \text{in} \quad \Omega \times (0,\infty), \\
\dfrac{\partial v}{\partial t} = \operatorname{div}[\nabla(a_2 v + \alpha_{21} uv + \alpha_{22} v^2) + b_2(\nabla\Phi(x))v] \\
\qquad\qquad + c_2 v - \gamma_{21} uv - \gamma_{22} v^2 \quad \text{in} \quad \Omega \times (0,\infty), \\
\dfrac{\partial}{\partial n}(a_1 u + \alpha_{11} u^2 + \alpha_{12} uv) + b_1 \dfrac{\partial \Phi(x)}{\partial n} u = 0 \quad \text{on} \quad \partial\Omega \times (0,\infty), \\
\dfrac{\partial}{\partial n}(a_2 v + \alpha_{21} uv + \alpha_{22} v^2) + b_2 \dfrac{\partial \Phi(x)}{\partial n} v = 0 \quad \text{on} \quad \partial\Omega \times (0,\infty), \\
u(0,x) = u_0(x) \quad \text{and} \quad v(0,x) = v_0(x) \quad \text{in} \quad \Omega.
\end{cases}
$$

Here, $\Omega \subset \mathbb{R}^2$ is a bounded region of C^2 class; $a_i > 0 (i = 1, 2)$ are positive constants; α_{ij}, b_i, c_i and $\gamma_{ij}(i,j = 1,2) \geq 0$ are non negative constants; $\Phi \in C^2(\bar{\Omega})$ is a given real function

This research was partly supported by Grant-in-Aid for Scientific Reseach (No. 03640176), Ministry of Education, Science and Culture.

on $\bar{\Omega}$; $\frac{\partial}{\partial n}$ denotes the outer normal derivative on the boundary $\partial\Omega$. u_0 and v_0 are initial functions which are assumed to satisfy:

$$\text{(I.F)} \qquad u_0, v_0 \in H^{1+\varepsilon}(\Omega) \quad \text{with some} \quad \varepsilon > 0 \quad \text{and} \quad u_0, v_0 \geq 0 \quad \text{in} \quad \Omega.$$

And, $u = u(t,x)$ and $v = v(t,x)$ are possibly non negative unknown functions.

This system has been introduced by Shigesada et al. [4] to describe the population dynamics in Ω of two competitive species which move under the influence of population pressure and of environmental potential $\Phi(x)$. The unknown functions $u = u(t,x)$ and $v = v(t,x)$ denote population densities of the two species. The boundary conditions show that the flow of individual is tangential on the boundary $\partial\Omega$.

We are concerned with the golbal existence of L^2 valued strict solution to (P.S). In some particular cases this problem has been already studied by several authors, Masuda and Mimura[3], Kim[2], Deuring[1], etc. In this Note we would like to handle general cases. In fact we shall show in the following two cases:

$$(\alpha.\text{i}) \qquad 0 < \alpha_{21} < 8\alpha_{11} \quad \text{and} \quad 0 < \alpha_{12} < 8\alpha_{22};$$

$$(\alpha.\text{ii}) \qquad \alpha_{11} > 0, \alpha_{21} = \alpha_{22} = 0 \quad \text{and} \quad \gamma_{11} > 0;$$

that there exists a global non negative solution to (P.S) for the initial functions u_0, v_0 satisfying (I.F).

This Note consists of four sections. In Section 2, we shall first verify that any real solution to (P.S) must be non negative by the truncation method. In Section 3, the existence and uniqueness of local solution will be obtained (without assuming (α.i) or (α.ii)) by applying the theory of abstract parabolic equations in Banach spaces (see [6]). In Section 4, a priori estimates for the solution u, v will be established under (α.i) or (α.ii) to obtain the global existence of solution.

2. Positivity of Solutions.

We shall observe that every real strict solution to (P.S) is non negative. To this end let us first consider an auxiliary problem:

$$(2.1) \qquad \begin{cases} \dfrac{\partial u}{\partial t} = \operatorname{div}[\nabla(a(t,x)u) + (\nabla\Phi(x))u] + c(t,x)u & \text{in} \quad \Omega \times (0,T] \\ \dfrac{\partial}{\partial n}(a(t,x)u) + \dfrac{\partial\Phi(x)}{\partial n}u = 0 \quad \text{on} \quad \partial\Omega \times (0,T] \\ u(0,x) = u_0(x) \quad \text{in} \quad \Omega, \end{cases}$$

in Ω. Here, $a(t,x) \geq \delta$ is a given positive function on $[0,T] \times \bar{\Omega}$ satisfying:

$$(2.2) \qquad \begin{cases} a \in \mathcal{C}([0,T]; H^{1+\varepsilon_1}(\Omega)) \cap \mathcal{C}((0,T]; H^2(\Omega)), \\ \qquad \|a(t)\|_{H^2} \leq At^{\frac{\varepsilon_2 - 1}{2}} \quad \text{near} \quad t = 0 \end{cases}$$

with some $\varepsilon_1, \varepsilon_2 > 0$ and a constant A. $\Phi \in \mathcal{C}^2(\bar{\Omega})$ is a given real function. And $c \in \mathcal{C}([0,T]; L^\infty(\Omega))$ is a given real valued function on $\Omega \times [0,T]$. Then,

PROPOSITION 2.1. *Let $u_0 \in L^2(\Omega)$ be non negative (≥ 0). If u is a real valued strict solution to (2.1) such that*

$$u \in \mathcal{C}([0,T]; L^2(\Omega)) \cap \mathcal{C}((0,T]; H^2(\Omega)) \cap \mathcal{C}^1((0,T]; L^2(\Omega)), \tag{2.3}$$

then $u(t)$ must be non negative for every $0 \leq t \leq T$.

PROOF: We use the truncation method. Set a decreasing function H_0 on $\mathbb{R}$ such that $H_0(\sigma) = 0$ for $\sigma \geq 0, H_0(\sigma) = \sigma^4$ for $-\frac{1}{2} \leq \sigma \leq 0, H_0(\sigma) = 1$ for $\sigma \leq -1$, and that $H_0 \in \mathcal{C}^\infty((-\infty, 0); \mathbb{R})$. And set $H_1(\sigma) = \int_0^\sigma H_0(\sigma)d\sigma$ and $H(\sigma) = H_2(\sigma) = \int_0^\sigma H_1(\sigma)d\sigma$. Clearly, H is a decreasing function such that $H(\sigma) > 0$ if $\sigma < 0$ and that $H(\sigma) = 0$ if $\sigma \geq 0$. We then consider a function $\phi(t)$ defined by

$$\phi(t) = \int_\Omega H(u(t,x))dx \quad \text{for} \quad 0 \leq t \leq T.$$

From (2.3), $\phi \in \mathcal{C}([0,T]; \mathbb{R}_+) \cap \mathcal{C}^1((0,T]; \mathbb{R})$. And, indeed, $\phi'(t)$ is given by

$$\phi'(t) = -\int_\Omega aH''(u)|\nabla u|^2 dx - \int_\Omega H''(u)u\nabla(a + \Phi) \cdot \nabla u dx + \int_\Omega cH'(u)u dx.$$

So that, noting that $H'(\sigma)\sigma, H''(\sigma)\sigma^2 \leq Const.H(\sigma), \sigma \in \mathbb{R}$, we obtain that

$$\phi'(t) \leq -\frac{\delta}{2}\int_\Omega H''(u)|\nabla u|^2 dx + \int_\Omega |\nabla a| H''(u)|u||\nabla u| dx$$
$$+ Const.\{\delta^{-1}\|\nabla\Phi\|_{\mathcal{C}}^2 + \|c\|_{L^\infty}\}\phi(t).$$

Further, since $|\frac{dH''^{\frac{1}{2}}}{d\sigma}(\sigma)\sigma| \leq Const.H''(\sigma)^{\frac{1}{2}}, \sigma \in \mathbb{R}$, it is observed that

$$\int_\Omega |\nabla a| H''(u)|u||\nabla u| dx \leq \frac{\delta}{2}\|H''(u)^{\frac{1}{2}}\nabla u\|_{L^2}^2$$
$$+ Const.\{\delta^{-\frac{p+2}{p-2}}\|\nabla a\|_{L^p}^{\frac{2p}{p-2}} + \delta^{-1}\|\nabla a\|_{L^p}^2\}\|H''(u)^{\frac{1}{2}}u\|_{L^2}^2,$$

where p is an arbitrary number such that $2 < p < \infty$ (note that $H''(u)^{\frac{1}{2}}u \in H^1(\Omega)$ and use the embedding: $H^1(\Omega) \subset L^q(\Omega)$ with $\frac{1}{p} + \frac{1}{q} = \frac{1}{2}$). Therefore, $\phi'(t) \leq h(t)\phi(t), 0 < t \leq T$, where

$$h(t) = Const.\{\delta^{-\frac{p+2}{p-2}}\|a(t)\|_{H^2}^{\frac{2p}{p-2}} + \delta^{-1}\|\nabla\Phi\|_{\mathcal{C}}^2 + \|c(t)\|_{L^\infty}\}.$$

Since p can be arbitrarily large, (2.2) implies that h can be: $h \in L^1(0,T;\mathbb{R})$. Hence, $\phi(t) \leq \phi(0)\exp(\int_0^t h(\tau)d\tau) = 0$. ∎

Now we can state:

THEOREM 2.2. *Let u_0 and v_0 satisfy (I.F). Assume that a real solution u, v to (P.S) on an interval $[0,T]$ satisfy:*

$$\begin{cases} u, v \in \mathcal{C}([0,T]; H^{1+\varepsilon_1}(\Omega)) \cap \mathcal{C}((0,T]; H^2(\Omega)) \cap \mathcal{C}^1((0,T]; L^2(\Omega)), \\ \|u(t)\|_{H^2}, \|v(t)\|_{H^2} \leq At^{\frac{\varepsilon_2 - 1}{2}} \quad \text{near} \quad t = 0 \end{cases} \tag{2.4}$$

with some $\varepsilon_1, \varepsilon_2 > 0$ and a constant A. Then $u(t)$ and $v(t)$ must be non negative for every $0 \le t \le T$.

PROOF: We use the contradictory. Let $T^* = \sup\{t \in [0, T]; u(s)$ and $v(s)$ are ≥ 0 for all $0 \le s \le t\}$ and suppose that $T^* < T$. Then, the Proposition 2.1 (with $a = a_1 + \alpha_{11}u + \alpha_{12}v$ and with $a = a_2 + \alpha_{21}u + \alpha_{22}v$) yields immediately a contradiction. ∎

3. Local Solutions.

Let u_0, v_0 satisfy (I.F). We shall show that there exists a unique local solution to (P.S).

We first have to rewrite (P.S) into another system which is almost symmetric near u_0, v_0 introducing suitable scale functions. Indeed, let $\varphi, \psi \in \mathcal{C}^2(\bar{\Omega})$ such that

$$0 < \varphi, \psi < \infty \quad \text{on} \quad \bar{\Omega} \quad \text{and} \quad \frac{\partial \varphi}{\partial n} = \frac{\partial \psi}{\partial n} \quad \text{on} \quad \partial\Omega$$

satisfy:

$$|\alpha_{12}\varphi(x)\psi(x)^{-1}u_0(x) - \alpha_{21}\varphi(x)^{-1}\psi(x)v_0(x)| < 2\sqrt{a_1 a_2} \quad \text{on} \quad \bar{\Omega}. \tag{3.1}$$

And let us change unknown functions from u, v to $u_\varphi = \varphi(x)u(t,x), v_\psi = \psi(x)v(t,x)$ respectively. Then, (P.S) becomes:

$$(P.S)' \quad \begin{cases} \dfrac{\partial u_\varphi}{\partial t} = \operatorname{div}[a_{11}(u_\varphi, v_\psi)\nabla u_\varphi + a_{12}(u_\varphi)\nabla v_\psi] + b_{11}(u_\varphi, v_\psi) \cdot \nabla u_\varphi \\ \qquad\qquad + b_{12}(u_\varphi, v_\psi) \cdot \nabla v_\psi + f(u_\varphi, v_\psi) \quad \text{in} \quad \Omega \times (0, \infty), \\ \dfrac{\partial v_\psi}{\partial t} = \operatorname{div}[a_{21}(v_\psi)\nabla u_\varphi + a_{22}(u_\varphi, v_\psi)\nabla v_\psi] + b_{21}(u_\varphi, v_\psi) \cdot \nabla u_\varphi \\ \qquad\qquad + b_{22}(u_\varphi, v_\psi) \cdot \nabla v_\psi + g(u_\varphi, v_\psi) \quad \text{in} \quad \Omega \times (0, \infty), \\ a_{11}(u_\varphi, v_\psi)\dfrac{\partial u_\varphi}{\partial n} + a_{12}(u_\varphi)\dfrac{\partial v_\psi}{\partial n} + h_1 u_\varphi = 0 \quad \text{on} \quad \partial\Omega \times (0, \infty), \\ a_{21}(v_\psi)\dfrac{\partial u_\varphi}{\partial n} + a_{22}(u_\varphi, v_\psi)\dfrac{\partial v_\psi}{\partial n} + h_2 v_\psi = 0 \quad \text{on} \quad \partial\Omega \times (0, \infty), \\ u_\varphi(0, x) = \varphi(x)u_0(x) \quad \text{and} \quad v_\psi(0, x) = \psi(x)v_0(x) \quad \text{in} \quad \Omega. \end{cases}$$

Here,

$$a_{11}(u_\varphi, v_\psi) = a_1 + 2\alpha_{11}\varphi^{-1}u_\varphi + \alpha_{12}\psi^{-1}v_\psi, a_{12}(u_\varphi) = \alpha_{12}\psi^{-1}u_\varphi$$
$$a_{21}(v_\psi) = \alpha_{21}\varphi^{-1}v_\psi, a_{22}(u_\varphi, v_\psi) = a_2 + \alpha_{21}\varphi^{-1}u_\varphi + 2\alpha_{22}\psi^{-1}v_\psi.$$

$b_{ij}(u_\varphi, v_\psi), 1 \le i, j \le 2$, are similarly affine functions with respect to u_φ and v_ψ; $f(u_\varphi, v_\psi)$ and $g(u_\varphi, v_\psi)$ are square functions with respect to u_φ and v_ψ; and $h_i = b_i \frac{\partial \Phi}{\partial n}, i = 1, 2$. All the coefficients of these functions are in $\mathcal{C}(\bar{\Omega})$ or $\mathcal{C}(\partial\Omega)$, which are determined by φ, ψ and Φ.

We then formulate (P.S)′, and therefore (P.S), as an abstract quasilinear equation:

$$\begin{cases} \dfrac{dU}{dt} + A(U)U = F(U), 0 < t < \infty, \\ U(0) = U_0 \end{cases} \tag{3.2}$$

in the product L^2-space $X = \{L^2(\Omega)\}^2$. Set also two other product spaces: $Z = \{H^{1+\varepsilon_1}(\Omega)\}^2$ and $W = \{H^{1+\varepsilon_2}(\Omega)\}^2$ with arbitrarily fixed $0 < \varepsilon_1 < \varepsilon_2 < \min\{\varepsilon, \frac{1}{2}\}$. Then (I.F) implies that the initial value $U_0 = \begin{pmatrix} \varphi u_0 \\ \psi v_0 \end{pmatrix}$ is in $W \subset Z$. For $U = \begin{pmatrix} u \\ v \end{pmatrix} \in K = \{U \in Z; \|U - U_0\|_Z < R\}, R > 0, A(U)$ denote linear operators defined by

$$\begin{cases} \mathcal{D}(A(U)) = \{\tilde{U} = \begin{pmatrix} \tilde{u} \\ \tilde{v} \end{pmatrix} \in \{H^2(\Omega)\}^2; \mathbb{B}(U;D)\tilde{U} = 0 \quad \text{on} \quad \partial\Omega\} \\ A(U)\tilde{U} = \mathbb{A}(U;D)\tilde{U}. \end{cases}$$

Here,

$$\mathbb{A}(U;D)\tilde{U} = -\begin{pmatrix} \operatorname{div}[a_{11}(\Re e u, \Re e v)\nabla\tilde{u} + a_{12}(\Re e u)\nabla\tilde{v}] \\ \operatorname{div}[a_{21}(\Re e v)\nabla\tilde{u} + a_{22}(\Re e u, \Re e v)\nabla\tilde{v}] \end{pmatrix} + \begin{pmatrix} b_{11}(u,v)\cdot\nabla\tilde{u} + b_{12}(u,v)\cdot\nabla\tilde{v} \\ b_{21}(u,v)\cdot\nabla\tilde{u} + b_{22}(u,v)\cdot\nabla\tilde{v} \end{pmatrix} + c\begin{pmatrix} \tilde{u} \\ \tilde{v} \end{pmatrix},$$

c being a sufficiently large constant specified below, and here,

$$\mathbb{B}(U;D)\tilde{U} = \begin{pmatrix} a_{11}(\Re e u, \Re e v) & a_{12}(\Re e u) \\ a_{21}(\Re e v) & a_{22}(\Re e u, \Re e v) \end{pmatrix} \frac{\partial\tilde{U}}{\partial n} + \begin{pmatrix} h_1 & 0 \\ 0 & h_2 \end{pmatrix} \tilde{U}.$$

$F(U) - cU + \begin{pmatrix} f(u,v) \\ g(u,v) \end{pmatrix}$ is a function on K with values in X. Finally, $U = \begin{pmatrix} u_\varphi(t) \\ v_\psi(t) \end{pmatrix}$, $0 \le t < \infty$, is an unknown function.

We may now apply the result on abstract parabolic equations in Banach spaces (actually in Hilbert spaces). In fact, according to [6], it is known that (3.2) admits a unique local strict solution provided that the following Conditions on $A(U), W \subset Z \subset X, F(U)$ and on U_0 are fulfilled.

(A.i) For $U \in K, \rho(A(U))$ (the resolvent sets of $A(U)$) $\supset \Sigma = \{\lambda \in \mathbb{C}; |\arg\lambda| \ge \theta_0\}, 0 < \theta_0 < \frac{\pi}{2}$; and an estimate

$$\|(\lambda - A(U))^{-1}\|_{\mathcal{L}(X)} \le \frac{M}{(|\lambda|+1)}, \lambda \in \Sigma, U \in K,$$

holds with some constant M.

(A.ii) For some $0 < \nu \le 1$, a Hölder condition

$$\|A(U)(\lambda - A(U))^{-1}\{A(U)^{-1} - A(V)^{-1}\}\|_{\mathcal{L}(X)} \le \frac{N\|U-V\|_X}{(|\lambda|+1)^\nu}, \lambda \in \Sigma, U, V \in K,$$

holds with some constant N.

(S.i) For some $0 < \gamma < 1, \|\cdot\|_Z \le D\|\cdot\|_W^\gamma\|\cdot\|_X^{1-\gamma}$ on W with some constant D.

(S.ii) The ball $\{U; \|U\|_Z \le 1\}$ of Z is a closed subset of X.

(S.iii) For some $0 < \beta < 1, \mathcal{D}(A(U)^\beta)$ (the domains of fractional powers $A(U)^\beta$) $\subset W, U \in K$, with continuous embedding $\|\cdot\|_W \le D_\beta\|A(U)^\beta\cdot\|_X$ with some constant D_β.

(Ex) A relation: $\beta + \nu > 1$ holds.
(F) $\|F(U) - F(V)\|_X \le L\|U - V\|_X, U, V \in K$, with some constant L.
(I) $U_0 \in \mathcal{D}(A(U_0)^\beta)(\subset W)$.

So that, let us check all these conditions. Fix $R > 0$ sufficiently small so that, on account of (3.1), every $U = \begin{pmatrix} u \\ v \end{pmatrix} \in K$, satisfies:

$$\{a_{12}(\Re e u) + a_{21}(\Re e v)\}^2 < 4a_{11}(\Re e u, \Re e v)a_{22}(\Re e u, \Re e v) \quad \text{on} \quad \bar{\Omega}. \tag{3.3}$$

Then we verify:

PROPOSITION 3.1. *$A(U), U \in K$, satisfy (A.i) if c is fixed sufficiently large.*

PROOF: For each $U \in K$, we consider a sesquilinear form $a(U;\cdot,\cdot)$ on $\{H^1(\Omega)\}^2$ defined by

$$a(U;\tilde{U}_1,\tilde{U}_2) = \int_\Omega \{[a_{11}(\Re e u, \Re e v)\nabla\tilde{u}_1 + a_{12}(\Re e u)\nabla\tilde{v}_1]\cdot\nabla\bar{\tilde{u}}_2 + [a_{21}(\Re e v)\nabla\tilde{u}_1 + a_{22}(\Re e u, \Re e v)\nabla\tilde{v}_1]\cdot\nabla\bar{\tilde{v}}_2\}dx$$

$$+ \int_\Omega \{[b_{11}(u,v)\cdot\nabla\tilde{u}_1 + b_{12}(u,v)\cdot\nabla\tilde{v}_1]\bar{\tilde{u}}_2 + [b_{21}(u,v)\cdot\nabla\tilde{u}_1 + b_{22}(u,v)\cdot\nabla\tilde{v}_1]\bar{\tilde{v}}_2\}dx + \int_\Omega c\{\tilde{u}_1\bar{\tilde{u}}_2 + \tilde{v}_1\bar{\tilde{v}}_2\}dx + \int_{\partial\Omega}\{h_1\tilde{u}_1\bar{\tilde{u}}_2 + h_2\tilde{v}_1\bar{\tilde{v}}_2\}d\sigma.$$

Since (3.3) yields that

$$a_{11}(\Re e u, \Re e v)\xi^2 + \{a_{12}(\Re e u) + a_{21}(\Re e v)\}\xi\eta + a_{22}(\Re e u, \Re e v)\eta^2 \ge \delta(\xi^2+\eta^2), \xi, \eta \in \mathbb{R},$$

with some $\delta > 0$, it follows that

$$\Re e\, a(U;\tilde{U},\tilde{U}) \ge \frac{\delta}{2}\|\tilde{U}\|^2_{H^1(\Omega)} + (c - Const.\delta^{-1})\|\tilde{U}\|^2_{L^2(\Omega)} \quad \text{on} \quad \{H^1(\Omega)\}^2.$$

Then, by virtue of the theory of sesquilinear forms, a linear operator $\mathcal{A}(U)$ in X is determined by each $a(U;\cdot,\cdot)$, and $\mathcal{A}(U)$ satisfies (A.i) with some uniform Σ and M provided c is sufficiently large. The domain $\mathcal{D}(\mathcal{A}(U))$ consists of $\tilde{U}$ such that $\mathbb{B}(U;D)\tilde{U} = 0$ in a certain weak sense and $\mathcal{A}(U)\tilde{U} = \mathbb{A}(U;D)\tilde{U}$ in the distribution sense. Therefore, we complete the proof if $\mathcal{D}(\mathcal{A}(U)) \subset \{H^2(\Omega)\}^2$. But this can be seen from the a priori estimate:

$$\|\tilde{U}\|_{H^2(\Omega)} \le Const.\{\|\mathbb{A}(U;D)\tilde{U}\|_{L^2(\Omega)} + \|\tilde{U}\|_{L^2(\Omega)} + \|\mathbb{B}(U;D)\tilde{U}\|_{H^{\frac{1}{2}}(\partial\Omega)}\}$$

on $\{H^2(\Omega)\}^2$. ∎

We next verify:

PROPOSITION 3.2. *$A(U), U \in K$, satisfy (A.ii) with any $0 < \nu < \frac{1}{2}$.*

PROOF: Let $U, V \in K$, and $\lambda \in \Sigma$. For any $\tilde{F}, \tilde{G} \in X$,

$$(A(U)(\lambda - A(U))^{-1}\{A(U)^{-1} - A(V)^{-1}\}\tilde{F}, \tilde{G})_X$$
$$= a(V; A(V)^{-1}\tilde{F}, (\bar{\lambda} - A(U)^*)^{-1}\tilde{G}) - a(U; A(V)^{-1}\tilde{F}, (\bar{\lambda} - A(U)^*)^{-1}\tilde{G}).$$

In view of the definition of $a(U; \cdot, \cdot)$, it is estimated by

$$|(A(U)(\lambda - A(U))^{-1}\{A(U)^{-1} - A(V)^{-1}\}\tilde{F}, \tilde{G})_X|$$
$$\leq Const. \|U - V\|_{L^2} \|A(V)^{-1}\tilde{F}\|_{H^2} \|(\bar{\lambda} - A(U)^*)^{-1}\tilde{G}\|_{H^2}^{1-\nu} \|(\bar{\lambda} - A(U)^*)^{-1}\tilde{G}\|_{L^2}^{\nu}$$

with any $0 < \nu < \frac{1}{2}$. Since $\mathcal{D}(A(U)^*) \subset \{H^2(\Omega)\}^2$ and $A(U)^*$ also satisfy (A.i), we obtain the result. ∎

In addition,

PROPOSITION 3.3. *For $0 \leq \theta < \frac{3}{4}, \mathcal{D}(A(U)^\theta) = \mathcal{D}(A(U)^{*\theta}) = \{H^{2\theta}(\Omega)\}^2, U \in K$, with norm equivalence.*

PROOF: Since $A(U)$ are maximal accretive operators in $X, \mathcal{D}(A(U)^\theta) = [X, \mathcal{D}(A(U))]_\theta$ and $\mathcal{D}(A(U)^{*\theta}) = [X, \mathcal{D}(A(U)^*)]_\theta$ for all $0 \leq \theta \leq 1$. So that, $\mathcal{D}(A(U)^\theta), \mathcal{D}(A(U)^{*\theta}) \subset \{H^{2\theta}(\Omega)\}^2, 0 \leq \theta \leq 1$. By the transposition, this implies that $A(U)$ is an isomorphism from $\mathcal{D}(A(U)^\theta)$ to $\mathcal{D}(A(U)^{*(1-\theta)})', 0 \leq \theta \leq 1$. On the other hand, by a direct calculation, it is seen that $A(U)$ is an isomorhism from $\{H^{2\theta}(\Omega)\}^2$ to $\{H^{2(1-\theta)}(\Omega)'\}^2$, for any $\frac{1}{4} < \theta < \frac{3}{4}$. Noting that $\mathcal{D}(A(U)^{*(1-\theta)})' \subset \{H^{2(1-\theta)}(\Omega)'\}^2$, we conclude the result. ∎

Other conditions are verified as follows. (S.i) is obvious from the interpolation. (S.ii) is also obvious from closedness of the closed unit ball of Z in the weak topology. (S.iii) is valid with $\beta = (1 + \varepsilon_2)/2$ by virtue of Proposition 3.3. (Ex) is also true since ν can be arbitrarily close to $\frac{1}{2}$. (F) is obvious since $F(U)$ is a square function. Finally, (I) follows from the Proposition 3.3. In this way we conclude that

THEOREM 3.4. *Let u_0, v_0 satisfy (I.F). Let $\varepsilon_1, \varepsilon_2, \nu$ and β be as above and take η arbitrarily in $1 - \nu < \eta < \beta$. Then, in the function space: $\mathcal{C}([0, \infty); H^{1+\varepsilon_1}(\Omega)) \cap \mathcal{C}^\eta([0, \infty); L^2(\Omega))$, there exists a unique local non negative solution u, v to (P.S) on an interval $[0, T], T > 0$, such that*

$$u, v \in \mathcal{C}((0, T]; H^2(\Omega)) \cap \mathcal{C}^1((0, T]; L^2(\Omega)). \tag{3.4}$$

Remark 3.5. If we use the regularity property of abstract parabolic equation (e.g. [7]), the local solution u, v obtained above is shown to be actually more regular than (3.4). For example, it is possible to verify that

$$u, v \in \mathcal{C}^1((0, T]; H^1(\Omega)) \cap \mathcal{C}^2((0, T]; H^1(\Omega)').$$

4. Global Solutions.

In the favorable cases that (α.i) or (α.ii) holds, we shall establish a priori estimates and show the global existence of solution.

In this section we use the notations:

$$P(u,v) = a_1u + \alpha_{11}u^2 + \alpha_{12}uv, \quad Q(u,v) = a_2v + \alpha_{21}uv + \alpha_{22}v^2$$
$$f(u,v) = c_1u - \gamma_{11}u^2 - \gamma_{12}uv, \quad g(u,v) = c_2v - \gamma_{21}uv - \gamma_{22}v^2.$$

Then,

$$\text{(P.S)} \qquad \begin{cases} \dfrac{\partial u}{\partial t} = \operatorname{div}[\nabla P + b_1(\nabla\Phi(x))u] + f & \text{in} \quad \Omega \times (0,\infty), \\ \dfrac{\partial v}{\partial t} = \operatorname{div}[\nabla Q + b_2(\nabla\Phi(x))v] + g & \text{in} \quad \Omega \times (0,\infty), \\ \dfrac{\partial P}{\partial n} + b_1\dfrac{\partial\Phi(x)}{\partial n}u = 0 & \text{on} \quad \partial\Omega \times (0,\infty), \\ \dfrac{\partial Q}{\partial n} + b_2\dfrac{\partial\Phi(x)}{\partial n}v = 0 & \text{on} \quad \partial\Omega \times (0,\infty), \\ u(0,x) = u_0(x) \quad \text{and} \quad v(0,x) = v_0(x) & \text{in} \quad \Omega. \end{cases}$$

We first prove:

PROPOSITION 4.1. *Let (α.i) be satisfied. For $0 < T < \infty$, assume that there exists a non negative strict solution u, v to (P.S) on $[0,T)$ such that*

$$u, v \in \mathcal{C}([0,T); L^2(\Omega)) \cap \mathcal{C}((0,T); H^2(\Omega)) \cap \mathcal{C}^1((0,T); L^2(\Omega)).$$

Then, the norms $\|u(t)\|_{H^2(\Omega)}$ and $\|v(t)\|_{H^2(\Omega)}$ remain being bounded as $t \to T$.

PROOF: Without loss of generality we can assume (in addition to (α.i)) that $\alpha_{12} = \alpha_{21}$. Because, if we change the unknown functions from u, v to $u_\varphi = \varphi u(t,x), v_\psi = \psi v(t,x)$ with two positive scale constants φ, ψ like (P.S)$'$, then it is easily observed that the Condition (α.i) is invariable by this change and that one can take some suitable φ and ψ such that $\alpha_{12} = \alpha_{21}$ in a reduced new system. Our a priori estimates consist of three steps.

Step 1. Multiply the first (resp. second) equation in (P.S) by u (resp. v) and integrate the product in $\Omega \times (0,t)$. And add the two equalities. Then, after some estimations,

$$\frac{1}{2}\int_\Omega \{u(t)^2 + v(t)^2\}dx + \int_0^t \int_\Omega (a_1|\nabla u|^2 + a_2|\nabla v|^2)dxdt$$
$$+ \int_0^t \int_\Omega \{(2\alpha_{11}u + \alpha_{12}v)|\nabla u|^2 + (\alpha_{12}u + \alpha_{21}v)\nabla u \cdot \nabla v + (\alpha_{21}u + 2\alpha_{22}v)|\nabla v|^2\}dxdt$$
$$\leq \varepsilon \int_0^t \int_\Omega (|\nabla u|^2 + |\nabla v|^2)dxdt + \varepsilon^{-1}Const.\left\{1 + \int_0^t \int_\Omega (u^2 + v^2)dxdt\right\}$$

with an arbitrary number $\varepsilon > 0$. While, (α.i) and $\alpha_{12} = \alpha_{21}$ imply here that

$$(4.1) \quad (2\alpha_{11}u + \alpha_{12}v)\xi^2 + (\alpha_{12}u + \alpha_{21}v)\xi\eta + (\alpha_{21}u + 2\alpha_{22}v)\eta^2$$
$$\geq \delta(u+v)(\xi^2 + \eta^2) \quad \text{for} \quad \xi, \eta \in \mathbb{R}; u, v \geq 0,$$

with some $\delta > 0$. Hence, for any $0 \leq t < T$,

$$(4.2) \qquad \int_\Omega \{u(t)^2 + v(t)^2\}dx \quad \text{and} \quad \int_0^t \int_\Omega (1+u+v)(|\nabla u|^2 + |\nabla v|^2)dxdt \leq Const..$$

Step 2. Multiply next the first (resp. second) equation in (P.S) by $\frac{\partial P}{\partial t}$ (resp. $\frac{\partial Q}{\partial t}$) and integrate the product in $\Omega \times (\tau, t)$ with a fixed $\tau, 0 < \tau < T$. (Note that from Remark 3.5 that $P, Q \in C^1([\tau, T); H^1(\Omega))$.) And add the two equalities. Then, after some estimation using (4.1), it is observed that

$$\int_\Omega \{|\nabla P(t)|^2 + |\nabla Q(t)|^2\}dx + \int_\tau^t \int_\Omega (1+u+v)(u_t^2 + v_t^2)dxdt$$
$$\leq Const. \left\{1 + \int_\tau^t \int_\Omega (|\nabla P|^2 + |\nabla Q|^2)dxdt + \int_\tau^t \int_\Omega (u^5 + v^5)dxdt\right\}.$$

While, since $|\nabla u| + |\nabla v| \leq Const.(|\nabla P| + |\nabla Q|)$ in Ω, it follows that

$$\int_\Omega (u^5 + v^5)dx \leq Const. \left\{1 + (\|\nabla u\|_{L^2} + \|\nabla v\|_{L^2}) \int_\Omega (|\nabla P|^2 + |\nabla Q|^2)dx\right\}.$$

Hence, in view of (4.2), for any $\tau \leq t < T$,

$$\int_\Omega \{|\nabla u(t)|^2 + |\nabla v(t)|^2\}dx, \int_\Omega \{|\nabla P(t)|^2 + |\nabla Q(t)|^2\}dx \quad \text{and}$$
$$\int_\tau^t \int_\Omega (1+u+v)(u_t^2 + v_t^2)dxdt \leq Const..$$

Step 3. In view of Remark 3.5, let us differentiate the first (second) equation in (P.S) in t, take the scalar product with $P_u u_t$ (resp. $Q_v v_t$) in $H^1(\Omega)' \times H^1(\Omega)$, and integrate the product on $[\tau, t]$. And afterwards add the two equalities. Then, after some estimation, we obtain that

$$(4.3) \quad \frac{1}{2}\int_\Omega \{P_u(t)u_t(t)^2 + Q_v(t)v_t(t)^2\}dx$$
$$+ \int_\tau^t \int_\Omega \{P_u^2|\nabla u_t|^2 + (P_uP_v + Q_uQ_v)\nabla u_t \cdot \nabla v_t + Q_v^2|\nabla v_t|^2\}dxdt$$
$$\leq \varepsilon \int_\tau^t \int_\Omega (1+u+v)^2(|\nabla u_t| + |\nabla v_t|)^2 dxdt$$
$$+ \varepsilon^{-1}Const. \left[1 + \int_\tau^t \int_\Omega \{(1+u+v+|\nabla u| + |\nabla v|)^2(u_t^2 + v_t^2) + |u_t^3| + |v_t^3|\}dxdt\right]$$

with an arbitrary $\varepsilon > 0$. While,

$$P_u^2\xi^2 + (P_uP_v + Q_uQ_v)\xi\eta + Q_v^2\eta^2 \geq \delta(1+u+v)^2(\xi^2 + \eta^2) \quad \text{for} \quad \xi, \eta \in \mathbb{R}; u, v \geq 0$$

with some $\delta > 0$. In addition,

$$\int_\Omega (|\nabla u| + |\nabla v|)^2(u_t^2 + v_t^2)dx \le \varepsilon^2(\|\nabla u_t\|_{L^2}^2 + \|\nabla v_t\|_{L^2}^2)$$
$$+ \varepsilon^{-2} Const.(1 + \|u\|_{H^2}^2 + \|v\|_{H^2}^2)(\|u_t\|_{L^2}^2 + \|v_t\|_{L^2}^2).$$

So that, we obtain that

$$\int_\Omega (|\nabla u| + |\nabla v|)^2(u_t^2 + v_t^2)dx$$
$$\le \varepsilon^2(\|\nabla u_t\|_{L^2}^2 + \|\nabla v_t\|_{L^2}^2) + \varepsilon^{-2} Const.(1 + \|u_t\|_{L^2}^4 + \|v_t\|_{L^2}^4),$$

if we notice that

LEMMA 4.2. *An estimate*

$$\|u\|_{H^2(\Omega)} + \|v\|_{H^2(\Omega)} \le Const.(1 + \|P\|_{H^1(\Omega)} + \|Q\|_{H^1(\Omega)})$$
$$\times (\|\Delta P\|_{L^2(\Omega)} + \|\Delta Q\|_{L^2(\Omega)} + \|P\|_{H^1(\Omega)} + \|Q\|_{H^1(\Omega)})$$

holds for all pairs of non negative functions $u, v \in H^2(\Omega)$ which satisfy boundary conditions: $\frac{\partial P}{\partial n} + b_1 \frac{\partial \Phi}{\partial n} u = 0$ and $\frac{\partial Q}{\partial n} + b_2 \frac{\partial \Phi}{\partial n} v = 0$ on $\partial\Omega$.

PROOF OF LEMMA: We shall use the well known estimate

$$\|u\|_{H^2(\Omega)} \le Const.(\|\Delta u\|_{L^2(\Omega)} + \|u\|_{L^2(\Omega)} + \|\frac{\partial u}{\partial n}\|_{H^{\frac{1}{2}}(\partial\Omega)}) \quad \text{for all} \quad u \in H^2(\Omega),$$

noting that

$$\begin{pmatrix} P_u & P_v \\ Q_u & Qv \end{pmatrix} \begin{pmatrix} \Delta u \\ \Delta v \end{pmatrix} = \begin{pmatrix} \Delta P \\ \Delta Q \end{pmatrix} - 2 \begin{pmatrix} \alpha_{11}|\nabla u|^2 + \alpha_{12} \nabla u \cdot \nabla v \\ \alpha_{21} \nabla u \cdot \nabla v + \alpha_{22}|\nabla v|^2 \end{pmatrix} \quad \text{in} \quad \Omega,$$

$$\begin{pmatrix} P_u & P_v \\ Q_u & Q_v \end{pmatrix} \begin{pmatrix} \frac{\partial u}{\partial n} \\ \frac{\partial v}{\partial n} \end{pmatrix} + \begin{pmatrix} b_1 & 0 \\ 0 & b_2 \end{pmatrix} \frac{\partial \Phi}{\partial n} \begin{pmatrix} u \\ v \end{pmatrix} = 0 \quad \text{on} \quad \partial\Omega. \quad ∎$$

Finally we also obseve that

$$\int_\Omega \{(1 + u + v)^2(u_t^2 + v_t^2) + |u_t^3| + |v_t^3|\}dx$$
$$\le \varepsilon^2(\|\nabla u_t\|_{L^2}^2 + \|\nabla v_t\|_{L^2}^2) + \varepsilon^{-2} Const.(1 + \|u_t\|_{L^2}^4 + \|v_t\|_{L^2}^4).$$

In this way (4.3) yields that, for any $\tau \le t < T$,

$$\int_\Omega \{P_u(t)u_t(t)^2 + Q_v(t)v_t(t)^2\}dx \quad \text{and}$$
$$\int_\tau^t \int_\Omega (1 + u + v)^2(|\nabla u_t|^2 + |\nabla v_t|^2)dxdt \le Const..$$

Then the desired result follows immediately from Lemma 4.2. ∎

We now handle the second case:

PROPOSITION 4.3. *Let (α.ii) be satisfied. Then the same assertion as the Proposition 4.1 is true.*

PROOF: The proof consists of five steps.

Step 1. Multiply the second equation in (P.S) by $pv^{p-1}, 2 \le p < \infty$, and integrate the product in $\Omega \times (0,t), 0 < t < T$. Then,

$$\int_\Omega v(t)^p dx + p(p-1)\int_0^t\int_\Omega a_2 v^{p-2}|\nabla v|^2 dxdt$$
$$\le \varepsilon\int_0^t\int_\Omega v^{p-2}|\nabla v|^2 dxdt + \varepsilon^{-1}Const.\left\{1+\int_0^t\int_\Omega v^p dxdt\right\}$$

with an arbitrary $\varepsilon > 0$. So that, for any $0 \le t < T$,

$$\int_\Omega v(t)^p dx \quad \text{and} \quad \int_0^t\int_\Omega v^{p-2}|\nabla v|^2 dxdt \le Const.. \tag{4.4}$$

Step 2. Integration of the first equation in (P.S) in $\Omega \times (0,t)$ yields that

$$\int_\Omega u(t)dx \le \int_\Omega u_0 dx + c_1\int_0^t\int_\Omega u dxdt - \gamma_{11}\int_0^t\int_\Omega u^2 dxdt.$$

So that, for any $0 \le t < T$,

$$\int_\Omega u(t)dx \quad \text{and} \quad \int_0^t\int_\Omega u^2 dxdt \le Const.. \tag{4.5}$$

Step 3. Multiply the first (resp. second) equation in (P.S) by u(resp. $\frac{\partial v}{\partial t}$) and integrate the product in $\Omega \times (\tau,t), 0 < \tau < T$. And add the two equalities. Then, after some estimation,

$$\int_\Omega\{u(t)^2 + |\nabla v(t)|^2\}dx + \int_\tau^t\int_\Omega(P_u|\nabla u|^2 + v_t^2)dxdt \le \varepsilon\left\{\int_\Omega|\nabla v(t)|^2 dx\right.$$
$$\left.+\int_\tau^t\int_\Omega(u|\nabla u|^2 + v_t^2)dxdt\right\} + \varepsilon^{-1}Const.\left\{1+\int_\tau^t\int_\Omega(u^3 + |\nabla v|^3)dxdt\right\}$$

with an arbitrary $\varepsilon > 0$. While,

$$\int_\Omega(u^3 + |\nabla v|^3)dx \le \varepsilon^2(\|u\|_{H^1}^2 + \|v_t\|_{L^2}^2) + \varepsilon^{-2}Const.(1 + \|u\|_{L^2}^4 + \|v\|_{H^1}^4).$$

Therefore we conclude in view of (4.4) and (4.5) that, for any $\tau \le t < T$,

$$\int_\Omega\{u(t)^2 + |\nabla v(t)|^2\}dx \quad \text{and} \quad \int_\tau^t\int_\Omega\{(1+u)|\nabla u|^2 + v_t^2\}dxdt \le Const.. \tag{4.6}$$

Step 4. Multiply the first equation in (P.S) by $\frac{\partial P}{\partial t}$ and inegrate the product in $\Omega \times (\tau,t), 0 < \tau < T$. On the other hand, differentiate the second equation in (P.S) in t, take the scalar product with $\frac{\partial v}{\partial t}$ in $H^1(\Omega)' \times H^1(\Omega)$, and integrate the product on $[\tau, t]$. And afterwards add the two equalities. Then some estimation yields that

$$\int_\Omega \{|\nabla P(t)|^2 + v_t(t)^2\}dx + \int_\tau^t \int_\Omega (P_u u_t^2 + |\nabla v_t|^2)dxdt$$
$$\leq \varepsilon \int_\tau^t \int_\Omega u_t^2 dxdt + \varepsilon^{-1} Const. \left\{1 + \int_\tau^t \int_\Omega (|\nabla P|^2 + u^5 + |v_t^3|)dxdt\right\}$$

with an arbitrary $\varepsilon > 0$. While,

$$\int_\Omega u^5 dx \leq Const.(1 + \|\nabla u\|_{L^2} \|\nabla P\|_{L^2}^2),$$

$$\int_\Omega |v_t^3| dx \leq \varepsilon \|\nabla v_t\|_{L^2}^2 + \varepsilon^{-1} Const.(1 + \|v_t\|_{L^2}^4) \quad \text{with any} \quad \varepsilon > 0.$$

Therefore we obtain in view of (4.6) that, for any $\tau \leq t < T$,

$$\int_\Omega \{|\nabla P(t)|^2 + v_t(t)^2\}dx \quad \text{and} \quad \int_\tau^t \int_\Omega \{(1+u)u_t^2 + |\nabla v_t|^2\}dxdt \leq Const.. \tag{4.7}$$

Step 5. Differentiate the first equation in (P.S) in t, take the scalar product with $P_u u_t$ in $H^1(\Omega)' \times H^1(\Omega)$, and integrate the product on $[\tau, t], 0 < \tau < T$. Then,

$$\int_\Omega P_u(t) u_t(t)^2 dx + \int_\tau^t \int_\Omega P_u^2 |\nabla u_t|^2 dxdt \leq Const. \Big[1 + \int_\tau^t \int_\Omega \{u^2 |\nabla v_t|^2$$
$$+ (|\nabla u| + |\nabla v|)^2 (u_t^2 + v_t^2) + (1 + u + v)^2 (u_t^2 + v_t^2) + |u_t^3| + |v_t^3|\}dxdt\Big].$$

We here notice that an estimate

$$\|u\|_{H^2(\Omega)} \leq Const.(1 + \|P\|_{H^1(\Omega)} + \|v\|_{H^1(\Omega)})(\|\Delta P\|_{L^2(\Omega)} + \|P\|_{H^1(\Omega)} + \|v\|_{H^2(\Omega)})$$

holds for all pairs of non negative functions $u, v \in H^2(\Omega)$ which satisfy boundary conditions: $\frac{\partial P}{\partial n} + b_1 \frac{\partial \Phi}{\partial n} u = 0$ and $a_2 \frac{\partial v}{\partial n} + b_2 \frac{\partial \Phi}{\partial n} v = 0$ on $\partial\Omega$, which is observed in an analogous way as for Lemma 4.2 by using

$$P_u \Delta u = \Delta P - P_v \Delta v - 2\alpha_{11} |\nabla u|^2 - 2\alpha_{12} \nabla u \cdot \nabla v \quad \text{in} \quad \Omega,$$
$$P_u \frac{\partial u}{\partial n} = (b_2 a_2^{-1} P_v v - b_1 u) \frac{\partial \Phi}{\partial n} \quad \text{on} \quad \partial\Omega.$$

Then we have:

$$\int_\Omega u^2 |\nabla v_t|^2 dx \leq Const.(1 + \|u_t\|_{L^2}^2) \|\nabla v_t\|_{L^2}^2,$$

$$\int_\Omega \{(|\nabla u| + |\nabla v|)^2 (u_t^2 + v_t^2) + (1 + u + v)^2 (u_t^2 + v_t^2) + |u_t^3| + |v_t^3|\}dx$$
$$\leq \varepsilon \|\nabla u_t\|_{L^2}^2 + \varepsilon^{-1} Const.(1 + \|u_t\|_{L^2}^4 + \|\nabla v_t\|_{L^2}^2)$$

with an arbitrary $\varepsilon > 0$. Therefore we conclude in view of (4.7) that, for any $\tau \leq t < T$,

$$\int_\Omega \{1 + u(t)\} u_t(t)^2 dx \quad \text{and} \quad \int_\tau^t \int_\Omega (1+u)^2 |\nabla u_t|^2 dx dt \leq Const.$$

This together with (4.7) completes the proof. ∎

It is not difficult to verify that, if the norms $\|u(t)\|_{H^2(\Omega)}, \|v(t)\|_{H^2(\Omega)}$ of a strict solution to (P.S) on an interval $[0, T), 0 < T < \infty$, do not blow up as $t \to T$, then the solution can be extended beyond the extreme point T. Thus we obtain:

THEOREM 4.4. *Let (α.i) or (α.ii) be satisfied. Then, for the initial functions u_0, v_0 satisfying (I.F), there exists a global non negative strict solution to (P.S) such that $u, v \in \mathcal{C}^1((0,\infty); L^2(\Omega)) \cap \mathcal{C}((0,\infty); H^2(\Omega))$.*

REFERENCES

1. P. Deuring, *An initial-boundary value problem for certain density-dependent diffusion system*, Math. Z. **194** (1987), 375–396.
2. J. U. Kim, *Smooth solutions to a quasi-linear system of diffusion equations for a certain population model*, Nonlinear Analysis **8** (1984), 1121–1144.
3. M. Mimura, *Stationary pattern of some density-dependent diffusion system with competitive dynamics*, Hiroshima Math. J. **11** (1981), 621–635.
4. N. Shigesada, K. Kawasaki and E. Teramoto, *Spatial segregation of interacting species*, J. theor. Biol. **79** (1979), 83–99.
5. A. Yagi, *Fractional powers of operators and evolution equations of parabolic type*, Proc. Japan Acad. **64 A** (1988), 227–230.
6. A. Yagi, *Abstract quasilinear evolution equations of parabolic type in Banach spaces*, Bollettino U. M. I. **5-B** (1991), 341–368.
7. A. Yagi, *Maximal regularity of abstract parabolic evolution equation, an alternative approach of using evolution operator*, Reports Fac. Sci. Himeji Insti. Tech. **2** (1991), 1–9.
8. H. Tanabe, "Equation of Evolution," Pitman, London, 1979.

Index